Nanotechnology
for
Telecommunications

Nanotechnology
for
Telecommunications

Edited by
Sohail Anwar
M. Yasin Akhtar Raja
Salahuddin Qazi
Mohammad Ilyas

CRC Press
Taylor & Francis Group
Boca Raton London New York

CRC Press is an imprint of the
Taylor & Francis Group, an **informa** business

CRC Press
Taylor & Francis Group
6000 Broken Sound Parkway NW, Suite 300
Boca Raton, FL 33487-2742

First issued in paperback 2017

© 2010 by Taylor and Francis Group, LLC
CRC Press is an imprint of Taylor & Francis Group, an Informa business

No claim to original U.S. Government works

ISBN-13: 978-1-4200-5325-8 (hbk)
ISBN-13: 978-1-138-11381-7 (pbk)

Library of Congress Cataloging-in-Publication Data

Nanotechnology for telecommunications / editors, Sohail Anwar, ... [et al.].
      p. cm.
    "A CRC title."
    Includes bibliographical references and index.
    ISBN 978-1-4200-5325-8 (hardcover : alk. paper)
      1. Telecommunication--Equipment and supplies. 2. Telecommunication--Technological innovations. 3. Nanotechnology. 4. Nanoelectronics. I. Anwar, Sohail. II. Title.

TK5103.N36 2010
621.382--dc22   2010006627

Visit the Taylor & Francis Web site at
http://www.taylorandfrancis.com

and the CRC Press Web site at
http://www.crcpress.com

Contents

Preface

Nanotechnology is the creation of functional materials, devices, and systems through the control of matter at the nanometer length scale, and the exploitation of novel properties and phenomena developed at that scale. Nanotechnology holds singular promise to revolutionize science, engineering, and technology. It has already had a significant impact on countless industries including communications, medicine, environmental cleanup, agriculture, and several others. Innovative materials, components, and systems based on nanotechnology are recognized as promising growth innovators in the years to come. It is expected that nanotechnology will eventually merge into a technology cluster offering a complete range of functionalities in information, energy, construction, environmental, and biomedical domains.

Current advances in nanotechnology have resulted in new approaches for improvements in telecommunications and information processing. Traditional electronic devices are increasingly being replaced by optoelectronic devices such as photonic crystals and quantum dots. Displays with low energy consumption can be produced using carbon nanotubes. Components based on the microelectromechanical system (MEMS) and nanoelectromechanical system (NEMS) hold significant promise for future developments in wireless communications.

This book provides a broad spectrum of technical information and research ideas regarding the use of nanotechnology in telecommunications and information processing. It describes current and emerging nanotechnology developments that hold promise for significant innovations in telecommunications. It also reviews a broad range of the nanoscience and nanotechnology concepts and their influence on every aspect of telecommunications.

This book is organized into 17 chapters and a glossary of technical terms. Chapters 1 and 2 provide background information for nanoscience and nanotechnology. Chapter 1 presents an overview of new emerging nanotechnology materials, processes, devices, and their applications. It also includes a discussion regarding the social implications of nanotechnology. Chapter 2 provides a basic understanding of the impact of nanotechnology on telecommunications. It also presents examples of the integration of voice and data to enhance existing telecommunication applications and to enable new applications.

Chapters 3 through 5 describe the specific applications of nanotechnology in telecommunications. Chapter 3 provides a description of the application of nanostructures in passive fiber-optic power control devices. It also summarizes the lifetime and performance testing of these devices and related international standardization. Chapter 4 presents the basic types, characteristics, fabrication, and applications of nanotubes in telecommunications. Nanotubes exhibit novel properties that make them potentially useful for many applications in nanotechnology, electronics, and optics. Chapter 5 examines the unique characteristics of silicon-based nanostructures that are fundamentally different from those of the bulk crystalline material. Some of these characteristics make nanoscale silicon attractive for use in many technical applications, such as light-emitting devices for optical communications, photovoltaic solar cells, and fully integrated optoelectronic systems.

Chapters 6 through 8 focus on nanostructured optoelectronic materials. Chapter 6 presents an overview of the research and development and the future prospects of III–V wide bandgap semiconductor (WBGS) nanostructures and optoelectronics devices. Some of the recent developments in Al_2O_3- and SiC-based devices are discussed. Chapter 7 highlights the recent developments in device fabrication processes, enabled by nanotechnology, to realize modern optical systems on chip and board levels. It focuses on the role of nanophotonics in on-board and on-chip communications. Chapter 8 describes crystalline colloidal array photonic crystals. These crystals constitute promising systems for optical switching and display or memory storage devices.

Chapters 9 through 11 describe MEMS, NEMS, and their applications in communication systems. Chapter 9 provides extensive details regarding the processing methods, such as silicon bulk and surface micromachining, microstereolithography, LIGA, and electrodeposition, for the fabrication of MEMS and NEMS. The systematic modeling of MEMS and NEMS from device level up to system level is covered in Chapter 10, which provides a detailed account regarding system-level modeling practices and issues. The challenges and potentials of the use of MEMS in wireless communications are presented in Chapter 11. This chapter also reviews the solutions to several MEMS technology problems, such as temperature drift, dielectric degradation, packaging, and stiction.

Quantum dot cellular automata (QCA) has become a very promising technology for designing the new generation of processors. Its potential for high density and regularity makes it very useful for manufacturing digital data processors for digital telecommunication systems and networks. Chapter 12 provides a good understanding of the QCA and its applications in telecommunication systems.

Information processing by transporting carriers in nanoelectronic devices and circuits are shown to be severely affected by the breakdown of Ohm's law when the applied voltage in a nanoscale device exceeds the critical voltage. This nonohmic behavior is the cause of the current saturation leading to a resistance blowup. In Chapter 13, we learn that this nonohmic nonlinear behavior affects both digital and analog signal processing.

The current trend is to miniaturize and make products multifunctional, smaller, lighter, and faster. This is leading to the use of nanoscale structures and materials in developing semiconductor chips. However, the scale of the physical systems that use these nanoscale electronic devices is still large and therefore results in serious challenges to the establishment of interconnections between nanoscale devices and the outside world. Chapter 14 describes the different levels of interconnections and presents details regarding the standardized assembly processes for a broad spectrum of semiconductor devices providing micro-, nano-, bio-, and optoelectronics functions.

Recent progress on quantum networks provides a new method for the transmission of messages. Due to significant interest in networks carrying quantum information, a switch-based networking technology for quantum networks is needed. Chapter 15 presents several concepts regarding quantum switching and its applications in quantum networks.

The creation of a fundamental nanotechnology knowledge base and the associated new innovations needs establishing a laboratory and human resource infrastructure that can effectively deal with the challenges of nanoscience and nanotechnology. Also, as worldwide demand for energy surges at an ever-increasing rate, there is a new urgency to improve the efficiency and sustainability of power-generation technologies. Chapter 16 discusses these two topics.

Being a platform technology, nanotechnology feeds its output into many industries, which use these inputs to improve their products. Thus, any effort to commercialize this technology has to be supported by scientific and engineering research coupled with an innovative well-funded product development and marketing program. Chapter 17 focuses on the issues and opportunities associated with the commercialization of nanotechnology. It also describes some of the key factors for success in the commercialization of nanotechnology.

Finally, a glossary of key terms used in the description of nanotechnology applications in telecommunications is provided. The glossary does not list each and every technical term used to describe nanotechnology applications in telecommunications. However, every effort has been made to include the important terms.

It is hoped that this book will serve as a useful source of technical and scientific information for professionals, researchers, and scientists who want to discover the challenges, issues, and opportunities associated with the current and future development of nanoscale telecommunication systems.

Sohail Anwar

MATLAB® and Simulink® are registered trademarks of The MathWorks, Inc. For product information, please contact:

The MathWorks, Inc.
3 Apple Hill Drive
Natick, MA 01760-2098 USA
Tel: 508 647 7000
Fax: 508-647-7001
E-mail: info@mathworks.com
Web: www.mathworks.com

Acknowledgments

I would like to thank my coeditors, Dr. Yasin Akhtar Raja, Dr. Salahuddin Qazi, and Dr. Mohammad Ilyas for their kind help. Special thanks are extended to Nora Konopka and Jennifer Ahringer at Taylor & Francis Group/CRC Press for their support and encouragement. I wish to add here that without the direction and creative insight of the management and staff at CRC Press, this book would not have taken its final form.

Editors

Sohail Anwar is an associate professor of engineering and the coordinator of Electrical Engineering Technology Program at the Altoona College of The Pennsylvania State University, Altoona, Pennsylvania. In addition, he is a professional associate of the management development programs and services at The Pennsylvania State University, University Park, Texas.

Dr. Anwar is currently serving as the chair of the Electronics Engineering Technology Consulting Faculty Committee of Excelsior College, Albany, New York. Since 1996, he has been serving as a visiting professor of electrical engineering at the Université d'Artois, Artois, France. He has also been serving as an invited professor of electrical engineering at the Shanghai Normal University, China, since 2009. He has been developing active research collaborations with his colleagues in China.

Dr. Anwar has been a program evaluator for the Accreditation Board for Engineering and Technology, Baltimore, Maryland. He is currently serving as the editor in chief of the *Journal of Engineering Technology*, the associate editor in chief of the *International Journal of Engineering Research and Innovation*, the executive editor of the *International Journal of Modern Engineering*, and the associate editor of the *Journal of the Pennsylvania Academy of Science*.

Dr. Anwar is the series editor for the Nanotechnology and Energy Series currently being developed by Taylor & Francis Group/CRC Press. Moreover, he is writing a textbook titled *Renewable Energy Systems* to be published by Taylor & Francis Group/CRC Press in late 2010. He is also the author of three guide books on renewable energy, solar energy, and biofuels.

Dr. Anwar has been teaching academic courses in the topical areas of digital systems design, microprocessors, circuit analysis, engineering design, signals and systems, quality control, programmable logic controllers (PLCs), and nanotechnology. In addition, he developed and conducted industrial training in the topical areas of PLCs, hydraulics, electrohydraulics, electrical machines, pneumatics, and electromechanical circuits. He has extensive experience in developing and teaching online courses in engineering and engineering technology.

Dr. Anwar is a senior member of IEEE and a member of the American Society for Engineering Education (ASEE); the Association of Technology, Management, and Applied Engineering (ATMAE); and the Pennsylvania Academy of Science.

Salahuddin Qazi received his PhD in electrical engineering from the University of Technology, Loughborough, England, and his BS in electronic engineering from the University College of Wales, Bangor, United Kingdom. He is currently a full professor in the School of Information Systems and

Engineering Technology at the State University of New York Institute of Technology, Utica, New York. Dr. Qazi has been a chair and coordinator of several programs including electrical engineering technology, photonics, and graduate programs in advanced technology, which he helped to develop. He has worked, conducted research, and taught in the United Kingdom and the Middle East before coming to the United States.

Dr. Qazi won several research awards and grants to conduct research and has published several articles and book chapters in the area of fiber-doped amplifiers, wireless security, optical wireless communications, and micro-electromechanial system (MEMS)–based wireless and optical communications. He is currently working on a National Science Foundation (NSF) project to develop an instructional laboratory for the "Visualization and Manipulation of Nanoscale Components." Dr. Qazi has participated as an invited speaker and presenter in several international conferences and workshops. He was a CO-PI for a U.S.–Pakistan conference on "High Capacity Optical Networking and Enabling Technology," for 2004 and 2005, which was jointly supported by the NSF of the United States and the Higher Education Commission of Pakistan.

Dr. Qazi is a recipient of several professional awards including the William Goddel award for research creativity at SUNYIT, and engineering professionalism and outstanding professional development by Mohawk Valley Engineering Executive Committee. He was also awarded several IEEE awards for outstanding support and service to the SUNYIT student chapter and forging closer relations with the IEEE Mohawk Valley section. Dr. Qazi is a senior member of IEEE and a member of the American Society of Engineering Education. He has been a member of the board of directors of Mohawk Valley Resource Center for Refugees for over 10 years.

Dr. M. Yasin Akhtar Raja is a professor of physics and optical science at the University of North Carolina (UNC), Charlotte, North Carolina. He joined UNC Charlotte in 1990 and has served since then in various faculty positions with a pioneering role in developing interdisciplinary new degree programs; he has also participated in establishing new academic units and research centers. During the past decade, he served as a founding member of the task force, and program and steering committees of PhD in electrical and computer engineering, biomedical sciences, and optical science and engineering. He also served on the planning and steering committee for establishing the College of Information Technology (now College of Computing and Informatics) at UNC Charlotte (www.uncc.edu).

In his home department, he played a leading role in establishing a new graduate degree (MS) in applied physics and optics. He also served as a graduate program director in applied physics and optics program for two years. During this period, he led the pioneering work for establishing the new PhD and the optics center (http://opticscenter.uncc.edu). He is the founding member of the Center for Optoelectronics and Optical Communications, and interdisciplinary graduate programs (PhD and MS) in optical science and engineering. Professor Raja has served and chaired high-level statuary and search committees. For example, he served five years as a member of the University Faculty Council; he chaired and served on committees for hiring center directors as well as department chairs and professors.

In the research area, he leads optical communication labs and coordinates the optical networks testbed initiative at UNC Charlotte. His research expertise and interests include the design, integration, and characterization of photonic components and novel nanophotonic devices (e.g., vertical cavity surface-emitting lasers [VCSELs] and LEDs) for optical communication and signal processing. Professor Raja has developed various new courses in optical communication and networks as well as in the field of nanophotonics; these courses serve optical science and engineering PhD and MS students in several colleges and departments. He is currently engaged in network design; simulations; testbed and experimental implementations, especially network topologies; challenges in optical layer; and passive optical networks for broadband access. He has several patents and has over 150 published articles in journals and conference proceedings to his credit. He has a wide range of collaborators from national and international academic as well as industrial institutions.

Dr. Raja received his PhD in 1988 in the area of optical physics/optoelectronics from the University of New Mexico (UNM), Albuquerque, New Mexico, where he coinvented RPG lasers and conducted a pioneering research (as a coinventor) on a special type of semiconductor lasers "VCSELs" at the Center for High Tech Materials (CHTM) (www.chtm.unm.edu). He also contributed in developing the high-tech laser research labs at CHTM, which was then a newly established center at UNM. He has organized six major international conferences (as general chair) sponsored by NSF (http:// honet-ict.org) and IEEE ComSoc, and has served on several national and regional symposia technical committees and NSF panels. Dr. Raja is a senior member of IEEE and ComSoc (serving as chair in greater Charlotte region); he is also a member of SPIE, OSA, and FTTH Council (www.ftthcouncil.org).

Mohammad Ilyas is associate dean for research and industry relation and professor of computer science and engineering in the College of Engineering and Computer Science at Florida Atlantic University, Boca Raton, Florida. He is also serving as interim chair of the Department of Mechanical and Ocean Engineering. He received his BSc in electrical engineering from the University of Engineering and Technology, Lahore, Pakistan, in 1976. From March 1977 to September 1978, he worked for the Water and Power Development Authority, Pakistan. In 1978, he was awarded a scholarship for his graduate studies and he completed his MS in electrical and electronic engineering in June 1980 at Shiraz University, Shiraz, Iran. In September 1980, he joined the doctoral program at Queen's University in Kingston, Ontario, Canada. He completed his PhD in 1983. His doctoral research was about switching and flow control techniques in computer communication networks. Since September 1983, he has been with the College of Engineering and Computer Science at Florida Atlantic University. From 1994 to 2000, he was chair of the Department of Computer Science and Engineering. From July 2004 to September 2005, he served as interim associate vice president for research and graduate studies. During the 1993–1994 academic year, he was on his sabbatical leave with the Department of Computer Engineering, King Saud University, Riyadh, Saudi Arabia.

Dr. Ilyas has conducted successful research in various areas including traffic management and congestion control in broadband/high-speed communication networks, traffic characterization, wireless communication networks, performance modeling, and simulation. He has published 1 book, 16 handbooks, and over 160 research articles. He has supervised 11 PhD dissertations and more than 38 MS theses to completion. He has been a consultant to several national and international organizations. Dr. Ilyas is an active participant in several IEEE technical committees and activities. He is also a senior member of IEEE and a member of ASEE.

Contributors

Kavan Acharya
Department of Physics and Optical
  Science
University of North Carolina
  at Charlotte
Charlotte, North Carolina

Seyed Allameh
Department of Physics
  and Geology
Northern Kentucky University
Highland Heights, Kentucky

Sohail Anwar
Altoona College
Penn State University
Altoona, Pennsylvania

Vijay K. Arora
Universiti Teknologi Malaysia
Skudai, Johor, Malaysia

and

Wilkes University
Wilkes-Barre, Pennsylvania

Emanuel F. Barros
Advanced Technology Innovation
Santa Cruz, California

Harry Efstathiadis
University at Albany
State Universtiy of New York
Albany, New York

William Eisenberg
Systems Integration Solutions
San Jose, California

Adam A. Filios
Farmingdale State College
State University of New York
Farmingdale, New York

Andre Girard
EXFO E.O. Engineering Inc.
Quebec, Canada

Ahmed S. Khan
College of Engineering
  and Information Sciences
Devry University
Addison, Illinois

M. Khizar
Department of Physics and Optical
  Science
University of North Carolina
  at Charlotte
Charlotte, North Carolina

Sy-Yen Kuo
Department of Electrical Engineering
National Taiwan University
Taipei, Taiwan
Republic of China

Marta K. Maurer
Penn State University
Altoona College
Altoona, Pennsylvania

Shahram Mohammad Nejad
Nanoptronics Research Center
School of Electrical Engineering
Iran University of Science
  and Technology
Tehran, Iran

Moshe Oron
KiloLambda Technologies Ltd.
Tel Aviv, Israel

Salahuddin Qazi
State University of New York
   Institute of Technology
Utica, New York

Ehsan Rahimi
Nanoptronics Research Center
School of Electrical Engineering
Iran University of Science and
   Technology
Tehran, Iran

M. Yasin Akhtar Raja
Department of Physics and Optical
   Science
University of North Carolina
   at Charlotte

and

Center for Optoelectronics and
   Optical Communications
University of North Carolina
   at Charlotte
Charlotte, North Carolina

S. Manian Ramkumar
Center for Electronics Manufacturing
   and Assembly
Manufacturing and Mechanical
   Engineering Technology/
   Packaging Science
Rochester Institute of Technology
Rochester, New York

Yeong S. Ryu
Farmingdale State College
State University of New York
Farmingdale, New York

Fidel M. Salinas
LearningSolutions.3rdPlanet
Santa Clara, California

Kamal Shahrabi
Farmingdale State College
State University of New York
Farmingdale, New York

Murad Shibli
Mechanical Engineering Department
United Arab Emirates University
Al-Ain, United Arab Emirates

Denise M. Smith
LearningSolutions.3rdPlanet
Santa Clara, California

I-Ming Tsai
Department of Electrical Engineering
National Taiwan University
Taipei, Taiwan
Republic of China

Raphael Tsu
Department of Electrical and
   Computer Engineering
University of North Carolina
   at Charlotte
Charlotte, North Carolina

Kashif Virk
GN Resound A/S
Ballerup, Denmark

Shekar Viswanathan
Department of Applied Engineering
National University
San Diego, California

1

An Overview of Nanotechnology and Nanoscience

Ahmed S. Khan

CONTENTS

1.1 Introduction

Nanotechnology is the builder's final frontier.

Richard Smalley
1996 Nobel Laureate in Chemistry

The accomplishments of the twentieth century are revolutionizing science and technology in the twenty-first century. Researchers have gained the ability to measure, manipulate, and organize matter on nanoscale—1 to 100 billionths of a meter. At the nanoscale, physics, chemistry, material science, biology, and engineering converge toward common principles, mechanisms, and tools. This convergence of multiple disciplines will lead to a significant impact on science, technology, and society.

Historically, every new technological advance and innovation remakes the world. The time to remake the world has become shorter with every new technological revolution. The industrial revolution took almost two centuries to reshape the world, the electronics revolution around 70 years, the information revolution two decades, and innovations in biotechnology and nanotechnology to reshape the world could be just a matter of less than a decade. Historically, the world was divided into the first world and the third world, but the information revolution revealed the "digital-divide," and advances in nanotechnology will divide the world into the nano-haves and nano-have-nots (Hjorth et al. 2008).

The nanoscale is not just another jump toward miniaturization, but a qualitatively new scale. The new behavior is dominated by quantum mechanics, material confinement in small structures, large interfacial volume fraction, and other unique properties, phenomena, and processes. Many current theories of matter at the microscale will be inadequate to describe the new phenomena at the nanoscale (Roco and Bainbridge 2001).

As the global economy continues to be transformed by new technology, an intense competition will grow for intellectual capital and intellectual property. Technology will continue to drive the global and domestic GDP (Khan 2006).

The National Science Foundation predicts that the global marketplace for goods and services using nanotechnologies will grow to $1 trillion by 2015 and employ 2 million workers. It is estimated that by 2015 nanotechnology will be a $3 trillion-a-year global industry. In 1997, the investment in nanotechnology stood at $430 million to more than $9 billion in 2004. There are more than 800 products in the marketplace that have been developed using nanotechnology (Ehrmann 2008).

1.1.1 What Is Nanotechnology?

A number of definitions exist in literature. According to the National Nanotechnology Initiative (NNI), nanotechnology is an area that encompasses the following traits or characteristics:

1. Research and technology development at the atomic, molecular, or macromolecular levels, in the length scale of approximately 1–100 nm range
2. Creating and using structures, devices, and systems that have novel properties and functions because of their small and/or intermediate size
3. Ability to control or manipulate matter on the atomic scale (Minoli 2006)

TABLE 1.1

Matters of Scale

Item	Dimensions in Nanometers (nm) [1 nm = 0.000000001 m]
Human hair	100,000
Diameter of bacterium cell	1,000–10,000
Range of visible light	400–700
Human immunodeficiency virus	90
Transistor dimensions on a CPU (2008)	43
Cell membrane	10
Drug molecule	~5–10
QD	~1–5
Diameter of DNA	2.5
Individual atom	0.1

According to Professor Stephen Fonash, nanotechnology is manipulating matter at the atomic and molecular scale, seeing matter at the atomic and molecular scale, and exploiting the unique capabilities and properties of structures fabricated at the atomic and molecular scale (Fonash 2008). In short, nanotechnology refers to the convergence of multiple disciplines and applied technologies dealing with particles and structures having dimensions in the range of a nanometer, i.e., one-billionth of a meter. The diameter of a human hair is ~100,000 nm. Presently PC notebooks employ CPU chips that contain transistors having dimensions of 43 nm. It is said that currently more nanoscale transistors are made in a year than grains of rice grown. Table 1.1 compares various entities and systems that occur in nature using a nanoscale.

1.2 Nanotechnology Evolution

1.2.1 How Did the Field of Nanotechnology Evolve?

Nature is full of nanostructures (Figures 1.1 through 1.7). Nanoparticles and structures have been used by humans for quite a long time. For example, in the fourth century, Romans used metal nanoparticles to make glass cups that changed their colors based on the transmission or reflection of light. An artifact from this period called the Lycurgus Cup resides in the British Museum in London. The cup, which depicts the death of King Lycurgus, is made from soda lime glass containing silver and gold nanoparticles (40 ppm gold nanoparticles and 300 ppm silver nanoparticles). The color of the cup changes from green to a deep red when a light source is placed inside it (Barber and Freestone 1990). The Irish manufactured stained-glass windows in AD 444 using nanoparticles. Italians also employed nanoparticles in creating sixteenth-century Renaissance pottery (Poole and Owens 2003). These were influenced by Ottoman techniques. In south Asia, traditional medicine also employed fine particles in the composition of *Tib-e-Unani* medicines.

In 1960, prominent physicist Richard Feynman presented a visionary and prophetic lecture at the meeting of the American Physical Society entitled, "There is plenty of room at the bottom," in which he speculated the possibility and potential of nano-sized materials. In 1974, the term "nanotechnology" was used for the first time by Nori Taniguchi in Tokyo, Japan, at the International Conference on Production Engineering (Minoli 2006). However, it was not until the 1980s, with the development of the appropriated methods of fabrication of nanostructures, that a notable increase in research activity occurred and a number of significant developments materialized (Poole and Owens 2003).

Nanoparticles have been used for ages, but the ability to understand the chemical and physical processes at the nanoscale have only been understood recently. Control and repeatability have been achieved, and the knowledge base and applications have been refined. Table 1.2 presents a summary of the applications of nanotechnology in the various domains of science. Table 1.3 presents a summary of the equipment and processes used in nanotechnology research and development.

TABLE 1.2

Applications of Nanotechnology

Domain/Area	Applications
Physics	Conductivity measurement of a single molecule
	Conduction through small junctions with few defects
	Observation of magnetic scattering of spin-polarized currents are possible
Optics	Fabrication of lasers and waveguides, optical switches, modulators, and photonic crystals
Electronics	Fabrication of less than 10 nm by integrating electronics with MEMS and optics
MEMS/NEMS	Fabrication of micro- and nanomechanical systems–based sensors and actuators
Life sciences	Use of nanostructures can be used to simulate biological structures, sort or detect cells or molecules, and control cell growth
	Fabrication of nanoprobe for in vitro applications
Chemistry	Fabrication of monolayers, dendrimers, functionalized nanotube structures
	Development of chemical sensors and chemical mixing systems using microfluids
	Development of new materials and processes for nanostructure fabrication

Source: National Nanotechnology Infrastructure Network, www.nnin.org

TABLE 1.3

Nanotechnology Equipment and Processes

Thin film deposition and growth	Evaporation, plasma chemical vapor deposition (CVD), sputtering, oxidation, molecular beam epitaxy (MBE), nanotube growth
Thin film etching	Ion mill, reactive ion etching, plasma etching
Lithography	Electron beam lithography, contact photolithography, projection photolithography
Synthesis, nanostructure coatings, and molecular assembly	Nanotube and nanowire growth, nanoparticle synthesis and characterization, molecular synthesis, electrophoretic particle deposition, porous materials, nanocomposites, self-assembly, cell culture/biological coatings/templates
Novel patterning, structuring, manipulation, and mesosopic/microscopic assembly	Nanoimprint lithography, hot embossing, microcontact printing, nanoxerography, interferometric lithography, nanopore formation scanned probe lithography, QD growth/patterning, optical tweezers, nanosphere assembly, electrofluidic assembly of particles, chemically directed assembly of particles/devices, organo/molecular coating
Inspection and characterization	Atomic force microscopy (AFM), fluorescence microscopy, confocal microscopy, ultrahigh resolution electron microscopy, focused ion beam, spectroscopic ellipsometery

Source: National Nanotechnology Infrastructure Network capabilities, www.nnin.org

---

BOX 1.1 NATURE AT NANOSCALE

Blue morpho butterfly

FIGURE 1.1
The iridescent colors of the blue morpho butterfly's wings are produced by nanostructures that reflect different wavelengths of light. The wingspan of this butterfly is about 10–15 cm. (Courtesy of Wikimedia Commons, http://www.nisenet.org/viz_lab/image-collection)

(continued)

BOX 1.1 (continued)   NATURE AT NANOSCALE

Blue morpho butterfly wing ridges

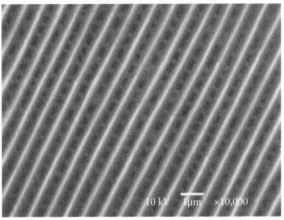

FIGURE 1.2

This scanning electron microscope (SEM) image shows ridges on a blue morpho butterfly wing scale. These ridges contain nanoscale structures that reflect light to create the morpho's iridescent colors. Each ridge is about 800 nm wide. (Courtesy of S. Yoshioka, Osaka University, Osaka, Japan, http://www.nisenet.org/viz_lab/image-collection)

Gecko foot

FIGURE 1.3

The feet of the gecko cling to virtually any surface. This image shows the sole of a gecko's foot. The adhesive lamellae on the sole have millions of branching hairs that nestle into nanoscale niches on the contact surface. (Courtesy of A. Dhinojwala, University of Akron, Akron, OH, http://www.nisenet.org/viz_lab/image-collection)

BOX 1.1 (continued)   NATURE AT NANOSCALE

Gecko foot (1600×)

FIGURE 1.4
The feet of the gecko cling to virtually any surface. This SEM image shows the branching hairs on the foot's adhesive lamellae. These hairs nestle into nanoscale niches on the contact surface. (Courtesy of C. Mathisen, FEI Company, Hillsboro, OR, http://www.nisenet.org/viz_lab/image-collection)

Nasturtium leaf

FIGURE 1.5
The lotus effect describes water droplets rolling off leaf surfaces, removing dirt and contaminants. This phenomenon can also be seen in the more common nasturtium. These leaves are covered with wax nanocrystal bundles that trap air and force water to bead and roll off. (Courtesy of A. Otten and S. Herminghaus, Göttingen, Germany, http://www.nisenet.org/viz_lab/image-collection)

(continued)

BOX 1.1 (continued)   NATURE AT NANOSCALE

Nasturtium leaf (2500×)

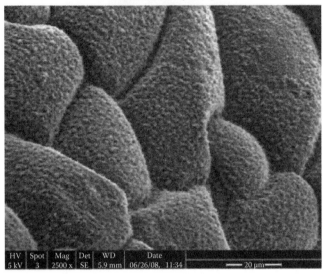

FIGURE 1.6
The lotus effect describes water droplets rolling off leaf surfaces, removing dirt and contaminants. This phenomenon can also be seen in the more common nasturtium. This image shows leaf sections covered with wax nanocrystal bundles that trap air and force water to bead and roll off. (Courtesy of A. Marshall, Stanford University, Stanford, CA, http://www.nisenet.org/viz_lab/image-collection)

1.3 Characteristics of Bulk and Nano-Sized Materials

1.3.1 What Is the Difference between Bulk and Nano-Sized Properties of Materials?

When the dimensions of solids are lowered to nanometers, the properties of these solids change—properties like strength, melting temperature, color, electrical conductivity, thermal conductivity, reactivity, and magnetism. The magnitude of such change is size dependent, i.e., the dimensions in nanometers determine the amount of change. When the dimensions of a material are reduced from large macroscopic values (meters or centimeters) to very small sizes, the properties remain the same at first, and then small changes start to take place until finally when the dimensions fall below 100 nm, significant changes in properties occur. Reduction in one dimension to nano size, but keeping other two dimensions large, results in a structure known as *quantum well*. Reduction in two dimensions to nano size, while one remains large, results in a structure called *quantum wire*; and the reduction of all three dimensions to nano size results in the structure known as a *quantum dot* (QD). At the nanoscale, the number of surface atoms is greater than that of the bulk material. The number of atoms in the nanocluster (surface-to-volume ratio) determines the change in the electronic, structural, optical, and acoustic properties of the material. This surface-to-volume ratio relationship allows the possibility of using size in nanometer regime to design and engineer materials with new and possibly technologically interesting properties (Poole and Owens 2006). Also, between the gas and condense phase of liquids and solids lies an intermediate state called clusters; these small agglomerations of atoms and molecules have properties not found in gases or condensed matter. This relatively new field of cluster science has the potential to make important advances in areas as far-reaching as material science, environmental chemistry, catalysis, and biochemistry (MRI 2007).

1.4 Applications of Nanotechnology

1.4.1 What Are the Applications of Nanotechnology?

At the nanoscale, materials exhibit novel electronic, optical, and magnetic properties, which have led to new applications of nanotechnology. Advances in nanoscience have enabled researchers to manipulate the behavior of a "single cell," reverse disease, and repair and grow human tissues. Nanotechnology is supplying improved services in the areas of energy, lighting, computing, printing, and water filtration. Nanotechnology innovations, such as QDs, semiconductor nanoparticles, carbon nanotubes, and nanoshells (see Table 1.4), have led to the fabrication of electronics hardware devices using the "bottom-up" approach in contrast to present "top-down" approach.

Nanotechnology has numerous other applications. In concrete manufacturing, the introduction of nanotubes increases the strength of cement. In plastics and polymers manufacturing, the use of nanoparticles allows precise control over color change and increases the strength of the materials. Nanocomposite materials are being used to manufacture various auto parts including body panels, instrument panel, bumper fascia, front-end module bolster, etc. In the medical field, nanotechnology has been used to study cancer cell structure, and nanoparticles have been used for the florescent imaging of tumors. Fluorescent nanoparticles have been employed for the diagnosis and treatment of cancer. Nanoparticles of gold and palladium have been used to remove toxic chemicals from water. Nanoparticles have been

TABLE 1.4

Potential Applications of Nanotechnology Innovations

Nanotechnology Example	Potential Applications
Buckyball: A soccer-ball-shaped molecule made of 60 carbon atoms	Composite reinforcement Drug delivery
Carbon nanotube: A sheet of graphite rolled into a tube	Fuel cells High-resolution displays Composite reinforcement
QD: A semiconductor nanocrystal whose electrons show discrete energy levels like an atom	Energy-efficient light bulbs Medical imaging
Nanoshell: A nanoparticle composed of silica surrounded by a gold coating	Medical imaging Cancer therapy

used as catalysts, cancer therapies, sunscreens, antibacterial composites, solar cells, and, more recently, advances in microfluidics and biosensors have made it possible to fabricate lab-on-a-chip devices.

Nanotechnology has enabled manufacturing processes to yield smaller, faster, and more energy-efficient electronic, photonic, and optoelectronic devices. The first generation of nanotechnology (1990s–early 2000s) dealt with improving the performance characteristics of existing micro materials; the second generation of nanomaterials (2006 onward) is leading to the fabrication of devices that are cleaner, stronger, lighter, and more precise (Minoli 2006).

Researchers at Penn State and the University of Southampton (United Kingdom) have created a breakthrough in optical electronics by developing a technique to fill an optical fiber with micro- and nano-semiconductor structures. This structure enables the optical fiber to carry multiple wavelengths of light, and thus increases the data transmission capability tremendously.

New developments in nanotechnology sensors have generated new interest in microelectromechanical (MEMS)-based systems. Such devices have applications in communications, medical diagnosis, commerce, the military, aerospace, and satellite systems. The extensive array of fabrication infrastructure developed for the manufacture of silicon integrated circuits has contributed to the development of systems having components of micrometer dimensions. Thin film deposition processes coupled with advanced lithographic techniques are used to make MEMS devices. The major advantages of MEMS devices are miniaturization, multiplicity, and the ability to directly integrate the devices into microelectronics (Poole and Owens 2003). The integration of MEMS and nanotechnology not only improves overall device performance, such as reliability, but also reduces weight, size, power consumption, and production costs. Hybrid systems using MEMS and nanotechnology allow the design and development of novel electro-optic sensors, lasers, and RF/millimeter-wave components such as phase shifters, switches, tunable filters, micromechanical

(MM) resonators, acoustic and infrared (IR) sensors, photonic devices, accelerometers, gyros, automobile-based control and safety devices, and unmanned aerial vehicles (UAVs) for battlefield applications (Jha 2008).

Following the recent discovery of the antimicrobial properties of silver metal, silver structure and particles are being used in the manufacture of various products. When silver is transformed into very small particles, it exhibits antimicrobial properties that are not present in the bulk material. This antimicrobial property of silver has been incorporated in various products. For example, a company manufactures socks with silver threads woven in with cotton and other synthetic fibers to prevent odor.

Historically, in the Indian subcontinent during the Mughal empire and in the princely estates, it was a common practice in the households to use metallic silver bowls for drinking water, silver rings were used to treat various ailments, silver leafs were used to decorate desserts, and silver particles were mixed in various medicines.

The use of QDs has revolutionized the optoelectronics area. QDs offer superior optical properties, high quantum efficiency (~95%), and size-tunable emission. The QDs fabricated through the colloidal synthesis of semiconductors have many applications such as optical sources and in flat-panel displays (Zhanao et al. 2008).

Another major example of the application of nanotechnology is food science. Advances in nanotechnology have enabled food science to improve food safety and quality, ingredient technologies, processing, and packaging. It is estimated that nanofood market will be worth $20.4 billion by 2010. Table 1.5 presents a summary of nanotechnology research activities and applications in various food science areas (Floros 2008).

TABLE 1.5

Nanotechnology Research and Applications in Food Science Areas

Food Science Area	Nanotechnology Research/Application
Food safety and quality	Development of sensors to detect a single molecule
	Research focus is to improve the detection capability of bimolecular by exploiting nanomaterials such as carbon nanotubes, silicon wires, and zinc oxide nanorods
	Biosensors are being developed that can detect changes in the environmental conditions (temperature, oxygen, chemical contaminants, and microbial contaminants)
	Biosensors would allow detection of microorganisms and toxins
Ingredient technology and systems	Nanoparticle utilization for flavor, antioxidants, antimicrobials, bioactives, etc.
Food processing	Development of new membrane separation systems and catalysts
Food packaging	Development of low permeability, high-strength plastics
	High-performance edible packaging

BOX 1.2   EXAMPLES OF NANOTECHNOLOGY
(FIGURES 1.8 THROUGH 1.14)

Quantum corral

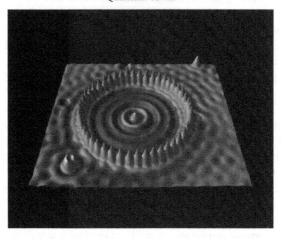

FIGURE 1.7
The corral is an artificial structure created from 48 iron atoms (the sharp peaks) on a copper surface. The wave patterns in this scanning tunneling microscope image are formed by copper electrons confined by the iron atoms. The radius of the corral is about 7 nm. (Courtesy of D. Eigler, IBM Almaden Research Center, San Jose, CA, http://www.nisenet.org/viz_lab/image-collection)

Nickel nanowires

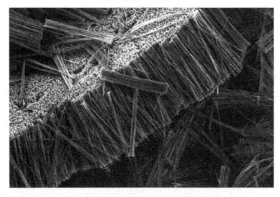

FIGURE 1.8
The orientation of the nickel nanowires shown in this SEM can be changed by altering the direction of an applied magnetic field. The nanowires are 100–200 nm in diameter and about 20 μm in length. (Courtesy of W. Crone, University of Wisconsin-Madison, Madison, WI, http://www.nisenet.org/viz_lab/image-collection)

BOX 1.2 (continued)   EXAMPLES OF NANOTECHNOLOGY

ZnO nanowires

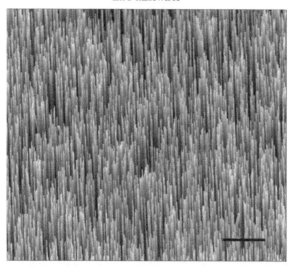

FIGURE 1.9
Vertical arrays of zinc oxide (ZnO) nanowires on a sapphire substrate. The sample displayed in the image is about 10 μm wide. (Courtesy of S. Dayeh, University of California at San Diego, San Diego, CA, http://www.nisenet.org/viz_lab/image-collection)

A nanowire resting on a human hair

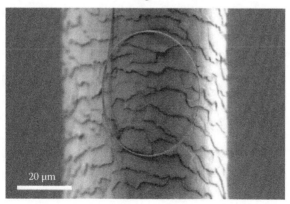

FIGURE 1.10
This SEM image compares the relative thickness of a typical nanowire in comparison with a strand of human hair. The hair is about 100 μm in diameter; the nanowire's diameter is about 100 nm. (Courtesy of http://www.nisenet.org/viz_lab/image-collection)

(continued)

BOX 1.2 (continued)   EXAMPLES OF NANOTECHNOLOGY

Nanotube yarn

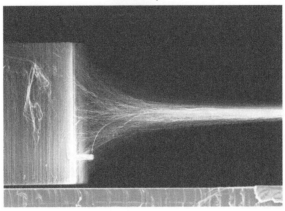

FIGURE 1.11
This SEM image shows nanotube yarn fibers drawn from a "nanotube forest." The yarn's diameter is about 1 μm. The nanotubes from which it is being drawn are each about 10 nm in diameter. (Courtesy of M. Zhang, UTD, Dallas, TX, http://www.nisenet. org/viz_lab/image-collection)

Silicon nanowire

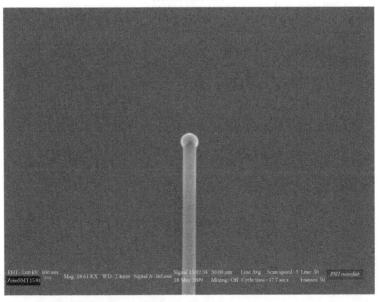

FIGURE 1.12
Silicon nanowire. (Courtesy of Fonash Research Group, Penn State University, Philadelphia, PA.)

BOX 1.2 (continued) EXAMPLES OF NANOTECHNOLOGY

ZnO nanowire

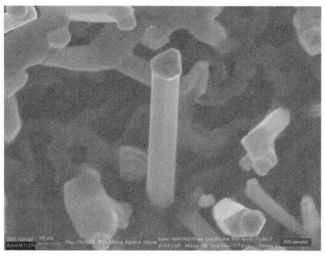

FIGURE 1.13
ZnO nanowire. (Courtesy of Fonash Research Group, Penn State University, Philadelphia, PA.)

FIGURE 1.14
ZnO nanowire growth. (Courtesy of Fonash Research Group, Penn State University, Philadelphia, PA.)

1.5 Nanotechnology and Education and Workforce of the Future

1.5.1 What Would Be the Impact of Nanotechnology on Education and Workforce of the Future?

It is estimated that by 2015, nanotechnology will be a $3 trillion-a-year global industry. In 1997, the investment in nanotechnology stood at $430 million, rising to more than $9 billion in 2004. Presently more than 800 products have been developed using nanotechnology.

The development of nanotechnology requires multidisciplinary teams of highly trained researchers with backgrounds in biology, medicine, mathematics, physics, chemistry, material science, electrical engineering, and mechanical engineering. For innovative advances in nanotechnology, the key is researchers with expertise in the multiple subsets of these disciplines since so many implications and fields are linked to a nano-"micro-revolution" (Khan 2006).

Education and training in nanotechnology require special laboratory facilities that can be quite expensive (see Table 1.3). The cost of creating and maintaining nanotechnology facilities is a major challenge for educational institutions. But by using innovative approaches such as interuniversity collaboration, academia–industry partnerships, and Web-based remote access to nanofabrication facilities, educational institutions can overcome the cost-related challenges and thus help students and faculty to become innovative nanotechnology researchers (Khan 2006).

To address these new demands of the global marketplace, a skilled workforce is required that can move from industry to industry without retraining. The new workforce will consist of researchers, technicians, and educators. To develop this workforce, new interdisciplinary educational programs need to be developed and revised. To promote the conditions for the development of nanotechnology workforce in the United States, the NNI program was established in 2001. The NNI (www.nni.org) provides a vision of the long-term opportunities and benefits of nanotechnology. The goals of NNI are as follows:

- Advance a world-class nanotechnology research and development program.
- Foster the transfer of new technologies into products for commercial and public benefit.
- Develop and sustain educational resources, a skilled workforce, and the supporting infrastructure and tools to advance nanotechnology.
- Support the responsible development of nanotechnology.

1.6 Social Implications of Nanotechnology

1.6.1 What Are the Social Implications of Nanotechnology?

The projected impact of nanotechnology has been touted as a second industrial revolution—not the third, fourth, or fifth, because despite similar predictions for technologies such as computers and robotics, nothing has yet eclipsed the first (Hall 2005).

Society is at the threshold of a revolution that will transform the ways in which materials and products are created. How will this revolution develop? The opportunities that will develop in the future will depend significantly upon the ways in which a number of challenges are met. As we design systems on a nanoscale, we develop the capability to redesign the structure of all materials—natural and synthetic—along with rethinking the new possibilities of the reconstruction of any and all materials. Such a change in our design power represents tremendous social and ethical questions. In order to enable our future leadership to make decisions for sustainable ethical, economic nanotechnological development, it is imperative that we educate all nanotechnology stakeholders about the short-term and long-term benefits, limitations, and risks of nanotechnology. The social implications of nanotechnology encompass many fundamental areas, such as ethics, privacy, environment, and security (www.epic.org).

Nanotechnology, like its predecessor technologies, will have an impact on all areas. For example, in health care it is very likely that nanotechnology in the area of medicine will include automated diagnosis. This in turn will translate into fewer patients requiring physical evaluation, less time needed to make a diagnosis, less human error, and a wider access to health-care facilities. However, with nanomedicines if the average life span of humans increases, it will create a large portion of elderly persons requiring medical attention, resulting in increased health expenditures (Moore 2004).

It is essential for the nanotechnology stakeholders to strive to achieve four social objectives: (1) developing a strong understanding of local and global forces and issues that affect people and societies, (2) guiding local–global societies to the appropriate uses of technology, (3) alerting societies to technological risks and failures, and (4) developing informed and ethical personal decision making and leadership to solve problems in a technological world (Khan 2006).

Advances in nanotechnology also present numerous challenges and risks in health and environmental areas. Nanotechnology risk assessment methods and protocols need to be developed and implemented by the regulatory bodies. Eric Drexler, author of *Engines of Creation*, has identified four challenges in dealing with the development, impact, and effects of nanotechnology on society (Drexler 1989):

1. The challenge of technological development (control over the structure of matter)
2. The challenge of technological foresight (the sense of the lower bounds of the future possibilities)
3. The challenge of credibility and understanding (a clearer understanding of these technological possibilities)
4. The challenge of formulating public policy (formulating polices based on understanding)

Table 1.6 presents a summary of the present and future impacts of nanotechnology. Table 1.7 lists nanotechnology application areas and potential benefits and risks.

TABLE 1.6

Present and Future Impact of Nanotechnology Area

Nanotechnology Area	Present Impact	Future Impact
Dispersions and coatings	Thermal barriers Optical barriers Image enhancement Ink-jet materials Coated abrasive slurries Information-recording layers	Target drug delivery/gene therapy multifunctional nanocoatings
High surface area materials	Molecular sieves Drug delivery Tailored catalysts Absorption/desorption materials	Molecule-specific sensors Large hydrocarbon or bacterial filters Energy storage Gratzel-type solar cells
Consolidated materials	Low-loss soft magnetic materials	Superplastic forming of ceramics
	High hardness, tough WC/Co- cutting tools	Ultrahigh strength, tough structural materials
	Nanocomposite cements	Magnetic refringements
		Nano-filled polymer composites
		Ductile cements
Nanodevices	GMR read heads	Terabit memory and microprocessing
		Single-molecule DNA sizing and sequencing
		Biomedical sensors
		Low noise, low threshold lasers
		Nanotubes for high brightness displays
Biological	Biocatalysis	Bioelectronics
		Bioinspired prostheses
		Single-molecule-sensitive biosensors
		Designer molecules

Source: Siegel, R.W., Technological impact: Present and potential.

TABLE 1.7

Nanotechnology: Benefits and Risks

Nanotechnology Application Area	Potential Benefits
Ecology	Nanoparticles have extremely high surface areas compared to their volume; this characteristic makes them ideal for the fabrication of
	New catalysts
	Heat reflection layers
	Aerogels for transparent damping layers in solar architectures
	Super thermal insulators
	Transparent layers showing resistance against wear and abrasion or anti-damping properties
Energy	Nanodevices will allow cleaner energy production and improved storage
	Small, compressed particles enable new photovoltaic cells, with simpler structure than conventional ones
	Plastics to be used as the electrode materials
Dematerialization	Nanocrystalline particles, with a monodisperse size distribution, to be formed into macroscopic parts with higher strength and resistance against mechanical and thermal load, despite the smaller amounts of material required. These parts can be hard and flexible in a unit and can replace scarce materials
	New processing techniques using remarkably lower temperatures offer possibilities for minimizing energy consumption during component fabrication
Health	More effective pharmaceuticals with reduced secondary effects due to improved basic understanding of the efficacy of natural human substances like insulin or hormones
	New form of localized drug delivery systems based on the potential of water-soluble, pharmacologically active substances when attached to nanometer-size particles
	External control and incorporation of target information by incorporating magnetic particles or antibodies into the drug delivery system
Electronics/telecommunications	Logical building blocks, for digital electronics, based on particles or molecules
	Nano-sized electronic data storage and processing systems

(continued)

TABLE 1.7 (continued)

Nanotechnology: Benefits and Risks

Nanotechnology Application Area	Potential Benefits
	Potential risks/detrimental effects
Genetics/medicine/health care	Artifacts based on nanotechnology incorporate genetic material or have genetic modification or repair as an objective
	If the artifact incorporates some kind of computing and sensing element, say for the controlled delivery of a drug, additional risks arise for the patient if these elements should malfunction
	Invasion of privacy and of human body through the planting and implanting of computing-cum-communication devices without the knowledge of those affected knowing this has been done
	Security and safety of the person, since it will be very difficult initially to detect the presence of nano-sized artifacts that are capable of breaching security and harming the individual
	In warfare, controlled distribution of biological and nerve agents may become feasible
Materials/composites	The general problem with composite materials is that they are more difficult to recycle and consume more energy during recycling than pure materials
	Wide-scale introduction of composite materials can increase environmental problems
Self-assembling and self-replicating nature of nanotechnological processes	In manufacturing area, many processes will need to be redesigned embodying new principles, particularly relating to containment of active or waste products

Source: Meyer, M., Socio-economic research and nanoscale science and technology, societal implications of nanoscience and nanotechnology, National Science Foundation, pp. 224–225, 2001, Available online: http://www.wtec.org/loyola/nano/NSET.Societal. Implications/

1.7 Conclusion

This chapter provided an overview of the state of the art in nanotechnology by exploring the topics of evolution, characteristics, equipment, and processes. It also discussed the applications, societal impact and ethical implications, benefits, and risks of nanotechnology. New demands on the curricula and workforce of the future were also explored. It is imperative for all nanotechnology stakeholders to guide society to the appropriate uses of nanotechnology, alert society to the risks and potential failures, and provide a vision in helping to solve societal problems that are related to nanotechnology. As the convergence of multiple disciplines occurs in the form of nanotechnology discoveries and developments, the time to remake the world will become shorter, and thus the nanotechnology stakeholders will be relied upon to guide society by making future decisions for the humane, just, and responsible utilization of this "invisible" technology.

References

Barber, D. J. and Freestone, I. C. 1990. An investigation of the origin of the color of the Lycurgus Cup by analytical transmission electron microscopy, *Archaeometry*, 32(1), 33–45.

Drexler, E. 1989. The challenge of nanotechnology, viewed April 28, 2009. http://www.halcyon.com/nanojbl/NanoConProc/nanocon1.html#anchor528648

Ehrmann, R. 2008. Hands-on nanofabrication workshop for educators, Center for Nanotechnology Education and Utilization, Penn State University, Philadelphia, PA, December 2–4.

Floros, J. 2008. Nanoscale science & technology for food. Hands-on nanofabrication workshop for educators, Center for Nanotechnology Education and Utilization, Penn State University, Philadelphia, PA, December 2–4.

Fonash, S. 2008. The overall picture: The world of nanotechnology. Hands-on nanofabrication workshop for educators, Center for Nanotechnology Education and Utilization, Penn State University, Philadelphia, PA, Tuesday, December 2–4.

Hall, S. 2005. *Nanofuture: What's Next for Nanotechnology*, Vol. 9. Amherst, New York: Prometheus Book.

Hjorth, L. et al. 2008. *Technology and Society: Issues for the 21st Century and Beyond*. Upper Saddle River, NJ: Pearson.

Jha, A. R. 2008. *MEMS and Nanotechnology-Based Sensors and Devices for Communications, Medical and Aerospace Applications*. New York: CRC Press.

Khan, A. S. 2006. Examining the impact of nanotechnologies for Science, Technology and Society (STS) students. *2006 ASEE Conference Proceedings*, Chicago, IL.

Meyer, M., Socio-economic research and nanoscale science and technology, societal implications of nanoscience and nanotechnology, National Science Foundation, pp. 224–225, Available online: http://www.wtec.org/loyola/nano/NSET.Societal. Implications/

Minoli, D. 2006. *Nanotechnology Applications to Telecommunications and Networking*, Vol. 2. Hoboken, NJ: John Wiley & Sons.

Moore, F. 2004. Implications of nanotechnology applications: Using genetics as a lesson. *Health Law Review*, 10(3), 9–15.

MRI, Focus on Materials. Summer 2007. Building nanomaterials one super atom at a time, Penn State, p. 10, http://www.nano.gov/html/about/home_about.html

National Nanotechnology Infrastructure Network. www.nnin.org

National Nanotechnology Infrastructure Network Capabilities. www.nnin.org

National Science Foundation. The societal implications of nanoscience and nanotechnology, NSET Workshop Report, Arlington, VA, Available online: http://www.wtec.org/loyola/nano/

Poole, C. and Owens, F. 2003. *Introduction to Nanotechnology*, pp. 1–7. Hoboken, NJ: John Wiley & Sons.

Privacy implications of nanotechnology, Electronic Privacy Information Center, viewed April 28, 2009. http://www.epic.org/privacy/nano/

Roco, M. C. and Bainbridge, W. (Eds.) 2001. The societal implications of nanoscience and nanotechnology, NSET Workshop Report. National Science Foundation, Arlington, VA, Available online: http://www.wtec.org/loyola/nano/societalimpact/ nanosi.pdf

Siegel, R. W. Technological impact: Present and potential.

Zhanao et al. 2008. Developing bright and color saturated quantum dot light emitting diodes—Towards next generation displays and solid state lighting. Hands-on nanofabrication workshop for educators, Center for Nanotechnology Education and Utilization, Penn State University, Philadelphia, PA, Tuesday, December 2–4.

2

Impact of Nanotechnology on Telecommunications

Fidel M. Salinas, Emanuel F. Barros, William Eisenberg,
Denise M. Smith, and Shekar Viswanathan

CONTENTS

2.1 Introduction

For over 100 years, telecommunications has brought people the ability to communicate remotely thanks to early pioneers like Alexander Graham Bell, Werner Siemens, Thomas Edison, Nikola Tesla, and many others. Today, there are hundreds of telecommunications companies throughout the globe.

However, with all this technology, it is hard to believe that in this modern age of information, with technology advancing inexorably and, seemingly, by the hour, roughly half of the planet's inhabitants have never made or received a phone call, according to the International Telecommunications Union in the United Nations. As Louise Frechette, Deputy Secretary-General of the United Nations, has pointed out, "Half of the people in the world have never made a phone call; half of the world's population doesn't even live within 3 miles of a telephone! The technology exists to provide virtually everyone on this planet with access to telecommunications. Yet only 40%

of the world's households have a telephone line and over half of the world's population has never made or received a telephone call" (Frechette 1999). It is hoped that nanotechnology will enable telecommunications to provide communications to more of the world's population, especially with the expanding capacity of wireless communications.

What is nanotechnology? Nanotechnology is the understanding and control of matter at dimensions between approximately 1–100 nm, where unique phenomena enable novel applications. This provides the opportunity to create materials, devices, and systems with fundamentally new properties and functions because of their small structure (http://nano.gov).

The National Nanotechnology Initiative (NNI) was established in the fiscal year 2001 to coordinate Federal nanotechnology research and development. The NNI Office of Science and Technology Policy (OSTP) was established by the National Science and Technology Policy, Organization, and Priorities Act of 1976. The NNI supplement to the President's FY 2010 Budget provides for an investment in nanotechnology research and development of $1.6 billion.

See Figure 2.1 for a comparison of the scale of "things natural" and "things man-made."

2.2 Dimensions: A Snapshot

A starting point for putting nanotechnology in perspective must begin with its dimensions, both negative and positive. The term "nano" comes from the Greek word νάνος which is pronounced nanos meaning dwarf. The nano dimension, 10^{-9}, is a nanometer, a value of 10 to the minus 9. To put this into perspective, our fingernails grow roughly at the rate of 1 nm/s. That's about a layer of atoms. As a further comparison, a gene is made up of a few atoms. However, this still might not put the dimension in perspective.

In order to understand a gene at the nanolevel, it would be useful to consider a more familiar positive dimension, namely, imagining two humans, the size of a gene in a human blood cell with the island of Manhattan roughly, being the size of the blood cell. The contrast between 10^{-9} and what we see with technology at this very small dimension certainly demonstrates a dramatic difference. Figure 2.2, the powers of 10, shows the negative and positive ranges of these dimensions.

The possibility of "seeing" and "moving" an atom seemed remote for most of human history until IBM Fellow Donald M. Eigler, PhD, using the scanning tunneling microscope (STM), which he designed and built, demonstrated the ability to manipulate individual atoms with atomic-scale precision. On September 29, 1989, IBM Fellow, Dr. Don Eigler, was the first to controllably manipulate individual atoms on a surface, using the STM to spell out "I-B-M" by positioning 35 xenon atoms, and in the process, perhaps

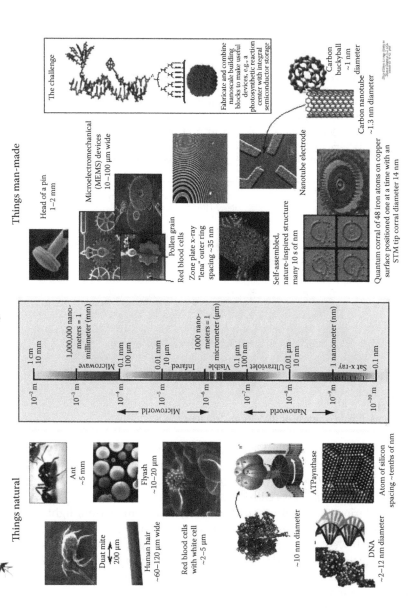

FIGURE 2.1

Scale of things natural and things man-made. (Courtesy of Office of Basic Energy Sciences, Office of Basic Energy Sciences, U.S. Department of Energy, Washington, DC.)

LearningSolutions.3rdPlanet:[TM] Dimensions of technology	

Powers of ten	Negative powers of ten
10^3 → kilobyte	10^{-3} → millitechnology
10^6 → megabyte	10^{-6} → microtechnology
10^9 → gigabyte	10^{-9} → nanotechnology
10^{12} → terabyte	10^{-12} → picotechnology
10^{15} → petabyte	10^{-15} → femtotechnology
10^{18} → exabyte	10^{-18} → attotechnology
10^{21} → zettabyte	10^{-21} → zeptotechnology
10^{24} → yottabyte	10^{-24} → yoctotechnology
10^{27} → xerabyte	10^{-27} → xeratechnology
10^{100} → googolbyte	10^{-100} → googoltechnology
10^n	10^{-n}

LearningSolutions.3rdPlanet Copyright © 2000–2009
Research and design by:
Dr. Fidel M. Salinas and Dr. Denise M. Smith

FIGURE 2.2
Dimensions of technology. (Courtesy of LearningSolutions.3rdPlanet.)

FIGURE 2.3
Xenon atoms. (Photo courtesy of IBM, Corp., San Jose, CA.)

created the world's smallest corporate logo (Figure 2.3). This event was likened to the Wright brothers' first flight at Kitty Hawk by Jack Uldrich in his book, *The Next Big Thing is Really Small: How Nanotechnology Will Change the Future of Your Business* (Uldrich 2003).

In considering the positive powers of 10, as a comparison, light velocity of 186,320 mps will go around the Earth at the equator about 7.5 times in one full second. If we capture a nanosecond of time, light travels approximately 11.75 in. On the other hand, the speed of sound in dry air at 68°F (20°C), at approximately 768 mph or a fifth of a mile per second, would take roughly 32 h to go around the Earth at the equator.

In general, the speed of sound c is given by

$$c = \sqrt{\frac{C}{\rho}}$$

where

C is a coefficient of stiffness (or the modulus of bulk elasticity for gas mediums)

ρ is the density

In continuing with the negative powers of 10, at 10^{-12}, in a picosecond, light travels about the length of a pepper grain and, at 10^{-15}, light travels about the distance across the diameter of a proton or neutron. To continue in the negative direction any further, we would have to follow the progress of the large hadron collider (LHC), the world's largest and highest energy particle accelerator, where it fills a circular tunnel 17 miles in circumference, and spans the border between Switzerland and France about 100 m underground.

On the positive powers of 10, the difference between a million (10^6) and a billion (10^9) can perhaps be shown with an analogy using money. For example, if one were to win the million dollar lottery and received the money at the rate of $1000 per day, it would take just under 3 years to collect the million dollars. If one were to win the billion dollar lottery, collecting the billion dollars at the same rate of $1000 per day, would take approximately 3000 years to collect the billion dollars. Following the same analogy, adding three more zeros, collecting the trillion (10^{12}) dollar lottery at the same rate, would take about 3 million years.

Parallel processing is a familiar axiom for computers: the computer is incredibly fast; however, it can only do one thing at a time. It is similar to a combine harvesting let us say, a 5000 acre wheat field. A single combine might take several days. However, if several combines were used in a row, the time taken to harvest the 5000 acre wheat field would perhaps be done in 1/20th or 1/40th the time. While computers are limited to the speed of light, a combine operates at a certain and constant speed.

A further example of parallel processing is the up and coming quantum computing wherein the same bits, called Q-bits in quantum computing, are processing multiple states at a time. This increasing potential computing power becomes exponential as you add Q-bits to the quantum processor. Some day, quantum computers will be able to factor any number to the nth degree and calculate solutions for intractable problems. The quantum processor would only require more Q-bits for larger calculations. Nanotechnology will provide the stable substrates to make quantum computing viable at substantially larger scales than today in the labs.

Of significant importance in accelerating applications for telecommunications through nanotechnology is the ongoing effort in developing faster supercomputers. IBM first broke the Petaflop barrier in May 2008 and that quest has continued with IBM's Research and Development work with universities. One such effort is its association with a Dublin lab in partnership with the Industrial Development Agency of Ireland for pursuing the Exaflop (10^{18}) barrier (Merritt and Rick 2009).

Similarly, a computer or processor is limited to a fixed amount of work while a bank of thousands of computers or (processors) would do the work of a single processor millions or billions of times faster than it would take a single processor to perform the same work, thus multiplying the work or processing throughput.

2.3 Global Standards

Global standards emerged from a multitude of government and proprietary standards to establish and unify international standards and protocols worldwide, permitting us to communicate across different architectures, computers, networks, and time zones. A sample of the major standards and protocols and organizations include the following:

- Internet protocol suite also known as transmission control protocol/Internet protocol (TCP/IP) has become the de facto international standard for connecting different vendors in a network. In the early 1970s, TCP/IP protocols were developed by Defense Advanced Research Projects Agency (DARPA), which later became ARPANET. This research project was to connect a number of different networks designed by different vendors into a network of networks (that eventually became the "Internet"). It was initially successful because it delivered a few basic services that everyone needs (file transfer, electronic mail, remote logon) across a very large number of client and server systems.

- Open systems integration (OSI), which began in 1977, was started by the International Organization for Standardization (ISO), a part of ITU, to coordinate the process of standardization of protocols for multi-vendor interoperability. However, the TCP/IP eclipsed OSI from being implemented worldwide.

- Computer teléphony integration (CTI), also called computer–telephone integration, is the technology that allows interactions coordinated with a telephone and a computer. As contact channels have expanded from voice to include e-mail, Web, and fax, the definition of CTI has expanded to include the integration of all customer contact channels (voice, e-mail, Web, fax, etc.) with computer systems. Section 2.5 gives an example of a customer merchant transaction using CTI and integrated voice and data technologies.

- Computer-supported telecommunications applications (CSTA) is an abstraction layer for telecommunications applications. It is independent of underlying protocols. It has a telephone device model that enables CTI applications to work with a wide range of telephone devices. Originally developed in 1992, it has continued to be developed and refined over the years. It is often the model that most CTI applications are built on and claim compliance with. It became an OSI standard in July 2000. It is currently being maintained by ECMA International. CSTA standard saw the introduction of CSTA, CSTA XML, and CSTA Object Model extensions. These extensions are in various states of completion but all extend the scope of CSTA.

- ECMA is the European Computer Manufacturers Association. ECMA International is an international, private (membership-based) nonprofit standards organization for information and communication systems. It acquired its name in 1994, when the ECMA changed its name to reflect the organization's international reach. As a consequence, the name is no longer considered an acronym and no longer uses full capitalization.

- The telephony application programming interface (TAPI) is a Microsoft application program interface (API) that provides CTI and enables PCs running Microsoft Windows to use telephone services. Different versions of TAPI are available on different versions of Windows.

2.4 Impact and Promise of Nanotechnology for Telecommunications

The impact of nanotechnology on telecommunications is manifesting itself in new applications in business, industry, and government. Such applications take advantage of integrating voice and data technologies over the Internet now with voice over the Internet protocol (VoIP), and help to reduce time and cost while at the same time, expanding applications by geometric proportions.

The combined technologies of voice and data are impacting societies across the globe, perhaps diminishing or at least reducing cultural dominance by using the common language of commerce, English, as an emerging business language. This combination is made possible through the blending of two technologies: *voice* and *data*. The technology is called computer teléphony (pronounced with emphases on the é) integration (CTI).

The impact of nanotechnology on telecommunications is, in many ways, enabling communications across not only physical borders but also across the cultural curtains of long-established economic, political, cultural, and even religious barriers. The relentless move toward *smaller-faster-cheaper* advancements in technology, especially, nanotechnology, the industrial revolution of the twenty-first century is accelerating the rate of this dynamic change.

Ray Kurzweil, PhD, entrepreneur, scientist, and futurist, and author of *The Singularity Is Near: When Humans Transcend Biology*, predicts that hardware and software will achieve human-level artificial intelligence (AI), with the broad suppleness of human intelligence including our emotional intelligence, by 2029. Ray's works include music synthesis, speech and character recognition, reading technology, virtual reality (VR), and cybernetic art. All of these pioneering technologies continue today as market leaders. Ray was the principal developer of the first Omni-font optical character recognition technology, the first print-to-speech reading machine for the blind, the first CCD flatbed scanner, the first text-to-speech synthesizer, the first music synthesizer capable of recreating the grand piano and other orchestral instruments, and the first commercially marketed large-vocabulary speech recognition system.

In his first book, written in the late 1980s, he talks of machine intelligence becoming indistinguishable from that of its human progenitors within the first half of the twenty-first century. The chess matches between IBM's Deep Blue computer and world chess champion, Garry Kasparov, were inconclusive as to who really won the various chess games, as AI had developed to the level where the computer program or a human was the opponent.

Dr. Kurzweil cofounded Singularity University with Peter Diamandis, CEO of the X Prize Foundation. Singularity University kicked off its first class at NASA Ames, Mt. View, CA on Monday, June 29, 2009. Dr. Kurzweil shows that technological change is exponential, contrary to the common sense "intuitive linear" view. He points out that we will not experience 100 years of progress in the twenty-first century; we are more likely to experience 20,000 years of progress if we continue to move at the current rates. For example, the "returns," such as chip speed and cost-effectiveness have also increased exponentially. Figure 2.4 shows the accelerating returns as we approach (the singularity).

In the chess matches between world champion, Garry Kasparov and IBM's Deep Blue, several of the matches were a draw. It became indistinguishable whether Garry Kasparov was playing against a human or a machine. The Deep Blue system was capable of examining 200 million moves per second or 50 billion positions in the 3 min nominally allotted for a single move in a chess game. Deep Blue used more than brute computing force. It combined the power of its processors and a highly refined evaluation function that captured human grandmaster chess knowledge—including Kasparov's.

It is this kind of power that will advance telecommunications applications through miniaturization. One such application has been around for several years, namely, *load balancing*, which is the balancing of calls between sites based on performance criteria. Load balancing involves the ability of a telecommunications network to route calls to the available agent, regardless of

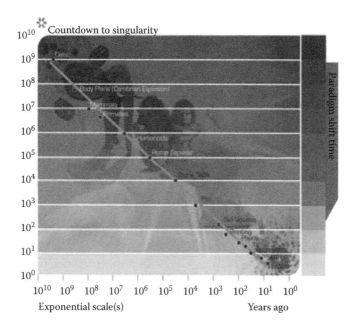

FIGURE 2.4
Exponential countdown to the singularity. (Courtesy of Dr. Ray Kurzweil.)

the physical location of the computers. In a Virtual Call Center, *load balancing* provides the ability to route to the available agent across an enterprise.

As the hardware is miniaturized further through nanotechnology advances, and takes less physical space, more automation will relieve the agent of repetitive tasks and ease communication throughout a particular transaction. See Figure 2.5, which shows the accelerated rate of the implementation of inventions.

Certainly, nanotechnology, with ongoing miniaturization at the atomic level, and ever-shrinking, lower cost creation of computer processors with greater power, is bound to enable telecommunications applications to a higher level of automation and, subsequently, become available to a greater population.

Technology, which is seemingly, changing "hourly" is driving this phenomenon and with this plethora of technologies, spawning applications also, by the hour, bringing with it the advantages, benefits, and, of course, the hazards and unintended consequences of new technologies. To paraphrase the late Professor, Dr. Melvin Kranzberg, Callaway professor of the History of Technology at Georgia Tech:

> Technology is neither good nor bad; nor is it neutral; it just is.

It is more than likely that the person who discovered fire (took a lot of heat) and criticism from his or her peers when he or she discovered fire. This is

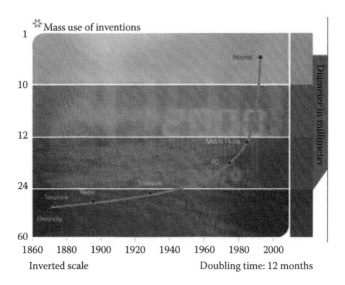

FIGURE 2.5
Accelerated use of inventions. (Courtesy of Dr. Ray Kurzweil.)

because they would have found out that though fire could keep one warm and could cook food, it could also burn down our houses, a negative side of this technology. It is this aspect that is discussed in Section 2.8. Technology has been part of being human ever since humans discovered fire or picked up a rock or stick to use as a tool.

Certainly, the global Internet benefits from developing technology and, in particular, nanotechnology, makes a smaller "footprint" of hardware installations possible and creates opportunities for new applications, which bring telecommunications to more of the planet's inhabitants. As hardware becomes smaller, and results in more affordable costs, it provides telecommunications applications to the have-nots of the planet, applications that have been enjoyed by the developed countries for past years. The cultural dominance of the "haves" will itself, level set the rest of the global population, so all can participate in the information highway in a true global economy.

Nanotechnology will further level the global playing field by localizing and massively reducing the costs of manufacturing anything. Communications will become the main conduit of civilization. Eventually there will only be in-home replication machines that will produce food, clothing, and all the other necessities and wants of humanity. Anything that cannot be replicated at home, say for instance a car, may be obtained from your local replication center. At this point, we will be a 100% digital economy, dealing only in information. Purchased products will pay royalties directly to the designers and innovators. The replication technology will use unprocessed raw materials that are relatively cheap right out of the ground.

The cost of the raw materials that make up the average car is about a couple hundred in today's dollars. This will dramatically reduce the costs of products by cutting out all the waste in production which will in turn, eliminate the existing messy processes of traditional manufacturing and the same technology will reverse all the environmental damage traditional manufacturing has caused. In addition, very little energy will be required for replication due to the elimination of all the steps in traditional manufacturing. Traditional manufacturing will go the way of the ugly cocoon that produces the beautiful butterfly. Nanotechnology and the communications infrastructure that will come with it will become the beautiful butterfly integrating all of humanity in a global village.

In today's world of technology, compared with technology of not that many years ago, machines rarely fail. A hard computer disk may have a mean-time-between-failure (MTBF) of tens, hundreds of millions of cycles, which can span several years. Reliability, availability, serviceability (RAS) have become less problematic. An additional benefit of improved reliability is lower maintenance service frequency and more predictable costs.

The ongoing miniaturization of disk storage continues to provide larger storage capacity with lower prices as technology advances into the twenty-first century. The impact on telecommunications in particular, expands the capacity of applications and brings telecommunications capability to more of the world.

Advancements in digital nanotechnology allow for more effective work-load balancing of call loads to be routed by additional computerized branch exchanges (CBXs). This means that telecommunications capability is made available to areas or countries to provide them with minimal and affordable telecommunications technology. While cell phones have become internationally ubiquitous, a telephone hard line remains a viable part of the telecommunications landscape.

The impact of nanotechnology on cell phone technology has made numerous applications possible and available for the first time. Its foremost benefit is that it has brought more telecommunications capability worldwide. In addition, an ever-increasing storage capability allows for pictures of higher resolution and full motion video with sound to be transmitted in real time rather than by data streaming. Nanotechnology advances enable higher storage capacity, coupled with faster transfer rates, and improve video and sound performance. This will further the integration of communication and expand virtual communities and their capabilities. VR technologies will become more prevalent in online communities like Second Life, where there are developing virtual economies. These communities will begin to thrive on a massive scale when the technology is cheaply available and is developed to the point of encompassing all our senses as in the movie *The Matrix*.

A major advantage of nanotechnology miniaturization is that it makes a smaller footprint possible. This provides additional storage space for remote shelf capability, so that equipment can not only be remote from the main branch exchange office, but also may be maintained in several locations and be managed by different central exchanges, also in different locations. Relational databases have benefited from this ever-shrinking technology, with the deployment of more computers or servers throughout the world.

In addition, with more interoperability implemented against the backdrop of international standards, local and remote equipment can be obtained from several telecommunications vendors, which helps to make the task of matching the organization's technical and budget requirements with available, off-the-shelf, commercial technology that adheres to established and committed enterprise architectures. In a major procurement, cost avoidance can be as an important factor as a projected actual cost savings in terms of current dollars compared to future dollars, which depreciate in time.

2.5 Transparent Transaction: A Scenario

In a transparent-to-the-client sales transaction, the West Coast agent takes over from the East Coast agent at quitting time; it is almost quitting time on the East Coast and, the West Coast is 3 h behind so the time zone is 3 h earlier. In a seamless transaction, as the East Coast agent prepares to go home, he or she transfers a voice call with the database linked to the voice transaction in a CTI application. This occurs because the data bits are also carried with the voice digital bits on the same transaction. This can be a voice conversation that is taking place using the technology (VoIP).

The ability to do this has certain benefits: The client experiences a personal attention transaction because his or her name, account reference number(s), address, billing/shipping information, product description, term and conditions of the transaction, potential merchandise return information, and, perhaps, other relevant information, is preserved in the data/voice transfer of the call.

The major advantage of this is that CTI technology has "preserved" the transaction information. Technology has helped to minimize cost and valuable time for both the merchant and the customer. This enables the merchant to complete the transaction, namely, to make the sale and help turn over the inventory. The customer is a happy camper because he or she has completed the purchase with one phone call and does not have to experience the frustration of making another call the next day, because it took place at an end of the day time zone. He or she would not have to repeat all of the same information in a new, repeated, and no doubt, frustrating call. As a result, the customer is unlikely to lose interest in the product or buy from another merchant be it online or onsite. So, this integration of voice and data technologies can indeed help avoid losing a sale, and can complete a transaction involving a completed phone call with a created, linked database and maybe retain and keep a customer happy to buy again.

2.6 Ongoing Research and Nanotechnology: Some Samples

The impact of nanotechnology on telecommunications opens up new and exciting worlds of applications. For example, industries such as Intel, IBM, Hewlett Packard, Siemens, and many other companies, organizations, from the public and private sector, are working on technology products that enable applications made possible by the use of nanotechnology. A sampling of emerging nanotechnology and applications for telecommunications includes the following:

- *Nanotechnology and storage at the atomic level*

  IBM is bringing single-atom data storage and molecular computers closer to reality, which can one day lead to new kinds of devices and structures built from a few atoms or molecules. The day may come when we can take a CBX in our bag, set up emergency telecommunications for 20,000 people with wireless transmission all in a matter of a few hours or minutes.

  Although still far from making their way into products, these breakthroughs will enable scientists at IBM and elsewhere to continue driving the field of nanotechnology, the exploration of building structures and devices out of ultratiny, atomic-scale components. Such devices might be used as future computer chips, storage devices, sensors, and for other applications nobody has imagined yet.

- *Nanophotonics*

  Nanophotonics has the potential to provide ultrasmall optoelectronic components, high speed and greater bandwidth. Current research into fabricating nanoelectronics could open the way for new methods of making nanophotonic devices, i.e., mass-producing light handling devices that may be only tens or hundreds of nanometers in size that will help drive this revolution. In molecular photonics, the aim is to achieve sufficient control over light–matter interactions through nanoscale physical, chemical, and structural modification interactions using a few or single molecules.

- *Self-assembly in nanotechnology*

  Future disk areal density is looking at potential to store the contents of 250 DVDs to fit onto a surface the size of a quarter. "I expect that the new method we developed will transform the microelectronic and storage industries, and open up vistas for entirely new applications," says Thomas Russell, director of the Materials Research Science and Engineering Center at UMass Amherst, visiting Miller Professor at UC Berkeley's Department of Chemistry, and one of the world's leading experts on the behavior of polymers. "This work could possibly be translated into the production of more energy-efficient photovoltaic cells."

2.7 The Promise and Future of Nanotechnology

As atomic miniaturization technology advances in the first decade of the twenty-first century, convergence subsequently accelerates to enable more applications. The power of nanotechnology is rooted in its potential to transform and revolutionize multiple technology and industry sectors, including aerospace, agriculture, biotechnology, homeland security and national defense, energy, environmental improvement, information technology, medicine, and

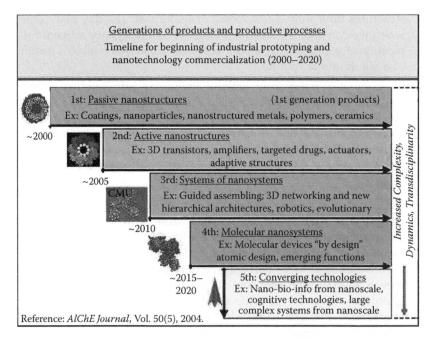

FIGURE 2.6
Generations of products and processes. (Courtesy of NSF, Arlington, VA; Roco, M.C., Nanoscale science and engineering education, Workshop "Partnership for Nanotechnology Education" Los Angeles, CA, April 26, 2009.)

transportation. Discovery in some of these areas has advanced to the point where it is now possible to identify applications that will impact the world we live in.

A major challenge and, at the same time, opportunity of characterizing nanotechnology, is that the field of nanotechnology crosses many boundaries of established academic disciplines. Any advances in nanotechnology for telecommunications have the potential to be advances for any other academic disciplines. Figure 2.6 shows the timeline for beginning of industrial prototyping and commercialization from 2000 through 2020.

2.8 Concerns about Nanotechnology

With the promise of new products, exciting new applications, and lower costs, nanotechnology also brings new hazards. Regulatory agencies do not yet exist to oversee the characterization of nanoproducts, or their action in a nanodimension. Neither have they thought of how to handle nanowaste. The city of Berkeley, California, amended its hazardous materials laws in 2006 to require nanotechnology researchers and manufacturers to report the nanomaterials with which they are working and how they are handling those materials, a move that is expected to draw international attention. As with all new technologies, we are treading in new territory.

Much research needs to be done to determine the hazards of nanotechnology. Existing regulations may not be effective in managing nanotechnology hazards. For example, nanoparticles already exist in consumer products such as sunscreens. However, due to manufacturer reticence and laissez-faire regulatory regimes, consumers often do not know what they are buying. Regulatory organizations will guide safety protocols with constant collaboration with industry and serve as a source for consumer confidence in the new technologies.

As with most technologies, there is often a propensity to reduce to practice, a destructive side of the technology. In *The Diamond Age*, sci-fi author Neal Stephenson posits a future in which street thugs use voice commands to fire nanoscale bullets from guns implanted in their skulls, while competing clans set loose clouds of deadly nanotech parasites, dubbed "nanosites," programmed to destroy anybody with the wrong DNA who happens to inhale the little beasties. Certainly, it is not too soon to start thinking through the implications of a world in which atoms and molecules become building blocks for nearly anything imaginable.

Just as in international telecommunications standards, international standards and regulations for the production, identification, transportation, distribution, and consumption of nanotechnology are also necessary. Whether such standards follow the U.S. and Asian models that favor a lighter regulation in the absence of proven danger or the more precautionary style of European Union regulation that tends to establish a product's safety before it is commercialized, there is potential for much debate.

2.9 Preparing Students for Nanotechnology

Research tells us that college students have chosen to major in math and science long before taking SATs and completing college admission essays. There is a strong correlation between math and science instruction during the school years: preschool through high school and the later career selection. Yes, teacher influence begins in preschool! For too long, math and science instruction emphasized the product, not the process. Detail about the process of solving math problems or scientific investigation was absent from instruction. In more recent years, scores on high stakes testing and other artificial measures of student learning and school accountability have significantly hampered both teacher creativity and student learning, especially with instructional content in math and science. Recently it has been found that the first 3 years of a teacher's tenure determines how effective the teacher is in teaching a particular subject. After all, not everyone is cut out to be a teacher. Increasing teacher pay and incentives to attract people who become these great teachers is imperative to bettering the educational system in general.

Young children are natural mathematicians; seeking order in their world through sorting objects by size or shape, to dividing boxes of crayons so everyone has an equal variety of colors. Children start learning about fractions when they cut their first birthday cake; they start learning about powers of 10 even faster when they receive their first money in a birthday card. As children mature, so does their curiosity. Instead of reading about magnets; children should be encouraged to complete experiments in small groups about magnetic fields or building a battery cell as these would not only teach them about the magnets or cells but also promote cooperative work that they could benefit from not only during their school years but also in their adult life. This can be an incentive and an early introduction to teamwork.

The foundation for building a "local" workforce strong in math and science starts in the young school years and continues across the life span. While teachers and school administration bear a great responsibility for school instruction, parents remain their child's first and most important teachers. Parents have a profound impact on every aspect of their child's life. Parents must take an active role and join teachers and administrators not only to create more interesting and innovating instructional methods in math and science but also to support their child's math and science learning at home.

2.10 Conclusion

Today, the advantage of emerging nanotechnologies will not only provide a smaller "footprint" on the computer or telecommunications floor, but also impact power consumption with an accompanying complimentary impact on the environment. As a comparison with past technologies, the computer storage systems in use today worldwide would most likely be unable to operate all at once using the vacuum tube technology of the past. In addition, the cooling requirements to maintain such systems would drain a tremendous amount of energy just to maintain a comfortable working environment.

The ongoing development and use of new technology in particular, nanotechnology, offers the promise of unlimited applications that can benefit society. Nanotechnology will no doubt encourage new applications in areas such as massive parallel computing and quantum computing, and facilitate the solving of complex problems and enable cryptography, simulation, and modeling, and environmental monitoring. The latter would enable meteorologists to make even more accurate weather predictions. Computers today still do not have the processing power, memory, or speed to process calculations that would facilitate the understanding of the fundamentals of matter; that effort is underway at the supercollider in France and Switzerland. Nanotechnology brings the promise to make possible quantum computing on a massive scale in this very area.

The technological singularity is what will be the ultimate culmination of rapidly advancing technology. Ray Kurzweil in his book, *The Singularity Is Near* identifies the trends of technology and where they will take us. The technological singularity will be when technology is advancing so rapidly that we can no longer make predictions about the next step without some sort of technological enhancement to allow us to understand and interact with the vast amount of information with which we will be bombarded with in real time. Some might say we have already started the integration of technology with the advent of smart phones like the iPhone and Blackberry.

People might wonder what our infrastructure will look like during this time in the future. As the technological singularity dictates, we do not really know where to begin, but what we can speculate on is that this infrastructure will have to integrate massive information bandwidth, massive parallel computing, localized in-home or on the fly manufacturing, and all of these systems will be out of sight, but not out of mind. So, no more ugly telephone poles, no more visible computer systems/servers to maintain, and no more polluting factories of any kind. The future is full automation of almost everything and 100% clean energy.

Imagine instead, an organically grown subterranean network, a biologically inspired root system that supplies unlimited real-time massive bandwidth, massive parallel computing that includes quantum computing, and a delivery system of raw feedstock materials for localized manufacturing using molecular nanotechnology. An infrastructure of that magnitude can grow organically and reach all its users, whether they are local or mobile. We can only get a glimpse of the infinite possibilities the technological singularity will bring us with today's rapidly advancing technology.

Appendix A: Nanotechnology Pioneers: A Snapshot

Francis Crick: (1917–2004) Nobel laureate, codiscovered DNA "double-helix" structure.

K. Eric Drexler: (1955–) K. Eric Drexler popularized the word molecular nanotechnology, had accusations of science fiction. [Engines of creation: The coming era of nanotechnology.]

Richard Feynman: (1918–1988) FeynmanOnline.com scientist, teacher, raconteur, and musician on. [There's plenty of room at the bottom: Feynman lectures on physics.]

Ray Kurzweil: (1948–) Inventor, futurist; speech recognition, synthesis, OCR, the law of accelerating returns.

Gordon Moore: (1929–) CEO Emeritus, Intel, Moore's law.

James Watson: (1928–) Nobel laureate, codiscovered DNA "double-helix structure with Francis Crick.

Richard Smalley: (1943–2005) Rice University professor, nobel laureate, codiscoverer of the "buckyballs/nanotubes" with Robert Curl and British Chemist Sir Harold Kroto.

Norio Taniguchi: (1912–1999) (谷口紀男) Tokyo University Science Professor, coined the term nanotechnology in 1974.

Nikola Tesla: (1856–1943) Wireless technology, Tesla coil, turbine, oscillator electric car, induction motor, rotating magnetic field, particle beam weapon and a most significant contribution, electricity.

References

Eigler, D. M. (Sept. 29, 1989), Used his scanning tunneling microscope (STM) to position individual atoms, IBM Corp. Almaden Research Laboratory, San Jose, CA. Retrieved July 14, 2009 from http://domino.research.ibm.com/comm/pr.nsf/pages/bio.eigler.html

European Organization for Nuclear Research (CERN, 2008). Retrieved July 14, 2009 from http://public.web.cern.ch/public/en/LHC/LHC-en.html

Frechette, L. (Oct. 25, 1999). Knowledge is the new global asset, the very premise of progress. ECA press release no. 90/1999. Retrieved July 14, 2009 from http://www.uneca.org/eca_resources/Press_Releases/1999_pressreleases/pressrelese9099.htm

Huawei India's R & D Center. (March 16, 2009). Technology News. Retrieved July 14, 2009 from http://www.siliconindia.com/shownews/Nanotubes_spun_into_threads_open_new_possibilities_in_communications-nid-53665.html

Kasparov, G. (1996). Garry Kasparov vs. IBM Deep Blue. Kasparov vs. Deep Blue. The first match, Philadelphia 1996. Retrieved July 14, 2009 from http://researchweb.watson.ibm.com/deepblue/meet/html/d.1.2.e.shtml

Kurzweil, R. (2006). *The Singularity is Near: When Humans Transcend Biology*. New York: Penguin Group.

Merritt, R. (June 23, 2009). Intel, IBM spar for lead in top 500 list. *EE Times*, Retrieved July 14, 2009 from http://www.pldesignline.com/news/218100774

Sotomayor, C. S. T. (March 18, 2004). Production nanophotonics—Dream or reality? Retrieved July 14, 2009 from http://www.innovations-report.com/html/reports/energy_engineering/report-27130.html

Yang, S. (Feb. 19, 2009). New method to assemble nanoscale elements could transform data storage industry. Retrieved July 14, 2009 from http://www.berkeley.edu/news/media/releases/2009/02/19_densechips.shtml

3

Nanotechnology in Fiber-Optic Telecommunications: Power-Control Applications

Andre Girard and Moshe Oron

CONTENTS

3.1 Introduction

Nanostructures are currently beginning to be used industrially as devices in fiber-optic (FO) telecommunications. Having a short history, these devices are designed to serve for 25 years in the environmental conditions of transmission systems. Consequently, they call for more than a physical proof of concept; they not only need dedicated packaging, new testing methods, and new standards for the devices, but also their nanomaterials. This chapter describes the state of the art of device development, lifetime and performance testing, standards in preparation, and some future directions in this novel industrial field.

Recently, a vast amount of research was directed toward the study of nanostructures and their interaction with light. This field opens new opportunities for the development of novel devices since it enables the creation of very high internal electric fields, enabling the realization of nonlinear effects using moderate light intensity inputs. Nanoparticles in a nanostructure have a large accumulative surface area that serves well for heat conduction. These two properties enable the creation of novel structures where, at the nanoscale, their optical properties differ very much from the bulk material.

This chapter describes in detail the application of nanostructures in passive, FO power control, showing two already-fielded examples, namely, the optical fuse and limiter.

3.2 Background

3.2.1 The Need for Optical Power-Control Devices

As telecommunication-network complexity increases and higher powers are transmitted through the optical fiber, allowing larger data transmission volumes, there is a greater need for power control and regulation. Today powers of +27 dBm (0.5 W) are used in FO networks. These power levels are very close to the fiber fusion breakdown, at about +30 dBm (1 W), for single-mode fibers (Davis et al. 1996, Dianov et al. 2006). Power control and regulation is consequently essential in today's FO networks. In some networks, namely, point-to-point (P2P), the optical peak power is well known and controlled. However, as network complexity increases, the peak power can be neither predicted nor easily controlled. Changes in the network configuration, such as bypassing or changing the transmission direction in a ring such as in reconfigurable optical add-drop multiplexer (ROADM), may significantly affect the power levels reaching the end points. Therefore, a wide dynamic range is required for allowable power handling. Moreover, networks are susceptible to undesirable power spikes that arise from amplifiers or external sources that are multiplexed into them. Currently, most, if not all, of the optical power control is performed by active electronic control loops using variable optical attenuators (VOAs) (Zhang et al. 2003). However, power control is now possible using passive optical devices based on nanostructures. The basic property of these components is their nonlinear interaction with the transmitted optical power, such as with optical fuses (similar to electrical circuit breakers) and optical power limiters (similar to electrical Zener diodes).

3.2.2 Optical Power-Control Devices

Passive optical power-control devices used in FO telecommunications include

- Optical fixed attenuator, in which the output power (P_{out}) increases linearly with input power (P_{in}).
- Optical fuse is an in-line device that is transparent under low power operation, but becomes permanently opaque when the input power reaches a predetermined power threshold level, at which point the output power drops irreversibly; as the fuse action is irreversible, optical fuses are designed to operate in emergencies, saving the network components from physical damage.
- Optical limiter is an in-line device that limits the output power to a certain predetermined level (namely, the limit power). At low input powers, the limiter is transparent, whereas at input powers higher than the limit power, the output power is and remains constant. As opposed to the optical fuse, the action of the optical power limiter is reversible and can then be repeated many times.
- Optical dynamic attenuator containing a fixed attenuator and a limiter in series, in which the output power increases linearly with input power up to the limit power, and then remains constant.
- Optical limiter fuse containing a limiter followed by a fuse, in which the output power is equal to input power up to the limit power, and then remains constant up to the threshold power, where the output power drops irreversibly.

Figure 3.1 shows the power profiles of ideal optical power-control devices.

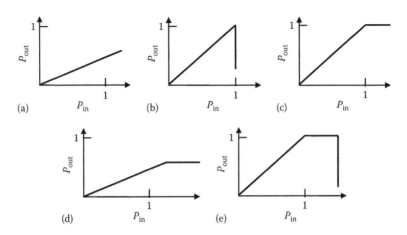

FIGURE 3.1
Output power versus input power for various power control devices. (a) Optical attenuator, (b) optical fuse, (c) optical power limiter, (d) optical dynamic attenuator, and (e) optical limiter-fuse.

3.2.3 Important Related Definitions (IEC 61931 1998)

Attenuator: An attenuator is a passive optical device that produces controlled signal attenuation in an optical fiber transmission line.

Attenuation: Attenuation is an optical property by which a reduction of optical power happens between the input and the output port of a passive optical component, a subsystem, or systems. Attenuation is expressed in decibel (dB) and defined as follows:

$$a = -10 \, \log\left(\frac{P_{out}}{P_{in}}\right) \tag{3.1}$$

where
P_{in} is the optical power launched into the input port
P_{out} is the optical power received from the output port, as shown in Figure 3.2

Attenuation has a positive sign. Insertion loss (with a negative sign) is often used to characterize the attenuation of a passive optical device.

Return loss: Return loss (RL) is an optical property by which a fraction of the input power is returned out of the same port of a passive optical component, a subsystem, or systems. RL is expressed in dB and defined as follows:

$$RL = -10 \, \log\left(\frac{P_{ref}}{P_{in}}\right) \tag{3.2}$$

FIGURE 3.2
Attenuation in a passive optical device.

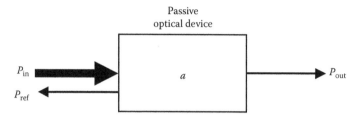

FIGURE 3.3
RL from a passive optical device.

where

P_{in} is the optical power launched into the input port
P_{out} is the power received out from the same port, as shown in Figure 3.3
P_{ref} is the power reflected from the same port, as shown in Figure 3.3

Bandpass: Bandpass is the specified range of wavelengths from $\lambda_{i\ min}$ to $\lambda_{i\ max}$ about a nominal operating wavelength λ_i, within which a passive optical device is designed to operate with the specified performance

Fuse reaction time: The reaction time of an optical fuse is the time period during which the fuse output power level is higher than fuse power threshold +1 dB, as shown in Figure 3.4.

Limiter reaction time: The limiter reaction time is the time period during which the limiter output power level is higher than limit power +1 dB, as shown in Figure 3.5.

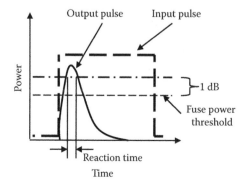

FIGURE 3.4
Optical fuse reaction time.

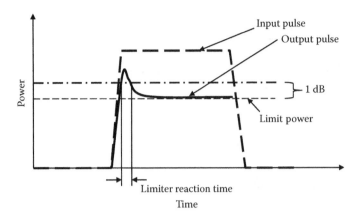

FIGURE 3.5
Optical limiter reaction time.

3.3 Nanostructures and Their Interaction with Light

There are quite a few similar definitions available on nanostructures:

- A material structure assembled from a layer or cluster of atoms with the size of the order of nanometers (http://www.answers.com/topic/nanostructure).

- An object of intermediate size between molecular and microscopic (micrometer-sized) structures (http://en.wikipedia.org/wiki/Nano structures).

- An atomic structure of a size between that of a molecule (of the order of a billionth of a meter, i.e., 1 nm) and a microscopic object (no larger than a bacterium, i.e., 100 nm) (http://www.cite-sciences.fr/lexique/defini-tion1.php?id_expo = 46&idmot = 481&radiob = &recho = &resultat = &num_page = 3&habillage = glp&lang = an&id_habillage = 79).

- An atomic, molecular, or macromolecular structure that has at least one physical dimension of approximately 1–100 nm and possesses a special property, provides a special function, or produces a special effect that is uniquely attributable to the structure's nanoscale physical size (United States Patent and Trademark Office, USPTO).

A few kinds of nanostructures relevant to FO communication shall be herein considered, such as

- Planar array, two dimensions
- Volume, three dimensions

TABLE 3.1

Description of Various Nanostructures

Structure No.	Type	Particle Material	Particle (Partially Absorbing)	Host Material (Transparent)
1	Planar	Metal and nonmetal	Spheres 2–10 nm diameter	Dielectric (glass, polymer, solgel, etc.)
2	Volume	Metal and nonmetal	Spheres 2–10 nm diameter	Dielectric (glass, polymer, solgel, etc.)
3	Planar	Metal and nonmetal	2–6 nm thick plates 20–100 nm diameter	Dielectric (glass, polymer, solgel, etc.)
4	Volume	Metal and nonmetal	2–6 nm thick plates 20–100 nm diameter	Dielectric (glass, polymer, solgel, etc.)

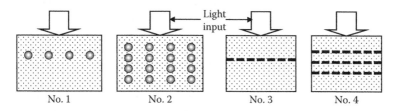

FIGURE 3.6
Illustration of nanostructures shown in Table 3.1.

The variation in the various typical parameters and types is shown in Table 3.1 and Figure 3.6.

The interaction of these kinds of nanostructures with light depends on the following parameters:

- Particle size and shape
- Distance between adjacent particles
- Particle material and its light-absorbing properties
- The dielectric material and its thermal, optical, and dielectric properties
- The nature of the impinging light, i.e., continuous wave (CW), quasi-CW, or pulsed

3.4 Single Nanoparticle

Light, when impinging on a single particle is partially absorbed by the particle and consequently heats it. At moderate particle temperatures, below its melting point, heat spreads out of the particle to the transparent dielectric matrix, mainly by conduction. This heat source creates a local temperature change ΔT, well known as the temperature profile around a point source (Green's function, http://www.greensfunction.unl.edu/home/whatisG/whatisG.html), creating an index of refraction Δn change (due to either positive or negative dn/dT). The matrix material, which is usually transparent, becomes more absorptive at elevated temperatures (Gillespie et al. 1965, Freeland 1988) creating a larger absorption area than the particle cross section. In addition to the temperature change, the electric field surrounding a metallic, conducting nanoparticle (e.g., gold, silver) in resonance conditions is changed (Quinten 2001, Kik et al. 2002, Kim and Kim 2008) and is orders of magnitude higher than the electric field of the impinging light. This process is reversible; no permanent damage is observed when the light is shut off.

When the absorbed light heats the particle to temperatures above its surroundings, melting point, the processes are different and these are non-reversible in most cases. At this point, when the process is nonreversible, the light fluence is called "damage threshold". When the absorbed light heats the particle to temperatures above its evaporation temperature, the process creates hot and dense plasma. The dielectric matrix is heated not only due to heat transfer by thermal conductivity (Donaldson 2001), but additional mechanisms as the injection of electrons through thermionic emission and UV light emission by the hot particle take over and change dramatically the nature of the surrounding matrix. This phenomenon results in light absorption by the matrix region that is now modified by energy transfer from the heated particle. It generates a much larger absorbing area, made of a much higher light-absorbing matter, namely, plasma. This process leads to crater formation, followed by the creation of permanent absorbing and scattering area in the dielectric matrix.

3.5 Nanostructure

Let us now look at a planar nanostructure made of particles and follow it through the processes described above. At moderate particle temperatures, below its melting point, heat spreads out of each particle to the transparent dielectric matrix by conduction. When the heat sources are ordered, at grating-like fixed distances, e.g., structure No. 1, each heat source creates a local temperature change in an area larger than the particle, corresponding to the index of refraction Δn change. We then get an optical grating of repeated low and high index of refraction n. This grating, when the grating spacing is in light wavelength sizes, is diffracting the light to different orders, thus reducing the forward intensity, the zero order, or the forward light transmission of the grating. In this way, the nanostructure becomes a non-linear optical filter, diffracting more energy out of the impinging light beam as light fluence is increased. This process is reversible; the grating contrast is linearly proportional to the light intensity, assuming dn/dT of the transparent dielectric matrix is linear. The onset of the nonlinearity is when the size of the light-absorbing area becomes much larger than the particle itself, and includes its heat-affected zone. As this nonlinear absorption in the heated dielectric is large, it diffracts more light by creating a high-contrast grating. This process is important for optical limiters.

In cases of volume nanostructure, e.g., structure No. 2, an additional effect is occurring. The collective contribution of the absorbed heat in many particles can be seen as absorbed by a volume absorber, cooled only at the side walls, located parallel to the light flow. If we look at structure No. 2 as part of an optical fiber, the thermal effect creates a temperature gradient: hot in the middle and cooler at the fiber perimeter, effectively making it a focusing lens. This process is also important for optical limiters.

When nonreversible processes are needed, such as in the case of optical fuses, one may prefer nanostructures as No. 3 and No. 4, where via between the plates is narrow, the strength of the electric field across via is large, and electrical breakdown and plasma can be created at lower light fluence than in structures No. 1 and No. 2.

3.6 Nanostructure Construction

Nanostructures, of the kinds described above, can be constructed using a variety of methods, some of which are described in this chapter.

The properties of many useful nanoparticles of spherical, needle-like, and other shapes, as well as their interaction with light, are well documented in the literature (Bohren and Huffman 1986, 1998, Kreibig and Vollmer 1995). No. 1 nanostructures, using gold nanoparticles embedded in glass, were used for the testing of laser damage thresholds as a function of particles size and separation. Their preparation methods are described by Papernov et al. (2001) and Donaldson (2001), who have used colloidal gold particles of selected size to build nanostructure No. 1 with different separations. An additional method (Donval et al. 2004b) consists of creating a No. 3 nanostructure using a very thin, 3–5 nm, deposited metallic layer on an insulator, letting the layer to "crack" into plates after deposition. There are many methods to create these "cracked" layers, in particular, a high degree of control of plate size and spacing has been achieved in recent years in the research of surface-enhanced Raman scattering spectroscopy and its substrate preparation methods (Haynes et al. 2005, Aroca 2006). In general, thin layers of conducting material are known to form a noncontinuous layer when deposited on dielectric substrates, either due to differences in the thermal expansion of the substrate and the deposited metal, or due to

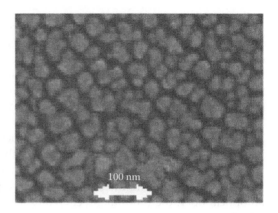

FIGURE 3.7
SEM picture of metallic 3–5 nm layer
on glass.

other mechanisms. Following the creation of the "cracked" layer, the layer is sandwiched between two dielectrics, filling vias as well (see Figure 3.7, where the island and via structure are seen, with 20–30 nm flake size and 10 nm via or tunnel). An additional way is starting with colloids of nanotubes in a dielectric material (Youichi et al. 2004) and creating No. 3 and No. 4 nanostructures.

3.7 Nanostructures as Optical Power-Control Devices

Two already-fielded examples of optical power-control devices are described in this section, showing a way to realize a device that takes advantage of the nonlinearity in nanostructures.

3.7.1 Optical Fuses

3.7.1.1 Market Needs

Fiber lasers, FO communication systems, and other systems designed for light delivery, such as in medical, industrial, and remote sensing applications, often handle high power levels or high fluence, up to 1 W or more in a single-mode fiber or waveguide. When these high specific fluences or powers per unit area are introduced into the systems, many thin film coatings, optical adhesives, and even bulk materials may be exposed to light beyond their damage thresholds and could eventually be damaged. Another issue of concern in high-power systems is laser eye safety, where well-defined upper limits are allowed for fluence out of fibers. These two issues can be resolved by using an optical fuse that will switch off the power propagation when the power exceeds the allowable fluence. Such an optical fuse is positioned either at the input of a sensitive optical device, at the output of a high-power device such as a laser or an optical amplifier, or integrated within an optical device.

In the past, there have been attempts to realize an optical fuse, mainly for high-power laser applications; special efforts were devoted to optical sights and eye-safety devices. The properties on which these solutions were based, included

- Self-focusing or self-defocusing, due to a high electric-field-induced index change through the Kerr nonlinearity.
- Reducing the optical quality of a gas, or a solid, transparent insert, positioned at the crossover spot of a telescope, by creating light-absorbing plasma in the crossover point. None of the past solutions have matured to a commercial product.

The optical fuse can be used as an add-on at the input port of, as shown in Figure 3.8a

- Power meters
- Optical switches
- Test equipment (e.g., spectrometers, optical spectrum analyzers, detectors)
- Receivers
- Multiplexers–demultiplexers

The optical fuse can be used as an add-on at the output port of the following, protecting the next device from damage by the light source, as shown in Figure 3.8b:

- Lasers
- Amplifiers
- Modulators

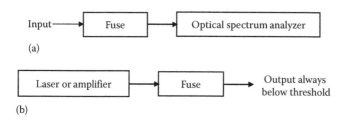

FIGURE 3.8
Examples of optical fuse placement at (a) input and (b) output of active devices.

3.7.1.2 Optical-Fuse Specifications for Optical Communication Networks

Optical fuses must comply with the following communication system requirements:

- Predetermined optical power transmission threshold-setting capability
- Unaffected, transmitted optical communication data, as long as the power is below the predetermined threshold power
- Wavelength independence in the optical communication bands
- Fast reaction time, offering fast response in protection against failure (which in case of detector failure is about 10 μs), but not too fast, affecting the transmitted data pulses (namely, not faster than 1 ns for 1 GHz modulation of the communication data)
- Single-mode, multimode, and polarization-maintaining fiber compatibility
- Stand-alone or integrated unit capability as an internal par into an optical subsystem
- Passive, no electrical power
- Operation in the needed environmental conditions, mainly temperature and humidity
- Orders of magnitude of opacity, or high insertion loss, when activated
- <40 dB RL
- Preferably bidirectional
- <1 dB threshold-level variability in time (years) and from fuse-to-fuse
- <1 dB attenuation
- Activation by CW, quasi-CW, and pulsed light input, according to its predetermined parameters

3.7.1.3 Optical Fuse: State of the Art

Some optical-fuse designs are active, using a control loop. These are (Levecq 2008) adaptive optics based or microelectromechanical system (MEMS) based (Hirata et al. 2005). Other designs are passive devices. These contain

- Heat-induced breakage of an optical-circuit glass bridge, linking silica glass fibers (Todoroki 2006)
- Two adjacent bulk materials, one is light-absorbing and the other heated and damaged by the temperature (Taneda et al. 1999)
- Semiconductor configuration having optical-fuse properties (Benedix 2001)

- Effect in a tunable liquid crystal waveguide with a Cr-grating coupler (Kamaga et al. 2007)
- Two nanostructure-based devices: (a) One using a light-transmitting medium containing carbon nanotubes (Youichi et al. 2004). (b) The other using a light-transmitting medium containing a layer of metallic nanoplates (Donval et al. 2004a,b, 2007). Up to now, only Donval-based device has been made commercially available.

3.7.1.4 How to Design and Produce a Fuse

The design steps of a nanostructure-based optical fuse or limiter start with a simulation of the temporal and spatial behavior of the light fluence passing through the device and its interaction with the nanostructure. A simplified model can be seen in Figure 3.9.

The light input is simulated as a continuous train of small temporal slices, each interacting with the nanostructure and creating a change (either due to temperature rise or electric field strength) in the nanostructure's absorption and index of refraction. The next temporal slice interacts with a modified nanostructure and so on. By carrying out this simulation, the whole interaction can be modeled, using different nanoparticle shape, material, spacing, and number of layers. The nanostructure's parameters such as size and boundary conditions, spectral transmittance, index of refraction, thermal properties, and dielectric strength are important. The light properties used are temporal modulation, spatial distribution, wavelength, line width, and intensity.

Simulation as described above led to device development (Donval et al. 2004a,b, 2007). Examples of fuse packaging are shown in Figure 3.10. The optical-fuse devices can be designed in two configurations, in-line and adaptor-like, as shown in Figure 3.11. The nanostructures, transparent dielectrics and particles, are placed between two fibers or two ferrules with fibers.

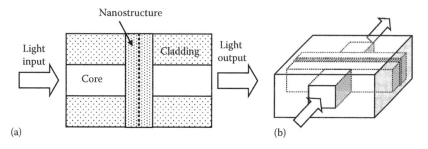

FIGURE 3.9
Design of a nanostructure-based optical fuse or limiter. (a) top view and (b) 3D view.

FIGURE 3.10
Optical fuse—in-line configuration with or without connectors. *Note*: Typical dimensions are 4 mm diameter by 40 mm length.

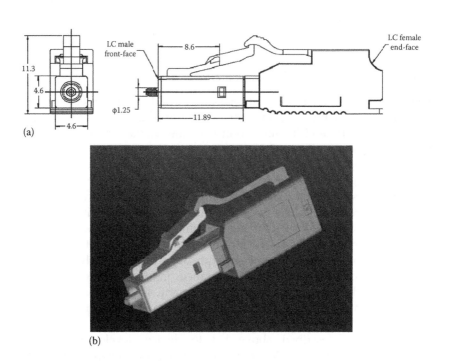

(a)

(b)

FIGURE 3.11
Optical fuse—plug-type configuration (LC plug). (a) Side views and (b) 3D view.

3.7.1.5 Fuse Design and Compliance to Market Requirements

The important fuse design parameters are discussed below, showing experimental data (courtesy of KiloLambda Technologies, Ltd.)

3.7.1.5.1 Predetermined Optical Power Transmission Threshold

For controlling the threshold power of the fuse, several techniques may be used. The first method uses different nanoparticle dimensions (Donaldson 2001). Generally, the threshold power decreases when using a larger particle. However, in this technique, the attenuation at the operating power also varies (the larger the particle, the higher the attenuation). Thus, if low attenuation is mandatory at the operating powers, this technique is rather limited in a dynamic range.

A second technique is to use fibers of different core size, or different mode field diameters (MFD). The commonly used fiber in optical communication networks is the single-mode, non-dispersion-shifted fiber (NDSF, zero dispersion in the 1310 nm region). This fiber has an MFD of approximately 10 μm at 1550 nm wavelength. Other fibers have either smaller or larger MFD. For example, high-numerical-aperture (HNA) fibers generally have smaller MFD. Thus, in HNA fibers, the light intensity or fluence (power per unit area) is larger than in NDSF, when operating with the same power through the fiber. Consequently, the power threshold in HNA-fiber optical fuse is lower than that in NDSF having the same nanostructure. Since there are several possible HNA fibers, with different MFD, one can control the threshold power using this technique. Moreover, the input and output fibers can still be NDSF, these are efficiently fusion spliced to HNA or other types of fibers. Hence, using the different types of fibers with the same nanostructures can lead to fuses having different thresholds with nearly the same attenuation at the operating powers. The same principle is used for multimode fibers (MMF) having various MFD. The predetermined optical power threshold can be reached by using a dedicated, specially designed nanostructure, for each power. Using combinations of the three techniques (Donval et al. 2004a,b, 2007) enables to design and manufacture fuses with thresholds from +17 to +31 dBm in devices using input and output NDSF.

Concerning testing the fuse performances, the first and most important measurements are the fuse functionality, operation mode, and repeatability of the threshold between different fuses, in the same production batch, having the same threshold rating. A typical test setup is shown in Figure 3.12.

The pass criterion is when the threshold limit is +20 dBm ± 0.5 dB. Measurement results are shown in Figure 3.13 for attenuation as well as threshold power. Ten samples of a batch of 100 were tested (the test is destructive). All samples passed the criteria, having threshold average and standard

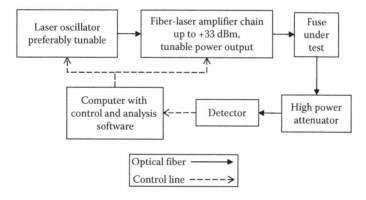

FIGURE 3.12
Typical test setup for measuring fuse performances.

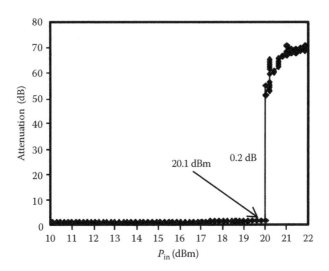

FIGURE 3.13
Typical attenuation of a 20 dBm optical fuse.

deviation of 20.1 dBm ± 0.2 dB. Measurements show that thresholds have no wavelength dependence: All thresholds were within +20 dBm ± 0.5 dB.

In Figure 3.13, we see the change in attenuation against the P_{in}. At threshold, the attenuation changes to 50 dB, and at higher powers to 70 dB giving five to seven orders of magnitude protection, or opacity.

As opposed to the desirable low attenuation, the fuse must have a high attenuation (low transmission) at high powers (above the threshold), and hold this high attenuation at input powers an order of magnitude more than the threshold, and not remelt and start transmitting again. This is obtained by a significant and permanent damage, which significantly increases the scattering and not the absorption of the impinging light.

Visual (microscope) inspection after damage (see Figure 3.14) reveals a cratered core, few microns deep, meaning that the volume of the material removed from the crater is few hundred times higher than the volume of the heated nanostructure.

Low attenuation is desirable at the operating powers. However, the metallic layer is generally absorbing and reflecting light. The reflection can be minimized by antireflective layers. The absorption of the metallic layer, however, is an intrinsic material property that cannot be completely eliminated. Therefore, the attenuation at the operating power is not negligible.

3.7.1.5.2 Fuse Reaction Time

The optical-fuse reaction time is the total time period during which the fuse output power level is higher than the fuse power threshold +1 dB. This time period shall offer reaction fast enough to protect against failure at

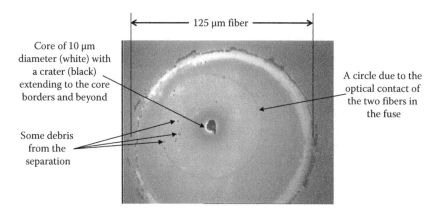

Core of 10 μm diameter (white) with a crater (black) extending to the core borders and beyond

125 μm fiber

A circle due to the optical contact of the two fibers in the fuse

Some debris from the separation

FIGURE 3.14
A separated, activated fuse made of a single-mode fiber.

the detector, the fiber, or the coating. A typical value may be about 10 μs, but should not be too fast, i.e., as fast as the modulation of the communication data, and should not be activated by the short data pulses at gigahertz and faster speeds. In this way, the transmitted data are unaffected by the fuse, as long as the power is below the predetermined threshold power.

The example illustrated in Figure 3.15 shows the reaction time in the case of nanostructures No. 1 and No. 3, and NDSF fibers with an input pulse

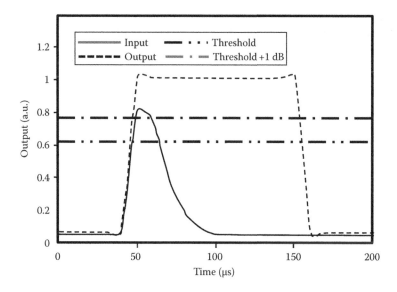

FIGURE 3.15
Optical-fuse reaction time. *Note*: 100 μs long, 10 μs rise time input pulse (fuse reaction time = ~10 μs).

of ~100 μs duration and 10 μs rise time. In this case, the threshold power is set at 0.62 (normalized units). Therefore, the measured fuse reaction time is about 10 μs. This reaction time is fast enough to prevent damage in fibers and detectors, but not fast enough to interfere with the GHz communication modulation.

3.7.1.5.3 Wavelength Independency in the Optical Communication Region

Testing wavelength dependence of the fusing threshold requires testing at different wavelengths. The purpose of this test is to ensure that the threshold does not change when using the fuse at difference wavelengths. A possible strategy is to perform the test at 1500 or 1600 nm, or using a tunable laser source (TLS). In the 1500–1600 nm window, the TLS can be replaced by using only the amplified spontaneous emission (ASE) of an amplifier, without the oscillator input, as a wide band source of radiation. Typical results are shown in Table 3.2.

The graph shown in Figure 3.16 is an example of attenuation measurement as a function of wavelength for an optical fuse. It can be seen that the attenuation variation is less than 0.1 dB within the wavelength range (measurement uncertainty is 0.01 dB). There is a periodic dependence due to the nanostructure dimensions along the beam propagation behaving like an

TABLE 3.2

Wavelength-Dependent Threshold Test

Wavelength Range (nm)	Low-Power Measurements before the Main Tests		Threshold Results (dBm)
	Attenuation (dB)	RL (dB)	
1500	1.4	42	20.1
1500–1600	0.9	43	20.4
1600	1.2	43	20.3

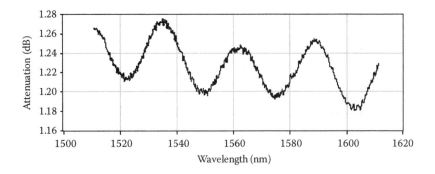

FIGURE 3.16
Attenuation as a function of wavelength for an optical fuse.

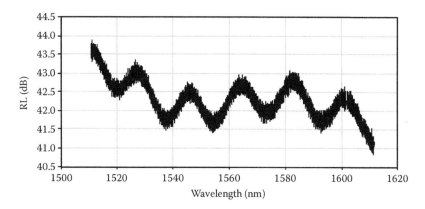

FIGURE 3.17
RL as a function of wavelength for an optical fuse.

interference layer, and this behavior is seen in the reflection as well but has a shifted phase, when attenuation is low the reflection (RL) is high.

An example of RL measurement as a function of wavelength for an optical fuse is shown in Figure 3.17. It can be seen that in this case, the RL change within the wavelength range is about 2 dB, while the device RL is ~40.5 dB (measurement uncertainty is 0.5 dB).

The polarization-dependent loss (PDL), i.e., the difference between the maximum and minimum attenuation as a function of all the states of polarization (SOP) of the light source, propagating through the device is an important parameter when lasers are concerned. Figure 3.18 shows an example of a PDL measurement of an optical fuse. It can be seen that in this case the PDL variation within the wavelength range is about 0.035 dB,

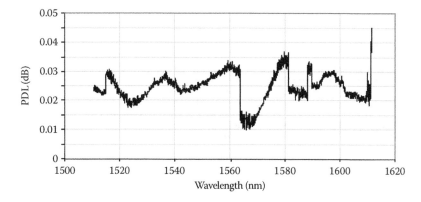

FIGURE 3.18
PDL as a function of wavelength for fuse.

while the PDL of the device is ~0.045 dB (measurement uncertainty is 0.005 dB). This kind of value for an optical fuse is considered as negligible PDL in optical fiber communications. The structure appears again in this measurement and may be explained as difference, due to polarization, of the interference in attenuation, and reflection.

3.7.1.5.4 Environmental Conditions, Mainly Temperature

An important parameter is the influence of the temperature of the fuse on the threshold power. Placing the fuse in an environmental chamber, as shown in Figure 3.19, having controlled temperature at −10 °C and at +70 °C, and feeding the fuse with slowly varying powers from +10 dBm up to +22 dBm ensures the fuse environmental qualification.

Difference in threshold average should not exceed 0.5 dB as shown in Table 3.3.

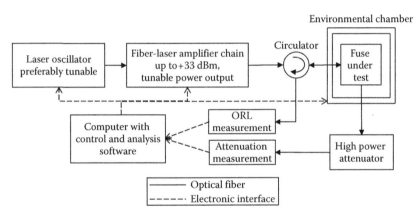

FIGURE 3.19
Typical test setup for qualifying a fuse environmental performances.

TABLE 3.3

Example of Test Results for Temperature-Dependent Threshold

| Sample Number | Temperature (°C) | Low-Power Measurements before the Main Tests | | Threshold Results (dBm) |
		Attenuation (dB)	RL (dB)	
1	−10	1.2	44	20.1
2	−10	1.5	47	20.5
3	70	1	43	19.5
4	70	1.2	43	20.1
5	70	0.9	41	20.1

3.7.1.5.5 Low Insertion and Reflection Loss

This is obtained by a combination of two methods. First, the nanostructure can be constructed in an angle to the fiber axis, namely, using an angled cleave or an angled fiber connector (or ferule). Second, coating layers are designed to have minimal reflections from the metallic plates.

Figure 3.20 shows the attenuation and RL power dependence for the +20 dBm fuse. The attenuation and RL are nearly constant up to +16 dBm input and from there and up one gets higher attenuation of up to 1.8 dB and

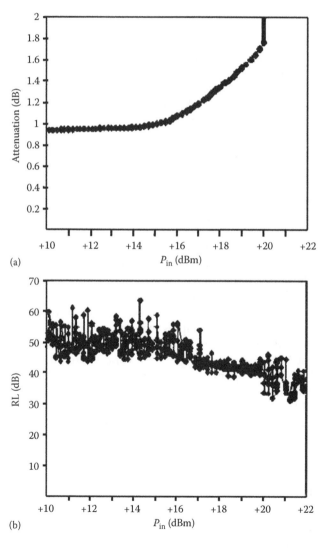

FIGURE 3.20
(a) Attenuation and (b) RL as a function of input power for a fuse.

more reflection (30 dB), compared to about 45 dB at the start. This is mainly due to heating of the nanostructure, changing the index of refraction of the nanostructure and its surroundings due to dn/dT, and reflection due to it. The parameters return to their original value when power goes down without reaching the threshold.

3.7.2 Optical Limiters

3.7.2.1 The Need

Fiber lasers, FO communication systems, and other systems for light delivery, such as in medical, industrial, and remote sensing applications, often handle high levels of optical power or high fluence, namely, up to 1 W or more in a SMF or waveguide. When these high specific fluence or power per unit area are introduced into the systems, many thin film coatings, optical adhesives and even bulk materials, and mainly detectors are exposed to light fluence beyond their damage thresholds and are eventually damaged. Another issue of concern in high-power systems is laser eye safety, where well-defined upper limits are allowed for fluence out of fibers. These two problems can be solved using a limiter that will limit the power propagation in a fiber or waveguide, when the power exceeds the allowed fluence. Such an optical limiter should be placed either at the input of a sensitive optical device or a detector, at the output of a high-power device such as a laser or an optical amplifier, or integrated within an optical device.

In the past, there have been few attempts to realize optical limiters. The optical limiter is a passive component that regulates the optical power in fibers. Under normal operation, when the input power is low, the limiter has no effect on the system. However, when the input power is high, the output power is limited to a predetermined level (P_{limit}). In a similar way to the optical fuse, the way to construct an optical limiter is based on the same principles. Various physical principles were proposed in the past for optical limiting (Tutt and Boggess 1993) using various materials like fullerenes (Tutt and Kost 1992, Sun and Riggs 1999), various organic materials (Perry et al. 1996, Perry 1997, Spangler 1999, Sun and Riggs 1999), liquids (Miller et al. 1998), and their specific interaction with light. None of these solutions have matured to a commercial product yet.

The limiter can be used as an add-on at the input port of the following (see Figure 3.21a):

- Power meters
- Optical switches
- Test equipment (e.g., spectrometers, optical spectrum analyzers, and detectors)
- Receivers or detectors
- MUX–DEMUX units

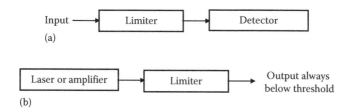

FIGURE 3.21
Examples of placement of an optical limiter at (a) input or (b) output of active devices.

The limiter can be used as an add-on at the output port of the following, protecting the next device from damage by the light source (see Figure 3.21b):

- Lasers
- Amplifiers
- Modulators

The commercially available (Oron et al. 2005, Choudhury et al. 2007, Donval et al. 2007) optical power limiter is based on nonlinear, absorption-induced diffraction and scattering. At low powers, there is only a residual absorption effect that results in relatively small attenuation (~1.5 dB). However, diffraction scattering becomes significant at high input powers, and allows only a fraction of the input power to propagate. The limiter configurations are similar to the fuse configurations.

3.7.2.1.1 Power Regulation

The limiter is used repeatedly, as shown in Figure 3.22, and ongoing long-life tests show its ability to work continuously for the needed 25 years of operation.

The optical power limiter is wideband. Power cycles performed for several wavelengths ranging from 1540 to 1625 nm reveal that the difference in output power, as a function of wavelength is less than ±0.3 dBm at every input power level, showing a slightly lower limit power at lower wavelength.

3.7.2.2 Optical Power Limiter Additional Applications

3.7.2.2.1 Equalization of Power between Channels in DWDM Systems

Equalization between different channels in dense wavelength division multiplexing (DWDM) systems is required in order to use optical amplifiers more efficiently, and to increase the signal-to-noise ratio. Power control in the optical level using an array of optical power limiters: a few sources of different wavelengths are multiplexed into a single fiber (see Figure 3.23). Each of these sources may have different power. The equalization is obtained by adding an optical power limiter on each line entering the multiplexer, so each of the input powers is limited by the limit power.

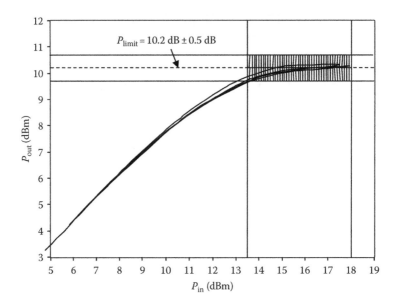

FIGURE 3.22
Optical limiter—output power versus input power. *Note*: 10 dBm, $\lambda = 1500$ nm, a few power cycles are shown (the input power goes up and down from +1 to +18 dBm).

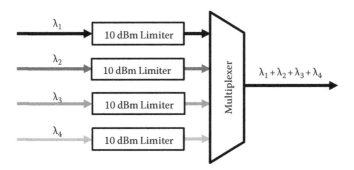

FIGURE 3.23
Power equalization of DWDM channels using a limiter array. *Note*: Four separate channels are shown, each varying between +12 and +17 dBm; they are injected into an array of limiters, equalized in power and multiplexed into a single fiber.

3.7.2.2.2 Power Flattening or "Noise Eater"

The optical power limiter can be used to remove power spikes as it truncates the excess power above the limit power. Also, if one has an unstable power source, one could use the limiter to flatten the output power profile, namely, a "noise eater." Figure 3.24 presents schematics of spike-removal and "noise-eater" applications.

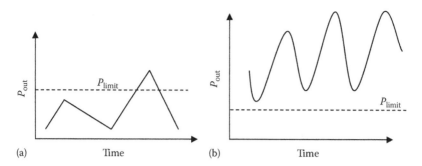

FIGURE 3.24
Input (full line) and output (dash line) of a limiter used for power flattening with (a) spike removal or (b) "noise eater."

3.7.2.2.3 Detector Protection from Optical Excess Power

When placed in front of a receiver or detector, the limiter function is twofold. First, it prevents damaging the detector from excess power. This may include prevention of reaching the damage threshold of certain devices or even prevention of catastrophic damage due to effects such as the fiber-fuse phenomenon. Second, it increases the dynamic range of the system beyond the original maximum power.

3.7.2.3 Optical Power Limiter Parameters

Some typical parameters of a limiter are presented as follows.

3.7.2.3.1 Time Response

Fast response is required in order to block the excess power. However, the response should be slower than the data rate, in order not to affect the transmitted data. The temporal response of an optical power limiter is shown in Figure 3.25. Limiter reaction time is defined as total time where the limiter output power level is higher than limit power +1 dB. The measurement is carried out using a long pulse, 5 ms to determine P_{limit}, +10.2 dBm in this case, and the reaction time is 117 μs.

3.8 Standardization

3.8.1 General

Two efforts of standardization are currently taking place: international standards development for nanostructure devices and related test methods. These activities are in progress in the International Electrotechnical Commission

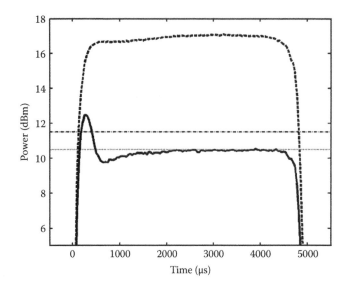

FIGURE 3.25
Long pulse for P_{limit} definition: input power (– – – –), output power (·········) where dotted line
(——) is the limit power and (— · — · —) is limit power +1 dB.

(IEC, www.iec.ch) Technical Committee TC86 (Fibre Optics*) and more
precisely in the working groups WG7 (standards and specifications for FO
passive components) and WG4 (standard tests and measurement methods
for FO interconnecting devices and passive components) of the subcommit-
tee SC86B (FO interconnecting devices and passive components), where the
following products are being looked at (all the following cited documents are
in progress—none of them have yet been published):

1. In-line fuse (IEC 61753-056-2 Ed. 1.0: Fibre optic interconnecting
   devices and passive components performance standard—Part
   056-2: Single mode fibre pigtailed style optical fuse for category
   C—Controlled environment)
2. In-line limiter (IEC 61753-058-2 Ed. 1.0: Fibre optic interconnecting
   devices and passive components performance standard—Part 058-2:
   Single mode fibre pigtailed style optical limiter for category C—
   Controlled environment)

---

* Scope: To prepare standards for fibre optic systems, modules, devices and components intended
  primarily for use with communications equipment. This activity covers terminology, characteristics, related tests, calibration and measurement methods, functional interfaces, optical,
  environmental and mechanical requirements to ensure reliable system performance.

3. Plug-type fuse (IEC 61753-057-2 Ed. 1.0: Fibre optic interconnecting devices and passive components performance standard—Part 057-2: Single mode fibre plug style optical fuse for category C—Controlled environment)

4. Plug-type limiter (IEC 61753-059-2 Ed.1.0: Fibre optic interconnecting devices and passive components performance standard—Part 059-2: Single mode fibre plug style optical limiter for category C—Controlled environment)

5. Fibre optic passive power-control devices (IEC 60869-1 Ed.4.0: Part 1: Generic specification)

The definitions of nanoparticles and nanostructures and handling methods are under study in the nanotechnology standardization committee TC113 (nanotechnology standardization for electrical and electronic products and systems) of the IEC.

3.8.2 Examples of Performance Results and Test Methods under Standardization Activities

3.8.2.1 Fuse

Fuses are manufactured in various threshold values, and it is recommended that continuous power applied to a fuse is 3 dB below the threshold values. Table 3.4 is intended to provide guidance on the power ranges of the various

TABLE 3.4

Single-Mode Threshold Powers for Optical Fuses

Sample Number	Threshold Power (dBm)	Recommended Power for Normal CW Work (dBm)
Item-18 SM	18	Up to 15
Item-19 SM	19	Up to 16
Item-20 SM	20	Up to 17
Item-21 SM	21	Up to 18
Item-22 SM	22	Up to 19
Item-23 SM	23	Up to 20
Item-24 SM	24	Up to 21
Item-25 SM	25	Up to 22
Item-26 SM	26	Up to 23
Item-27 SM	27	Up to 24
Item-28 SM	28	Up to 25
Item-29 SM	29	Up to 26
Item-30 SM	30	Up to 27

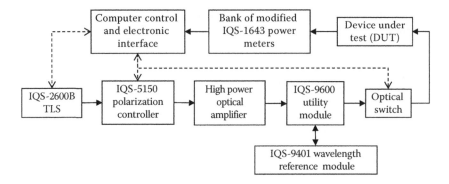

FIGURE 3.26

System setup design. (Courtesy of EXFO E.O. Engineering, Inc., Quebec, Canada, namely, the IQS-12004B high power WDM test system.)

fuses. It is not intended for specification. The values of operating energies used in performance verification shall be specified between the customer and supplier or shall be as defined in the manufacturer's specifications.

Testing of attenuation, RL, PDL, and functionality requires a setup providing the ability to measure these parameters under high power, e.g., up to +32 dBm. A dedicated setup, including a high-power amplifier and the ability to measure powers that are five and more orders of magnitude smaller than regularly measured, was designed specifically for this application. The setup is shown in Figure 3.26. Most of the data shown in this chapter were measured using this test system.

Fuses present few more problems in testing and performance verification, mainly due to the destructive performance testing and due to the nonlinear or threshold operation. The destructive performance testing needs statistical analysis to select the tested sample of each batch. The threshold operation of the fuse does not allow simple time-accelerated tests, where fuses are overpowered for short times simulating low power operation for long times, e.g., few days of testing in powers close to threshold simulates few years in the field at lower powers. This problem is not yet fully resolved and accelerated optical tests are an ongoing issue in the standardization committee.

3.8.2.2 Limiter

Limiter lifetime testing and performance verification present a problem due to its nonlinear operation. The physical model for accelerated optical tests is an ongoing issue in the standardization committee. At this point of time, lifetime is tested using real time, namely, years, and results are compared to accelerated tests in order to find the right correlation.

Table 3.5 is intended to provide guidance on the power ranges of the various limiters. It is not intended for specifications. Values of operating

TABLE 3.5

Powers for Single-Mode Optical Limiters

Sample Number	Limit Power, P_{limit} (dBm)	Sustained Input Power for CW Work (dBm)	Sustained Input Power for 1 s Exposure Every Minute (dBm)
Item L-0 SM	0	Up to 5	Up to 8
Item L-1 SM	1	Up to 6	Up to 9
Item L-2 SM	2	Up to 7	Up to 10
Item L-3 SM	3	Up to 8	Up to 11
Item L-4 SM	4	Up to 9	Up to 12
Item L-5 SM	5	Up to 10	Up to 13
Item L-6 SM	6	Up to 11	Up to 14
Item L-7 SM	7	Up to 12	Up to 15
Item L-8 SM	8	Up to 13	Up to 16
Item L-9 SM	9	Up to 14	Up to 17
Item L10 SM	10	Up to 15	Up to 18
Item L-11 SM	11	Up to 16	Up to 19
Item L-12 SM	12	Up to 17	Up to 20
Item L-13 SM	13	Up to 18	Up to 21
Item L-14 SM	14	Up to 19	Up to 22
Item L-15 SM	15	Up to 20	Up to 23
Item L-16 SM	16	Up to 21	Up to 24
Item L-17 SM	17	Up to 22	Up to 25

powers used in performance verification shall be specified between the customer and supplier or shall be as defined in the manufacturer's specifications.

3.9 Conclusion

We presented a newly developed family of nanostructure-based devices, namely, optical power fuses, optical power limiters, and their hybrids. We think this is the first practical device application to FO telecommunications. These devices can serve as both power-regulating and power-protecting devices in optical networks. Moreover, one can use a limiter as a wide-spectrum eye-safety device, in both fiber and free-space systems.

Near-future developments of devices are expected in nanostructure-based terminations, having very low reflectance (≤ 70 dB). Another near-future development, using nanostructures, is in fiber splicing, performed today by an electrical arc for fusion splicing and using adhesives for fiber to the bulk splicing. Absorbing thin nanostructure can be melted locally by in-fiber light, using very low powers. Using these nanostructures may enable fusion splicing of fiber to fiber and, more importantly, fiber to bulk-optic devices at energies as low as 1 µJ pulse, delivered by the spliced fiber itself.

Many advantages will be seen when nanostructure-based devices will find their way into short-distance communications, e.g., optical interconnects from chip to chip or intra-chip communications. Here the advantage of small dimensions and the ability to accommodate a nanostructure in an etched via on a chip is an asset that will certainly attract a lot of interest from the marketplace.

Acknowledgments

Dr. Andre Girard acknowledges the support of EXFO E.O. Engineering, Inc. Dr. Moshe Oron acknowledges the support of KiloLambda Technologies Ltd., making available the relevant data, and the BIRD foundation for supporting the joint KiloLambda–Molex research program. Thanks are also extended to Ariela Donval and Ram Oron for data contribution and comments. Special thanks go to Dan Oron for having reviewed the manuscript and for his helpful remarks.

References   AC

Aroca, R. 2006. *Surface Enhanced Vibrational Spectroscopy* (Wiley & Sons, New York).
Benedix, A. 2001. Semiconductor configuration having an optical fuse, United States Patent 6483166.
Bohren, C.F. and Huffman, D.R. 1986. Absorption and scattering of light by small particles. *Appl. Opt.* 25(18), 3166.
Bohren, C.F. and Huffman, D.R. 1998. *Absorption and Scattering of Light by Small Particles* (Wiley-VCH, Berlin, Germany).
Choudhury, A.N.M., Grzegorzewska, B., Hanrahan, T.S. et al. 2007. Dynamic attenuator—A new passive device to control optical power levels in networks, in *National Fiber Optic Engineers Conference* (NFOEC), Anaheim, CA, pp. 1–8, paper JThA88.
Davis, D.D., Mettler, S.C., and DiGiovanni, D.J. 1996. Experimental data on the fiber fuse, in *Laser-Induced Damage in Optical Materials: Proceedings of SPIE*, Vol. 2714, SPIE International Society for Optical Engineering, Bellingham, WA.
Dianov, E.M., Fortov, V.E., Bufetov, I.A. et al. 2006. Detonation-like mode of the destruction of optical fibers under intense laser radiation. *JETP Lett.* 83(2), 75–78.
Donaldson, W. R., ed. 2001. *Establishing Links between Single Gold Nanoparticles Buried inside SiO₂ Thin Film and 351-nm Pulsed-Laser-Damage Morphology*, Vol. 89, LLE review (Laboratory for Laser Energetics, Rochester, NY).

Donval, A., Goldstein, S., McIlroy, P., Oron, R., and Patlakh, A. 2004a. Passive components for high power networks, in *Optical Components and Devices*, Fafard, S. ed., *Proceedings of SPIE* 5577, International Society for Optical Engineering, Bellingham, WA.

Donval, A., Nevo, D., Oron, M., and Oron, R. 2004b. Optical energy switching device and method, United States Patent application 2005 0111782.

Donval, A., Nemet, B., Oron, M., Oron, R., and Shvartzer, R. 2007. Nanotechnology based optical power control devices, in *Technical Proceedings of the 2007 Nanotechnology Conference and Trade Show*, Nanotech Volume 1, Nano Science & Technology Institute, Cambridge, MA.

Freeland, C.M. 1988. High temperature transmission measurements of IR window materials. *Proc. SPIE*, 929, 79–86.

Gillespie, D.T., Olsen, A.L., and Nichols, L.W. 1965. Transmission of optical materials at high temperatures in the 1 μm to 12 μm range. *Appl. Opt.* 4, 1488–1493.

Haynes, C.L., McFarland, A.D., and Van Duyne, R.P. 2005. Surface enhanced Raman scattering. *Anal. Chem.* 77, 338A–346A.

Hirata, T., Mitama, I., Abe, M., Makita, K., Shiba, K., Hane, K., and Sasaki, M. 2005. Development of MEMS-based optical surge suppressor, in *OFC/NFOEC*, Vol. 4, pp. 3.

International Electrotechnical Commission (IEC) 61931, 1998. *Fibre Optic-Terminology*. pp. 86.

Kamaga, C., Segawa, Y., Tikhodeev, S., and Ishihara, T. 2007. Optical fuse effect in a tunable liquid crystal waveguide with a Cr grating coupler. *Appl. Phys. Lett.* 91(17), 173119-1–173119-3.

Kik, P.G., Martin, A.L., Maier, S.A., and Atwater H.A. 2002. Metal nanoparticle arrays for near field optical lithography. *Proc. SPIE* 4810, 7–13.

Kim, D.S. and Kim, Z.H. 2008. Polarization-selective imaging of the enhanced local field at gold nanoparticle junctions. *J. Korean Phys. Soc.* 52(1), 17–20.

Kreibig, U. and Vollmer, M. 1995. *Optical Properties of Metal Clusters*, Springer Series in Materials Science, Vol. 25 (Springer-Verlag, Berlin, Germany).

Levecq, X. 2008. Adaptive optics: Optical fuse protects intracavity laser components. *Laser Focus World*, March 2008.

Miller, M.J., Mott, A.G., and Ketchel, B.P. 1998. General optical limiting requirements, in *Proceedings on Nonlinear Optical Liquids for Power Limiting and Imaging*, San Diego, CA, C.M. Lawson, ed., pp. 24–29 (SPIE, Bellingham, WA).

Oron, R., Donval, A., Goldstein, S. et al. 2005. Optical power control components in networks, in *National Fiber Optic Engineers Conference* (NFOEC), paper JWA75.

Papernov, S., Schmid, A.W., Krishnan, R., and Tsybeskov, L. 2001. Using colloidal gold nanoparticles for studies of laser interaction with defects in thin films, in *Laser-Induced Damage in Optical Materials, Proceedings of SPIE*, Vol. 4347, pp. 146–154.

Perry, J.W. 1997. Organic and metal-containing reverse saturable absorbers for optical limiters, in *Nonlinear Optics of Organic Molecules and Polymers*, H.S. Nalwa, and S. Miyata, eds., p. 813 (CRC Press, New York).

Perry, J.W., Mansour, K., Lee, I.Y.S. et al. 1996. Organic optical limiter with a strong nonlinear absorptive response. *Science* 273(5281), 1533–1536.

Quinten, M. 2001. Local fields close to the surface of nanoparticles and aggregates of nanoparticles. *Appl. Phys. B* 73, 245–255.

Spangler, C.W. 1999. Recent development in the design of organic materials for optical power limiting. *J. Mater. Chem.* 9(9), 2013–2020.

Sun, Y.P. and Riggs, J.E. 1999. Organic and inorganic optical limiting materials: From fullerenes to nanoparticles. *Int. Rev. Phys. Chem.* 18(1), 43–90.

Taneda, Y., Ogata, T., Nagata, H., Ichikawa, J., and Higuma, K. 1999, Optical fuse, (two materials), United States Patent 6218658.

Todoroki, S. 2006. Heat induced breakage of an optical circuit glass bridge linking silica glass fibers. *J. Ceram. Soc. Jpn.* 114(8), 709–712.

Tutt, L.W. and Boggess, T.F. 1993. A review of optical limiting mechanisms and devices using organics, fullerenes, semiconductors and other materials. *Prog. Quantum Electron.* 17, 299–338.

Tutt, L.W. and Kost, A. 1992. Optical limiting performance of C-60 and C-70 solutions. *Nature* 356(6366), 225–226.

Youichi, S., Madoka, T., Yohji, T. et al. 2004. Light transmitting medium, US patent application 2005 0254760.

Zhang, X.M., Liu, A.Q., Chan, C.W., Thian, C.S., Cai, H., and Hao, J.Z. 2003. Optical power regulator using closed controlled MEMS variable optical attenuator (VOA), in *TRANSDUCERS '03, The 12th International Conference on Solid State Sensors, Actuators and Microsystems*, Boston, MA.

4

Nanotubes and Their Applications in Telecommunications

Murad Shibli

CONTENTS

4.1 Introduction

Since nanotubes exhibit extraordinary strength and unique electrical properties and are efficient conductors of heat, this chapter is dedicated to introducing the basic characteristics, properties, types, fabrication, and applications of nanotubes in the telecommunication industry. These cylindrical carbon molecules have novel properties that make them potentially useful for many applications in nanotechnology, electronics, and optics. Carbon nanotubes (CNTs) have many properties that make them ideal components of electrical circuits. For example, they have shown to exhibit strong electron–phonon resonances. These resonances can be used to make terahertz sources or sensors. Nanotube-based transistors have been made that operate at room temperature and that are capable of digital switching using a single electron. CNTs have already been used as wires to carry electricity and transistors to control electric current.

Carbon tubes, only nanometers or billionths of a meter in diameter, could serve as ultrabright light sources for telecommunications. Light-emitting nanotubes could be used to form efficient communication devices and, eventually, all-optical computer chips. Researchers found that if they inject electrons that carry negative charges into one end of a nanotube and holes, or positive charges, into the other end, the two combine to emit light whose wavelength is inversely proportional to the tube's diameter. Nanotechnology scientists have found a way to make the microscopic tubes emit light and have fashioned a nanotube transistor that emits 1.5 micron infrared light, a wavelength widely used in telecommunications.

Single-walled carbon nanotubes (SWCNTs) are excellent saturable absorbers because of their subpicosecond recovery time, low saturation intensity, polarization insensitivity, and mechanical and environmental robustness. In principle, different diameters and chiralities of nanotubes could be combined to enable compact, mode-locked fiber lasers that are tunable over a much broader range of wavelengths than other systems.

Recently CNTs are proving their usefulness in cutting-edge optoelectronic devices, where their unique combination of tunable electronic properties and nanoscale dimensions opens up a wide range of possibilities. Engineers have successfully demonstrated that improvements in the polymer matrix surrounding CNTs can help enable their use in laser signaling and detection for telecommunication applications. SWCNTs have the potential to produce and detect femtosecond (10^{-15} s) pulses, allowing for high data transmission rates through fiber optics.

Spacecraft and satellite communications rely on microwave devices such as microwave diodes that use a cold-cathode electron source consisting of CNTs and operate at a high frequency and at high current densities. Because it weighs little, responds instantaneously, and has no need for heating, this miniaturized electron source should prove valuable for microwave devices used in telecommunications.

The chapter is organized as follows: Section 4.2 presents a basic background on CNTs in terms of its classifications and methods of fabrication. Section 4.3 describes the basic categorization of CNTs and their unique electrical and electronic characteristics. The characterization of CNTs using Raman microspectroscopy is discussed in Section 4.4. This section introduces the IEEE standard test methods for the measurement of the electrical properties of CNTs. Moreover, the dynamical analysis of a single particle motion is analyzed in Section 4.6 using Schrödinger–Hamiltonian dynamics. Furthermore, Section 4.7 presents some of the promising applications of nanotubes in the telecommunications field. Finally, the conclusions are discussed in Section 4.8.

4.2 Background

CNTs have many unique physical, mechanical, and electronic properties. These distinct properties may be exploited so that they can be used for numerous applications ranging from sensors and actuators to composites. As a result, in a very short duration, CNTs appear to have drawn the attention of both the industry and academia. Since their discovery in 1991, CNTs have been intensively studied, and a number of new applications have been identified. Applications range from nanoelectronics to hydrogen absorption for battery electrodes and fuel cells.

CNTs were first discovered in 1991 by scientist Sumio Iijima of Japan's NEC Corporation. One decade later, he was awarded the Benjamin Franklin Medal in Physics, "for the discovery and elucidation of the atomic structure and helical character of multiwall and single-wall carbon nanotubes, which have had an enormous impact on the rapidly growing condensed matter and materials science field of nanoscale science and electronics" (Iijima 1991, Iijima and Ichihashi 1992).

CNTs are defined as molecular structures composed of cylindrical graphene shapes. These cylindrical structures are closed at either end with caps containing pentagonal-like rings due to carbon atoms' bond. Amazingly, at the microscopic structure, the carbon molecules are organized in a perfect matrix of hexagonal graphite sheets rolled up onto itself to form a hollow tube. The diameter of a nanotube is in the nanometer range and its length varies from microns to millimeters . Thus, it is 100,000 times less than the diameter of a sewing needle. CNTs can be fabricated in a single-wall or multiwalled structure which can be visualized by using microscopic nondestructive test (Figure 4.1).

There are two types of CNTs: (1) SWCNTs, as shown in Figure 4.2, which have only one single layer and (2) multiwall carbon nanotubes (MWCNTs), which have many cylindrical layers, as shown in Figure 4.3. SWCNTs and

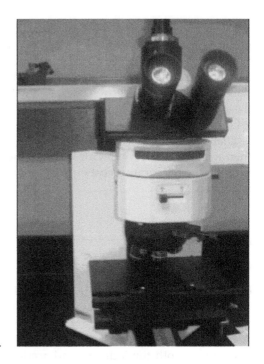

FIGURE 4.1
Visual microscopic nondestructive test.

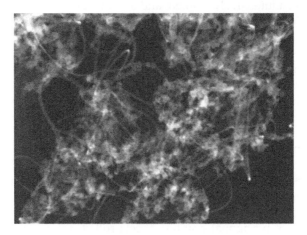

FIGURE 4.2
SWCNTs.

MWCNTs are different in their graphene cylinder arrangements (Iijima 1991, Iijima and Ajayan 1992, Iijima and Ichihashi 1992, Journnet et al. 1997).

CNTs are fabricated by vaporizing carbon graphite with an electric arc under an inert environment. The standard arc-evaporation fabrication method can produce only multilayered tubes. Meanwhile, the single layer

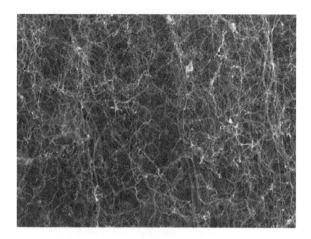

FIGURE 4.3
MWCNTs.

uniform nanotubes (constant diameter) were synthesized only lately in the 1990s. SWCNTs are manufactured using laser vaporization, arc technology, vapor growth, as well as other methods. Figure 4.4 shows two cases of uniform nanotubes and random nanotubes, respectively. Meanwhile, Figure 4.5 shows a microscopic image amplification of the nanotube structure.

CNTs can be filled with any media, including biological materials and fluids. Additionally, CNTs are known to act as conductors or insulators based on their molecular structure. Moreover, they are considered better than silver. The diverse potential application of such nanotubes, formed with a few carbon atoms in diameter, provides the possibility to fabricate and design devices at the atomic and molecular scale. Furthermore, CNTs are also known for their high mechanical strength, being much stronger than steel wires assuming that the diameter is the same, and have a better thermal conductivity than diamond.

CNTs have been used widely in microelectromechanical systems (MEMS) and nanoelectromechanical systems. To mention some applications but not all, two slightly nanotubes can be twisted, joined end to end, to act as the diode. Moreover, nanotube transistors can be manufactured using different alignments. It has been found that there are strong relationships between the

FIGURE 4.4
Aligned CNTs and random CNTs.

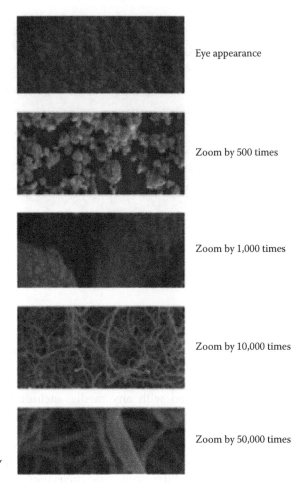

Eye appearance

Zoom by 500 times

Zoom by 1,000 times

Zoom by 10,000 times

Zoom by 50,000 times

FIGURE 4.5
MWNT: 10–40 nm diameter, 1 μm length.

nanotube's electromagnetic properties and its diameter and the degree of the molecule twist. Research on nanotechnology shows that the electromagnetic properties of CNTs depend on the molecule's twist. More specifically, if the graphite sheet forming the SWCNT is rolled up perfectly so that all its hexagons line up along the molecules axis, then the nanotube is a perfect conductor. On the other hand, if the graphite sheet rolls up at a twisted angle, the nanotube exhibits as a semiconductor. Additionally, since CNTs are stronger than steel wire, they can be added to the plastic to make the conductive composite materials.

CNTs can be implemented as nanostuctures for high-frequency operations, such as robust high bandwidth low power compatible logic gates, switches, and analog to digital pipeline converters. These logic gates, switches, and converters can have a direct application in nanocomputers, wireless communication, and networks.

4.2.1 Fabrication Methods of Nanotubes

4.2.1.1 Chemical Vapor Deposition

Chemical vapor deposition (CVD) has been found to be more beneficial over other nanostructure synthesis methods. In this fabrication method, nano-tubes are obtained at relatively low temperatures with high quality and purity. Moreover, the comparatively simple process can provide nanotubes of long length and controllable diameter. The catalyst deposition on Si substrates allows for formation of novel structures. Furthermore, the CVD method is scalable to industrial production levels.

4.2.1.2 Arc Discharge

Using the arc discharge method, nanotubes are found in soot produced in arc discharge with catalytic metals such as Fe, Ni, and Co as shown in Figure 4.6. Two graphite rods, separated by a few millimeters, are connected to a power supply. At 100 A, carbon vaporizes and forms a hot plasma. Contrary to CVD, in arc discharge, it is generally difficult to control the location and alignment of nanotubes. The resultant nanotubes are produced in small quantities, often short with random sizes and orientations, and typically tangled, making some applications difficult. The method requires evapor-ation of carbon atoms from solid targets at very high temperatures and the products generally require purification.

4.2.1.3 High Pressure Carbon Monoxide Conversion

In the high pressure carbon monoxide conversion (HiPco) method, high pressure carbon monoxide (CO) is heated with a volatile catalyst precursor

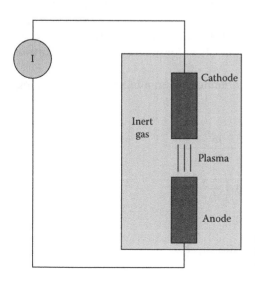

FIGURE 4.6
Schematic diagram of carbon arc dis-charge to fabricate CNTs.

such as $Fe(CO)_5$ at high temperatures. The single-walled nanotubes (SWNTs) are formed in the gas phase and are removed by filter from flowing, recirculating CO. Different from CVD, the HiPco method requires pressurized CO, very high process temperatures, and a metal catalyst that is difficult to remove at the end of the process.

4.2.1.4 Laser Ablation or Vaporization

In the laser ablation method, nanotubes are produced by pulsed YAG laser ablation of a graphite target in a furnace at temperatures near 1200°C. The target is hit with intense laser pulses, generating carbon gas from which the nanotubes are formed. Unlike CVD, in the laser ablation method, it is difficult to control the location and alignment of the nanotubes produced. Small quantities of nanotubes result from this process and they are generally tangled, making some applications difficult. The method also requires the evaporation of carbon atoms from solid targets at very high temperatures.

4.3 Characteristics of Carbon Nanotubes

Since the applications of CNTs are related to some of their specific properties, it is important to realize these properties in detail. As an example, CNTs can be used to store hydrogen because they have the ability to elastically sustain loads at high deflection angles. They are capable of storing and absorbing considerable energy (Sinha and Yeow 2005, Wong et al. 1997). SWNTs are structurally similar to a single layer of graphite: a semiconductor with zero bandgap. Moreover, SWNTs can be either metallic or semiconducting depending upon the tube diameter and the chirality. Chirality is defined as the sheet direction in which the graphite sheet (graphene) is rolled to form a nanotube cylinder.

The direction and diameter can be obtained by an integer pair (n, m) using (Terrones 2003)

$$d = \frac{k\sqrt{n^2 + m^2 + mn}}{\pi}$$

$$\alpha = \arctan\left(\frac{-\sqrt{3}n}{n + 2m}\right)$$

where
k is the lattice constant in the graphite sheet
α is the chiral angle

There are three defined categories of CNTs based on the relation between *m* and *n* (Dresselhaus et al. 1998):

1. Armchair: $n = m$ and chiral angle α equal to 30
2. Zigzag: $n = 0$ or $m = 0$ and chiral angle $\alpha = 0$
3. Chiral: other values of *n* and *m* and chiral angles α between 0 and 30

Figures 4.7 through 4.11 show the three categories of CNTs obtained by using free online nanotube Applet 2000. Meanwhile, the *xyz* coordinates of several cases of nanotubes are generated in Figures 4.12 through 4.19 by using the free version of the nanotube modeler of Jcrystalsoft 2005–2009. Moreover, the categories of electron diffraction of CNTs are displayed in Figures 4.20 through 4.22 via the free simulator software of the University of North Carolina of Chapel Hill.

All armchair SWNTs are metals. Meanwhile, those with $n - m = 3j$ with *j* as a nonzero integer are semiconductors with a tiny bandgap. All others are semiconductors, which have the bandgap that is inversely related to the diameters of the nanotubes.

The dielectric responses of the CNTs are found to be highly anisotropic. The electronic transport in metallic SWNTs and multiwalled nanotubes (MWNTs) occurs ballistically (without scattering) over long lengths owing to the nearly one-dimensional electronic structure. This enables nanotubes to

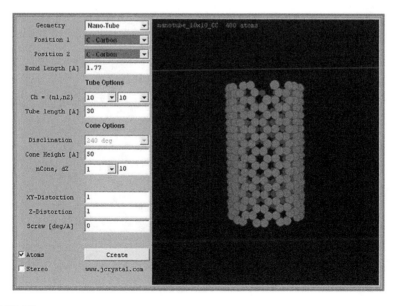

FIGURE 4.7
Armchair $n = m = 10$, with α equal to 30.

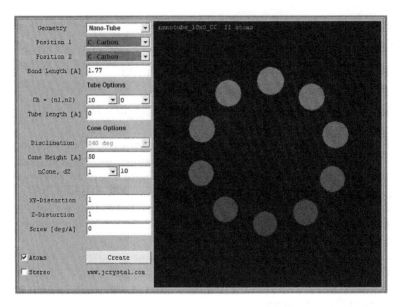

FIGURE 4.8
Zigzag $n = 10$, $m = 0$, with α equal to 0.

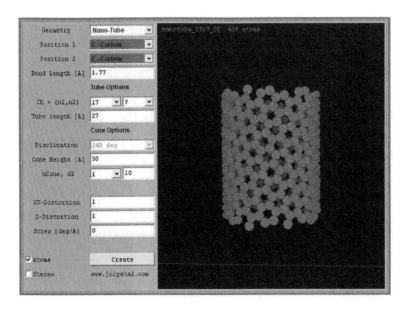

FIGURE 4.9
Chiral $n = 17$, $m = 7$, with α less than 30.

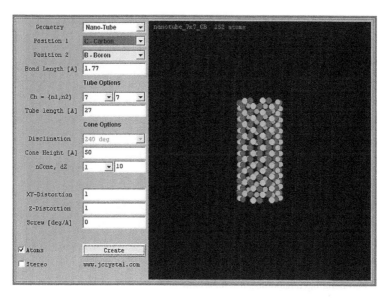

FIGURE 4.10
Armchair carbon–boron nanotube, $n = 7$, $m = 7$.

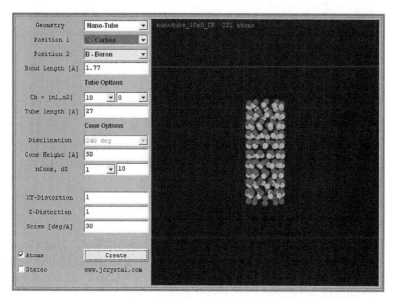

FIGURE 4.11
Zigzag carbon–boron nanotube, $n = 10$, $m = 0$, screw $30°/A$.

FIGURE 4.12
Armchair boron–nitrogen nanotube, $n = 10$, $m = 10$.

FIGURE 4.13
Armchair CNT, $n = 10$, $m = 10$, length $= 15$ Å.

carry high currents with negligible heating. The electrical and electronic properties of nanotubes are affected by distortions like bending and twisting. Bending introduces a pentagon–heptagon pair in CNTs, which results in metal–metal and semiconductor–metal nanoscale junctions that can be used for nanoswitches. When bending angles are more than 45, the effect of bending becomes very crucial. The ultimate effect of bending is a reduction

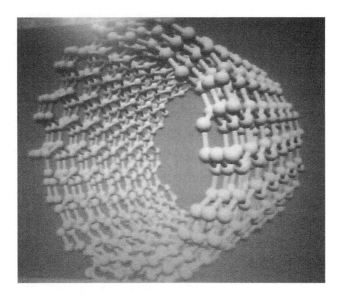

FIGURE 4.14
Armchair multiwall ($=3$) CNT, $n=10$, $m=10$, length $=15$ Å.

FIGURE 4.15
Zigzag single CNT, $n=10$, $m=0$, length $=15$ Å.

FIGURE 4.16
Zigzag multiwall $(=3)$ CNT, $n = 10, m = 0$,
length $= 15$ Å.

FIGURE 4.17
Armchair carbon nanosheet, $n = 10$, $m = 10$, length $= 7$ Å.

FIGURE 4.18
Chiral carbon nanosheet, $n = 10$, $m = 0$ length $= 7$ Å.

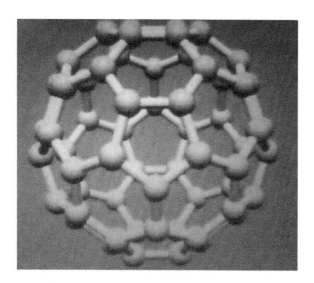

FIGURE 4.19
Bucky ball.

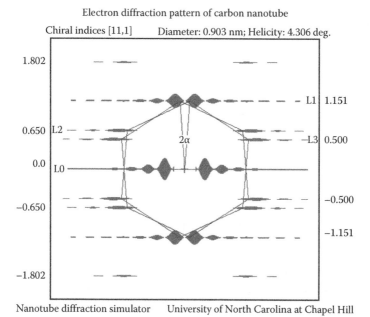

FIGURE 4.20
Nanotube diffraction by free online UNC simulator.

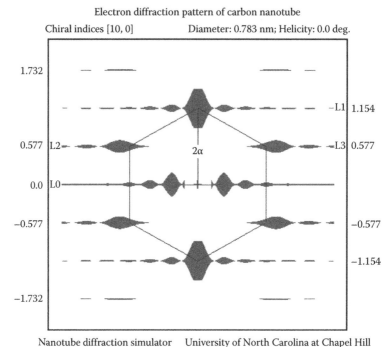

Electron diffraction pattern of carbon nanotube

Chiral indices [10, 0]   Diameter: 0.783 nm; Helicity: 0.0 deg.

Nanotube diffraction simulator   University of North Carolina at Chapel Hill

FIGURE 4.21
Nanotube diffraction by free online UNC simulator.

in the conductivity of CNTs. Upon twisting, a bandgap opens up that turns metallic CNTs to semiconducting CNTs. Twisting above a certain angle results in the collapse of CNT structures. Also, superconductivity in SWNTs has been observed, but only at low temperatures (Sinha and Yeow 2005).

Conductivity in MWCNTs is quite complex. Some types of armchair-structured CNTs appear to conduct better than other metallic CNTs. Furthermore, interwall reactions within MWCNTs have been found to redistribute the current over individual tubes nonuniformly. However, there is no change in current across different parts of metallic SWCNTs. The behavior of the ropes of semiconducting SWCNTs is different in that the transport current changes abruptly at various positions on the CNTs.

The conductivity and resistivity of ropes of SWCNTs has been measured by placing electrodes at different parts of the CNTs. The resistivity of the SWCNTs ropes was of the order of 10^{-4} Ω-cm at 27°C. This means that SWCNTs ropes are the most conductive carbon fibers known. The current density that was possible to achieve was 10^{-7} A/cm^2; however, in theory the SWCNTs ropes should be able to sustain much higher stable

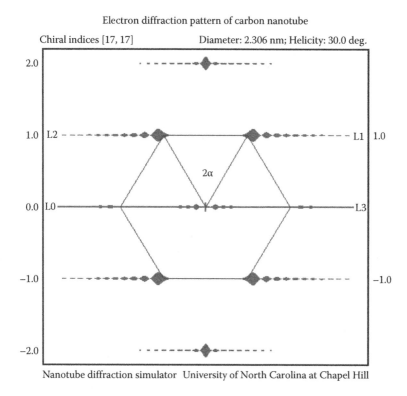

Electron diffraction pattern of carbon nanotube

Chiral indices [17, 17]   Diameter: 2.306 nm; Helicity: 30.0 deg.

Nanotube diffraction simulator University of North Carolina at Chapel Hill

FIGURE 4.22
Nanotube diffraction by free online UNC simulator.

current densities, as high as 10^{-13} A/cm^2. It has been reported that individual SWCNTs may contain defects. Fortuitously, these defects allow the SWCNTs to act as transistors. Likewise, joining CNTs together may form transistor-like devices. A CNT with a natural junction (where a straight metallic section is joined to a chiral semiconducting section) behaves as a rectifying diode, i.e., a half-transistor in a single molecule. It has also recently been reported that SWCNTs can route electrical signals at speeds up to 10 GHz when used as interconnects on semiconducting devices.

Field emission results from the tunneling of electrons from a metal tip into a vacuum, under the application of a strong electric field. The small diameter and high aspect ratio of CNTs is very favorable for field emission. Even for moderate voltages, a strong electric field develops at the free end of supported CNTs because of their sharpness. It has been realized that these field emitters must be superior to conventional electron sources and might find their way into all kinds of applications, most importantly flat-panel displays. It is remarkable that after only 5 years, Samsung actually realized a very bright color display, which will be shortly commercialized using this technology. While studying the field emission properties of MWCNTs, it was observed that together with electrons, light is emitted as well. This luminescence is induced by the electron field emission, since it is not detected without applied potential. This light emission occurs in the visible part of the spectrum, and can sometimes be seen with the naked eye.

It is true that the basic technology that produces x-rays has remained essentially almost the same for a century, but new nanotechnology experiments at the University of North Carolina at Chapel Hill and Applied Nanotechnologies have shown a significant improvement (Yue et al. 2009). The nanoscientist team has conducted experiments using CNTs to produce x-rays. It is demonstrated that the cold-cathode device can generate sufficient x-ray flux to create images of extremities such as the human hand and fish that are as good as standard x-rays.

Using CNTs is so advantageous since the machines associated with them can work at room temperature rather than the approximate 1500°C that conventionally x-ray machines now require and produce. Miniaturizing x-ray devices could have more major benefits such as allowing technicians to take x-rays inside or outside ambulances and manufacturing large-scale x-ray scanning machines for industrial inspections, airport security screening, and customs inspections. This might be considered a major breakthrough in x-ray technology.

CNTs have the intrinsic characteristics desired in material used as electrodes in batteries and capacitors, two technologies of rapidly increasing importance. CNTs have a tremendously high surface area, good electrical conductivity, and very importantly, their linear geometry makes their surface highly accessible to the electrolyte.

In any electronic circuit, but particularly as dimensions shrink to the nanoscale, the interconnections between switches and other active devices become increasingly important. Their geometry, electrical conductivity, and ability to be precisely derived make CNTs the ideal candidates for the connections in molecular electronics. In addition, they have been demonstrated as switches themselves.

The fabrication and analysis of the electrochemical behavior of vertically aligned CNT electrodes is engineered in Poh et al. (2009). Vertically aligned single-walled carbon nanotubes (VASWCNTs) are becoming increasingly recognized due to their fast electron transfer rates. However, the chemistry available for further functionalizing these electrodes is limited. In the former work a new approach to the fabrication of VASWCNTs is described. SWNTs were covalently attached to a p-type silicon (100) (Si) wafer surface using a thioester linkage in which the nanotubes were first acid treated and then, in the presence of a hydroxylated Si wafer surface, reacted with phosphorus pentasulfide (a mild electrophilic catalyst). The novel nanostructure was characterized using atomic force microscopy (AFM) showing vertical alignment with Fourier transform infrared (FTIR) spectroscopy indicating pendant thiocarboxylic acid groups for further reaction. In addition, electro-chemical properties using cyclic voltammetry (CV) indicate that the electrodes have excellent electrochemical properties with an electron transfer rate of 2.98×10^{-3} cm/s.

The electrochemical properties of an MnO_2/activated CNT (A-CNT) composite as an electrode material for supercapacitor is investigated in Ko and Kim (2009). MWCNTs were chemically activated using KOH in order to improve their specific surface area, electrical conductivity, and specific capacitance. Using such an A-CNT, a composite of MnO_2/A-CNT was prepared by the coprecipitation method, and its physical and electrochemical properties were evaluated for use as an electrode material in supercapacitors. For comparison, a composite of MnO_2/CNT was also prepared using an inactivated CNT and characterized in an aqueous solution of 1.0 M Na_2SO_4. The specific capacitances of the MnO_2/A-CNT composite electrode, measured using CV at scan rates of 10 and 100 mV/s, were found to be 250 and 184 F/g, respectively, compared with 215 and 138 F/g, respectively, for the MnO_2/CNT composite electrode. Because of CNT activation, the MnO_2/A-CNT composite electrode showed an improved performance in both the capacitance and cycle performance, due to the alleviation of the accumulated stress during charge–discharge cycling.

A study of the magnetic and electromagnetic properties of gamma-Fe_2O_3–MWCNTs and Fe/Fe_3C–MWCNT composites is reported in Xu et al. (2009). A ferromagnetic gamma-Fe_2O_3–MWCNT composite was prepared by a reverse microemulsion technique and was transformed to a Fe/Fe_3C–MWCNT composite by further heat treatment in an H_2 atmosphere at 950°C. Transmission electron microscopy (TEM) results suggest that MWCNTs are both surface decorated and filled with gamma-Fe_2O_3 or Fe/Fe_3C nanoparticles. The saturation magnetization and anisotropy field are particularly strong for Fe/Fe_3C–MWCNTs. The reflection loss of Fe/Fe_3C–MWCNTs is better than that of gamma-Fe_2O_3–MWCNTs at all frequencies between 2 and 18 GHz, resulting from enhanced magnetic loss and better matched characteristic impedance, rather than electric loss, as shown by the complex relative permeability. The microwave absorbing properties can be modulated simply by manipulating the thickness of the prepared Fe/Fe_3C–MWCNT composite for application in different frequency bands.

Spark plasma sintering of double-walled carbon nanotube–magnesia nanocomposites is presented in Legorreta Garcia et al. (2009). A double-walled carbon nanotube–MgO powder is prepared without any mixing. The applied pressure is the main parameter acting on densification. Increasing the maximum temperature and holding time is marginally beneficial. The nanotubes are blocking the matrix grain growth. The nanocomposite prepared using the most severe spark plasma sintering conditions (1700°C; 150 MPa) shows mostly undamaged nanotubes and a higher microhardness than the other materials, reflecting a better bonding between the nano-tubes and the matrix. The electrical conductivity of all nanocomposites is over 12 S/cm.

Investigating the dc voltammetry of ionic liquid (IL)-based capacitors, effects of faradaic reactions, electrolyte resistance, and voltage scan speed is implemented in Zheng et al. (2009) using an electrode of CNTs in EMIM-EtSO$_4$. CNT electrodes in combination with IL electrolytes are potentially important for energy storage systems. Such a study concentrates on the analytical aspects of CV to probe the double layer capacitance of these relatively unconventional systems (that involve rather large charge–discharge time constants).

New research analyzing the effect of CNT size on superconductivity properties of MgB$_2$ was implemented in Yeoh et al. (2005). Experimental results were presented for the incorporation of CNT in a polycrystalline MgB$_2$ superconductor based on x-ray diffraction and TEM measurements. Electron microscopy studies show that nanotubes are embedded into the MgB$_2$ matrix with a fraction of nanotubes found to be unreacted and entan-gled. In contrast, magnetization measurements indicate a change in the critical current density with the length of nanotubes and not with their outside diameter. This implies that longer nanotubes tend to entangle, pre-venting their homogenous mixing with MgB$_2$ and dispersion. It was found that CNT doping of MgB$_2$ enhanced the critical density and depressed the critical temperature.

The postgrowth processing of CNT arrays is reported in Yin et al. (2004). Highly ordered CNT arrays were fabricated by pyrolysis of acetylene using anodic–aluminum–oxide templates. To avoid the natural tendency of the nanotubes sticking together and forming haystack-like bundles when expos-ing the nanotubes from the growth template, a new postgrowth treatment process using a mixture of 6 wt% phosphoric acid and 1.8 wt% chromium oxide as the etchant and 0.1 wt% gum arabic or 5 wt% polymethacrylic acid as the dispersant was developed yielding for the first time well-aligned and spatially free-standing CNT arrays. The as-prepared CNT arrays, which are vertically aligned and well separated, could be used for many applications such as mechanical oscillators, field emission, and sensors and the exposed nanotubes offer a good platform for the study of the collective behavior of electrical and magnetic nanoarrays.

The building and testing of the organized architecture of CNTs is presented in Vajtai et al. (2003). The research focuses on the directed assembly of MWCNTs on various substrates into highly organized structures that include vertically and horizontally oriented arrays, ordered fibers, and porous mem-branes. The concept of growing such an architecture is based on the growth selectivity on certain surfaces compared with others. The selective placement of ordered nanotube arrays is achieved on patterned templates prepared by lithography or oxide templates with well-defined pores. The growth of nanotubes is achieved by CVD using hydrocarbon precursors and vapor phase catalyst delivery. Efforts in creating nanotube circuits selectively and controllably and on the spatially resolved electronic properties of nanotubes is reported as well.

Investigating the impact of electron–phonon scattering on the perform-ance of CNT interconnects for GSI is reported in Naeemi and Meindl (2005). The electron mean-free path in CNTs can be as large as several micrometers for small bias voltages. For large biases, electrons get backscattered by optical and zone-boundary phonons and nanotube resistance can increase by more than 100 times. This pair reveals that this kind of backscattering has a small impact (error 25%) in most interconnect applications of CNTs in which adequate numbers of nanotubes are connected in parallel. This is mainly due to relatively small electric fields along nanotubes when they are used as interconnects. This is in sharp contrast with transistor applica-tions of CNTs in which transconductance degrades considerably by elec-tron–phonon scatterings unless their channels are made ultrashort (10 nm).

The fabrication of regularly coiled CNTs was investigated by Biró et al. (2003). A new family of Haeckelite nanotubes was generated in a systematic way by rolling up a two-dimensional threefold coordinated carbon network composed of pentagon–heptagon pairs and hexagons in a proportion of 2:3. In this model, the n-Hx rings are treated like regular building blocks of the structure. A cohesion energy calculation shows that the stability of the generated three-dimensional Haeckelite structures falls between that of straight CNTs and that of 60. The electronic density of states of the Haeckelite computed with a tight binding Hamiltonian that includes the orbital only shows that the structures are semiconductors.

Experimental observation and quantum modeling of electron irradiation on SWNTs was implemented in Charlier et al. (2003). Experimental-based electron irradiation at high temperatures in a TEM is used to investigate isolated, packed, and crossing SWNTs. During continuous, uniform atom removal, the surfaces of isolated SWNTs heavily reconstruct leading to drastic dimensional changes. In bundles, the coalescence of SWNTs is observed and induced by vacancies via a zipper-like mechanism. "X," "Y," and "T" carbon nanostructures are also fabricated by covalently connecting crossed SWNTs in order to pave the way toward the controlled fabrication of nanotube-based molecular junctions and network architectures exhibiting exciting electronic and mechanical behavior. Each experiment is followed by quantum modeling in order to investigate the effect of the irradiation process at the atomic level.

The electronic properties of SWCNTs and their dependence on synthetic methods were investigated in Buzatu et al. (2004). Because of their high electrical conductivity and strength, high-sensitivity AFMs already use CNTs for their tips, and carbon nanostructures are also used as electron beam emitters for medical and scientific equipment. Electron emission is directly correlated with the work function and the ionization potential of CNTs. Gaussian 98 software was used to perform theoretical quantum

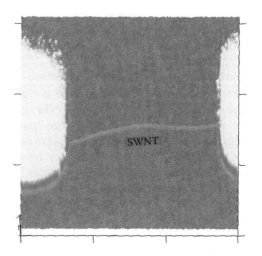

FIGURE 4.23
Nanotransistor.

calculations on a limited set of HyperChem 5.01 simulated metallic SWCNTs. These initial sets of calculations show that the bandgaps and work functions of these small carbon nanostructures are dependent upon the diameter of the tubes, and to a lesser degree so is the ionization potential. In addition, how the manufacturing methods can directly affect the diameter of the nanotubes produced is demonstrated, and therefore directly influences the electrical properties of the nanotubes.

Although today's state-of-the-art transistors are characterized by a physical gate length of approximately 50 nm, research groups in the industry and academia have already demonstrated devices with a gate length close to 30 nm. The trend of the transistor physical gate length is shown in Figure 4.23. In Chau et al. (2003), the performance and energy delay trends for research devices down to 10 nm and also discusses the 10 nm barrier and potential ways to break it were explored.

It was reported recently that a group of Korean scientists revealed they have a prototype of what they claim is the world's fastest nanoscale transistor. Figure 4.24 shows a simple nanotube-based transistor.

The team, led by Seoul National University engineering professor Seo Kwang-seok, reported on December 11, 2008 that they succeeded in creating a 15 nm high electron mobility transistor (HEMT) with a maximum frequency speed of 610 GHz. The team used its self-developed technology called the "slope etching process" to materialize hyperfine electrodes of the transistors. It was previously reported that Fujitsu Ltd. of Japan has developed the most advanced nanotransistor with a size of 25 nm and a speed of 562 GHz.

The synthesis of CNT by localized resistive heating of a MEMS structure in a room temperature chamber is fabricated in Christensen et al. (2003). It was reported that this was the first known vapor-deposition CNT growth method that does not need globally evaluated temperatures. The localized, selective, and scalable process is compatible with on-chip microelectronics

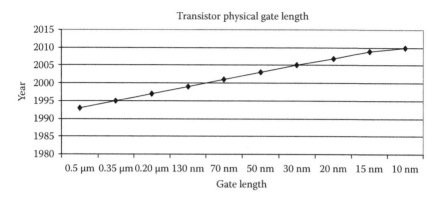

FIGURE 4.24
Miniaturization of physical gate length of nanotubes-based transistors.

and removes the necessity for post-synthesis assembly of nanostructures to form integrated circuits. This approach makes the direct integration of CNT devices on-chip transudation, readout, processing, and communication circuitry possible.

Utilization of CNTs as building blocks for nanodevices can only be realized when their electronic properties are preserved in the device configuration. It has been shown that the bending or overlapping of CNTs results in a decrease in their conductance. In this research of Ravindran et al. (2004), a significant refinement is reported to integrate quantum dots at the end of MWCNTs and to report a MWCNT–QD–MWCNT heterostructure characterized via FTIR.

Based on the intrinsic ballistics of electrons in nanotubes, a novel design of antenna with a nanotube is proposed in Qi and Rui (2004). Its radiation efficiency and radiation directivity have been investigated. The numerical results show that nanotube antennas possess lower loss and better radiation directivity and higher gain. Furthermore, because it has significantly decreased the area of the antenna at certain gains, the nanotube may be applied to broad areas.

An analysis of CNT intermolecular p–n tunnel junction transistors based on mesoscopic material parameters is presented in Richardson (2003). The tunneling rate across each junction was evaluated using Matsubara–Green's function technique. Characteristics of transistors such as gain, the input voltage versus output voltage, and the output voltage versus current is calculated.

Vertically aligned CNTs were grown by DC-PECVD on (100) cleaned Si using a mixture of methane and hydrogen as feed gases (Karamdel et al. 2006). The refinement process by means of a seed layer is patterned by a standard lithography method. This technique has been applied on P-type 00(100) silicon substrates for the formation of the gate region of N-MOSFET devices.

The resulting transistor has a drive current of 286 $\mu A/\mu m$ in VGs = 250 mV and a maximum g_m equal to 2100 mS/mm with a subthreshold slope of 100 mV/decade.

An extraction of the parameters of the high permittivity ultrathin (0.5–2.0 nm) gate dielectric technique is presented in Kar (2004). Such a technique has been proposed for the extraction of the parameters of MOS devices with ultrathin high-K gate dielectrics from the capacitance data. Experimental data confirms that barring the changeover regions, the space charge capacitance is an exponential function of the surface potential in accumulation and in strong inversion. It is a parabolic function of the same in depletion and weak inversion. This permits easy extraction of the gate capacitance, the doping density, quantization index, the flat band voltage, surface potential, and the dielectric potential.

The technological peculiarities of the production and construction of modern field silicon nanotransistors (FSNT) are considered in Lobanova et al. (2006) including teracycle transistors (TCT). Their performance capabilities and limit properties are analyzed.

The ballistic nanotransistors are investigated in Timp et al. (1999). It was achieved extremely high drive current performance and ballistic (T larger than 0.8) transport using ultrathin (less than 2 nm) gate oxides in sub-30 nm effective channel length in nano-MOSFETs. The peak drive performance in a nano-MOSFET was observed at 1.3 nm for a 1.5 V voltage power supply.

A simple and efficient method for the growth of highly aligned and densely packed CNTs under a wide range of growth parameters was successfully demonstrated in Zhu et al. (2005). The developed CVD deposition of A-CNTs is fully compatible with the current microelectronics fabrication process and has great potential in various applications of photonic devices, microelectronics, interconnect, and packaging. A U.S. provisional patent has been filed.

4.4 Characterization of Carbon Nanotubes Using Raman Microspectroscopy

One primary method for the analysis of carbon nanostructures is Raman microspectroscopy. Research studies have predicted that nanotubes can be used as conducting or insulating devices depending on their molecular structure, which leads to applications in electronics, transistors, and photonics. It is also suggested that nanotubes should be immensely strong, becoming the ultimate carbon fiber. Moreover, SWNTs have remarkable adsorptive and catalytic properties and can be used as basic components of a new generation of fuel cells. This molecular nature is unprecedented for devices of this size. The challenge exists to find a method capable of finding these tiny particles and nondestructively analyze them for structural composition. Raman microspectroscopy is capable of doing that and more.

Nanotubes were synthesized by various laboratories (Figure 4.25). They were imaged as shown in Figure 4.26. Several areas of each sample were evaluated. The instrument is equipped with a nanodiode laser. The integration times for spectral collection takes up to 20 s per acquisition. The spectra indicate that the purity and structure of the nanotubes can be determined using the Raman spectra. The results of two samples are shown in Figures 4.27 and 4.28. Raman is an effective way to screen nanotubes that is reliable, nondestructive, and requires little sample preparation.

FIGURE 4.25
Raman testing apparatus.

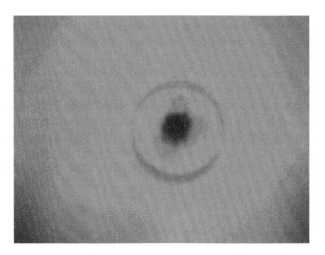

FIGURE 4.26
SWCNT sample to be tested by Raman system.

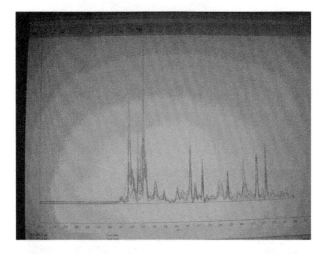

FIGURE 4.27
Raman results of two different samples of SWCNTs.

4.5 Standard Test Methods for Measurement of Electrical Properties of Carbon Nanotubes [IEEE 1650]

This standard provides methods for the electrical characterization of CNTs. The methods are independent of processing routes used to fabricate the CNTs. This standard covers the recommended methods and standardized

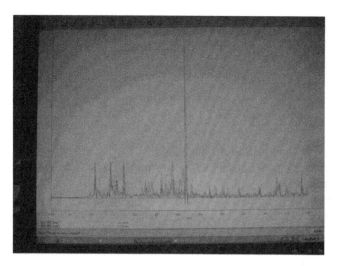

FIGURE 4.28
Raman results of two different samples of SWCNTs.

reporting practices for the electrical characterization of CNTs. Due to the nature of CNTs, significant measurement errors can be introduced if not properly addressed. This standard describes the most common sources of measurement error and gives recommended practices in order to minimize and/or characterize the effect of each error.

Standard reporting practices are included in order to minimize confusion in analyzing reported data. A disclosure of environmental conditions and sample size is included so that results can be appropriately assessed by the research community. These reporting practices also support the repeatability of results, so that new discoveries may be confirmed more efficiently. The practices in this standard were compiled from scientists and engineers from the CNT field. These practices were based on standard operating procedures utilized in facilities worldwide. Standardized characterization methods and reporting practices create a means of effective comparison of information and a foundation for manufacturing readiness.

4.6 Schrödinger–Hamiltonian Dynamical of Particles Motion along Nanotubes

Since CNTs are used as a media of moving particles, analyzing the particles' motion is so critical. The motion of a particle moving inside a nanotube can be analyzed using the governing Schrödinger equation to describe how the wave function of the particle changes in time. Assume that a particle moves along a single axis, i.e., the z-direction as a transnational motion. Then the Schrödinger equation that describes the motion can be represented in the second-order differential form:

$$-\frac{h^2}{2m}\frac{d^2\Phi(z)}{dz^2} + V(z)\Phi(z) = E\Phi(z) \qquad (4.1)$$

where
  z is the particle's position on the z-axis
  $\Phi(z)$ is the wave function, which is the amplitude for the particle to have a given position
  $V(z)$ represents the potential energy of the particle
  E is the sum of the particle's kinetic energy and potential energy

Now define the Hamiltonian function of the particle kinetic and potential total energies as

$$H(z, p) = \frac{p^2(z)}{2m} + V(z) = -\frac{h^2}{2m}\frac{d^2\Phi(z)}{dz^2} + V(z) \qquad (4.2)$$

in which $p = kh$ and h is the reduced Planck's constant. Assume that the particle moves from $z = 0$ to $z = z_f$, and the potential energy is

$$k = \sqrt{\frac{2mE}{h^2}} \qquad (4.3)$$

Thus, the motion of the particle is bounded in the "potential wall," and

$$\Phi(z) = \begin{cases} \text{continuous,} & \text{if } 0 \le z \le z_f \\ 0, & \text{if } z < 0 \text{ and } z < z_f \end{cases} \qquad (4.4)$$

If the particle moves such that $0 \le z \le z_f$, then the potential energy is zero and yields to

$$-\frac{h^2}{2m}\frac{d^2\Phi(z)}{dz^2} = E\Phi(z) \qquad (4.5)$$

The solution of the resulting second-order differential equation

$$\frac{d^2\Phi(z)}{dz^2} + k^2\Phi(z) = 0 \qquad (4.6)$$

where the solution for Equation 4.6 can be easily verified as

$$\Phi(z) = A\sin kz + B\cos kz \qquad (4.7)$$

The solution can only be easily verified by plugging in the solution for Equation 4.7 in the left hand side of the differential equation (4.5).

It is obvious that one must use the boundary conditions. We have $\Phi(z)|_{z=0} = \Phi(0) = 0$, and therefore $A = 0$. From $\Phi(z)|_{z=z_f} = \Phi(z_f) = 0$ using $B \sin kz_f = 0$, one must find the constant B and expression for kz_f. Assuming that $B \neq 0$ from $B \sin kz_f = 0$, we have $kz_f = n\pi$, where n is the positive integer (if $n = 0$, the wave function vanishes everywhere, and thus, $n \neq 0$). From $k = \sqrt{2mE/\bar{h}^2}$ and making use of $kz_f = n\pi$, we have the expression for the energy (discrete values of the energy that allow a solution of the Schrödinger equation) as

$$E_n = \frac{\bar{h}^2\pi^2}{2mc^2}n^2, \quad n = 1, 2, 3, \ldots \tag{4.8}$$

where the integer $n = 1$ designates the allowed energy level (n is called the quantum number). For example, if $n = 1$ and $n = 2$, we have $E_1 = \bar{h}^2\pi^2/mc^2$ (the lowest possible energy that is called the ground state) and $E_1 = 2\bar{h}^2\pi^2/mc^2$. This equation quantizes the energy of moving particles.

Figure 4.29 shows how the system behaves as a harmonic oscillator and shows the sinusoidal response. Then, this simplified nanotube dynamics can be used in microtransistors, gates, or switches analysis.

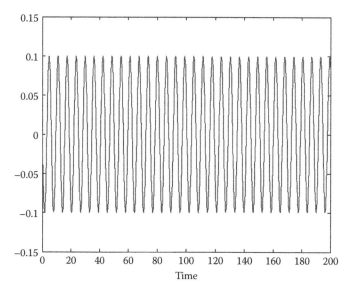

FIGURE 4.29
Free electron moving inside nanotube as quantum harmonic oscillator.

4.7 Carbon Nanotubes in Telecommunications

This section presents some of the promising applications of nanotubes in the telecommunications field ranging from tunable resistors, neural commentators, spacecrafts and satellites microwave diodes, fiber-optics transistors, to wireless and radio transmission as well as the IEEE fabrication standardization.

4.7.1 Resistivity of Nanotubes

A nanoscientists team at the University of North Carolina at Chapel Hill has discovered that the electrical resistance between nanotubes and graphite surfaces changes according to the orientation of the nanotubes. Such results might have a fruitful impact on the future of telecommunications.

More specifically, the researchers found that electrical resistance between a nanotube and a graphite surface increased by six times as the end of the nanotube is rotated by 360°. This happens because the atoms in the nanotube and the graphite are arranged in a hexagonal shape.

This highlights the usefulness of nanotubes as ultrasmall circuits. The preferred orientations of the nanotubes would necessarily be aligned when the devices are assembled to manipulate electrical resistance.

This discovery can be added to previous findings that the nanotubes have an incredible capability to bend and rebend. A carbon-13 nuclear magnetic resonance (NMR) is tested to measure the number of electrons in SWNTs as well as the tubes' electrical properties.

More research has been dedicated to recognizing the properties of metallic and semiconducting nanotubes and revealing the potential to fabricate either all-metallic or all-semiconducting CNTs with tunable structures. It has been found that the nanotubes need more significant energy to roll across some surfaces than to slide over these same surfaces due to electrical interactions as believed.

4.7.2 Carbon Nanotubes as Neural Communicators

CNTs characterized by their highly mechanical strength and electrical conductivity devices have been added to batteries to increase their surface area and are being used to develop light-emitting devices for telecommunications.

A group of researchers at the University of Texas have successfully demonstrated that mats of SWCNTs can communicate electrical signals to neurons. These results suggest that SWCN could be used as an electrical interface between neural prosthetics and the human body. Prosthetics are devices used to replace damaged or missing nerves in the human body.

Multifunctional and hybrid nanostructured materials for neural interfaces that are soft, have low impedance, high charge capacity density, and are capable of controlled drug release were designed in Abidian and

Martin (2009). The fabrication process includes the electrospinning of anti-inflammatory drug-incorporated biodegradable nanofibers, encapsulation of these nanofibers by an alginate hydrogel layer, then the electrochemical polymerization of conducting polymers around the electrospun drug-loaded nanofibers to form nanotubes and within the alginate hydrogel scaffold to form cloud-like nanostructures.

Rat neurons were grown on thick mats of CNTs seeded on flexible plastic sheets. Surprisingly, instead of treating the mats like a foreign surface, the nanotubes absorbed an important neural protein and formed a roughly textured carpet on which nerves grow rapidly. When an electrical charge signal was sent across the sheet, the neurons responded with an electrical signal of their own, called an action potential, indicating that the neuron did receive the message signal.

The cochlear implant is a common example of a neural prosthetic in which electrodes are used so that it responds to sound waves and sends electrical signals directly into the brain of a person suffering from severe hearing loss. Similarly, retinal prosthetics are being developed by neuroscientists hopefully to restore vision in blind people. Importantly, long-term neural prosthetic implants may possibly cause soreness and scarring, since the body treats them as foreign materials. This advantage can be avoided by covering the CNT with molecules that look friendly to human cells.

4.7.3 Nanotubes as Microwave Diodes in Spacecrafts and Satellites

Communication with spacecrafts and satellites depends on microwave devices. The currently used microwave devices are based on relatively inefficient thermionic electron devices that require heating and cannot be switched on instantaneously.

A microwave diode that uses the CNTs' cold-cathode electron source and operates at high frequency and high current densities has been developed by Teo et al. (2005). Since such a nanotube-based diode is very light, it can be switched on instantaneously and does not need a heating source, it can be very useful in microwave devices used in telecommunications.

The microwave diode was constructed so that the CNT field-emission source was directly driven at gigahertz (GHz) frequencies. It was composed in such a way that arrays of vertically aligned CNTs were integrated on a coaxial post in a resonant cavity. In the diode device, a high oscillating electric field at the end of the coaxial post was induced by radiofrequency electromagnetic radiation at the input. Furthermore, this electric field was amplified by the CNT array.

Each single CNT array consists of uniform individual CNTs located at distances as much as twice their height so as to minimize the electrostatic-field shield from neighboring emitters. Each cathode has an active area equal to $1/2 \times 1/2$ mm^2 (corresponds to 2500 CNTs) and each of the 16 cathodes can be formed at the same time.

The microwave diode device was operated at 1.5 GHz via various radio-frequency power inputs to generate different macroscopic electric fields at the array of CNT emitters. When the cavity walls and emitters are grounded, the radiofrequency electric field exists only inside the cavity. A spectrum analyzer connected to the output antenna shows the presence of the fundamental 1.5 GHz peak in the diode cavity. More specifically, cathodes were operated at 1 mA and 1.5 GHz for 40 h without decreasing the current output. When a 29 MV/m radiofrequency electric field is applied, the output at the anode reaches 3.2 mA, with a current density average of 1.3 A/cm^2. These values are corresponding to the amplified peak current of 30 mA and a current density of 12 A/cm^2 in the output waveform.

The importance of such results is shown as follows. Spacecraft and satellite thermionic communications are based on sources operated by direct current or at low frequency; their electron beam is usually modulated downstream in an extended interaction line, leading to physically long devices.

Meanwhile, utilizing CNT cold cathodes that have a cavity of a few hundred micrometers or less have low capacitances and can be operated at very high frequencies. Therefore, such efficient cold diodes can be used directly as the input stage of a microwave amplifier. Not to mention that CNT emitters are robust and do not suffer from electromigration because of their strong C–C covalent bonding. On the other hand, metal emitters show frequent thermal failure owing to field-induced sharpening.

CNT cathodes hold promise for a new generation of lightweight, compact, and efficient microwave devices for satellites or spacecraft telecommunications because of their tiny size and ability to generate and modulate the beam simultaneously without the need for high temperatures.

Using nanotube diode technology will help significantly in replacing conventional heavy, bulky, high temperature, microwave amplifiers. Typically, conventional satellites carry 50 microwave amplifiers on board, each weighing about 1 kg and measuring about 30 cm in length.

On average, it costs about 10,000 pounds sterling to send a single kilogram of payload into space. This cost will be dropped dramatically if the weight and size of the microwave devices are reduced. With a potential weight and size reduction up to 50%, the cost will drop by half and the capability of conventional satellite systems would be increased. Additionally, such a technology will drive the industry toward very low cost microsatellites that weigh about 10 kg.

Currently, most long-range telecommunication satellite systems are based on microwave links including high power transmitters on ground stations and on satellites. It is expected in the near future that these satellites will be equipped with 50 to a 100 traveling wave tubes (TWTs). As a result of the rapid growth of telecommunications and the saturation of the present frequency bands, new bands are needed to be allocated at higher frequencies (30–100 GHz) bringing the present technology close to its limit.

Space satellite manufacturers are looking for a technological breakthrough to satisfy the rapid need for low-cost microwave amplifiers that operate at higher frequencies. For this reason, cold cathodes using CNTs are considered as an ideal start. The overall advantage of the CNT-based cold cathodes usage will lead to a new generation of high frequency (30–100 GHz), compact, and low-cost vacuum microwave amplifiers.

It has been demonstrated for the first time at the University of California in Irvine that CNTs can send an electrical signal on a chip faster than traditional copper or aluminum wires, at speeds of up to 10 GHz. Such a breakthrough could lead to faster and more efficient computers and improved wireless network and cellular phone systems.

Modern semiconductor chips that are dedicated to interconnect signals via wiring system have shown a limitation in terms of processing speed and memory. For this fact, the semiconductor industry has recently shifted from using aluminum to copper as interconnectors because copper carries electrical signals much faster than aluminum.

Recently, it has become a new trend to shift the industry from copper to nanotubes since it will provide an even larger performance advantage in terms of speed. But it should be mentioned that before such a shift could occur, nanotube technology would need to show that it is more economical and requires precise assembly and packaging.

4.7.4 Carbon Nanotubes in Fiber-Optics Telecommunications

Recent research on CNTs has reported that they can be used as saturable absorbers to control laser output pulses over a much wider range of frequency bands than the equivalent dominant traditional semiconductors (Wang et al. 2008). Although the results have shown that CNT technology still has a relative inability to precisely control the matrix growth of CNTs, it has turned into an advantageous application for telecommunications.

Traditionally, fiber lasers are used in telecommunications by a standard laser diode attached to a fiber-optics array that includes a highly doped section of fiber. The commonly used dopant is the erbium (Er). The latter diode injects the Er-doped fiber that results into a range of 1500 nm wavelength commonly used in the telecommunications industry. Diverse research on CNTs has shown their effectiveness as a saturable absorber, which motivates us to evaluate whether they could eliminate the need for traditional expensive semiconductors.

Although CNTs vary in diameter and are not precise enough in optics and electronics applications, such variations can contribute to a broad spectrum of absorption. A band of 40 nm was achievable with CNTs in one order of magnitude higher than that observed in silicon semiconductor materials. Using CNT diodes would allow for wideband tunable laser telecommunications. Orientation manipulation of the nanotubes and eliminating more impurities in the CNTs and matrix material could improve the telecommu-nication pulse frequency.

Applications in spectroscopy, biomedical research, and the telecommuni-cations field show the need for ultrashort pulse lasers with spectral tuning capability. Mode-locked fiber lasers are the most powerful and convenient sources of ultrashort pulses. Including a broadband saturable absorber as a passive optical switch inside the laser diode cavity may offer tunability over a range of wavelengths.

Although semiconductor saturable absorber mirrors are widely used in fiber lasers, they have a limited operation range (typically limited to a few tens of nanometers); in addition to the fact that their fabrication is noticeably challenging in the wavelength range of 1.3–1.5 μm.

Comparably, SWCNTs characterized by their subpicosecond recovery time, low saturation intensity, polarization insensitivity, and mechanical and environmental robustness can be used as excellent saturable absorbers.

A nanotube–polycarbonate film was engineered in Wang et al. (2008) with a wide bandwidth (>300 nm) around 1.55 μm and then was used to dem-onstrate a 2.4 ps Er^{3+}-doped fiber laser that is tunable from 1518 to 1558 nm. Principally, different diameters of nanotubes could be combined to enable compact, mode-locked fiber lasers that are tunable over a much broader range of wavelengths than other traditional semiconductor systems. In add-ition to the properties CNTs are known for, such as strength, small diameter, and stunning electronic properties, they may play a significant role in the memory and processing chips of tomorrow in breaking through the major bottleneck in the microelectronics industry.

For computational and memory reason, the transistors are getting smaller and smaller. One of the challenges that should be overcome is the speed limiting factor due to using metal wires to send signals through the systems.

This challenge has motivated researchers to pursue alternative intercon-nections by using light. However, the electrical signals from the transistor must be converted into optical signals. That is to convert the electronic signal into a photonic signal. This was achievable at IBM's labs by using new nanotube devices. A simplified description of the optical nanotransistor is shown in Figure 4.30.

Light-emitting nanotubes may find applications in telecommunications. Currently, emitting light with nanotubes at the same wavelengths used by the telecom industry to send information through optical fibers is achieved. This frequency along nanotubes varies with different tube diameters, so that the wavelengths they emit can be tuned by changing the structure and without changing its chemicals.

IBM has recently reported in *Science* that its researchers have successfully fabricated nanotube devices that are 1000 times more efficient than its own previous ones in terms of emitting light. The IBM research team has success-fully fabricated a nanotube transistor that emits a 1.5 micron infrared light, a wavelength widely used in telecommunications.

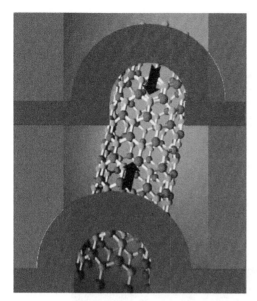

FIGURE 4.30
Nanotube transistor.

Using such light-emitting nanotubes may lead to the formation of efficient communications devices as well as all-optical computer chips. Astonishingly, the researchers found that if they inject electrons at one end of a nanotube and positive charges at the other end, they emit pulses of light or photons whose wavelength is inversely proportional to the tube's diameter.

4.7.5 Carbon Nanotubes for Wireless Communications and Radio Transmission

The systems and methods in which a CNT is used as a demodulator of amplitude-modulated (AM) signals are described in Rutherglen and Burke (2008). Due to the nonlinear current–voltage (I–V) characteristics of a CNT, the CNT induces the rectification of an applied radio frequency (RF) signal enabling the CNT to function as a demodulator of an AM RF signal. By properly biasing the CNT so that the operating point is centered on the maximum portion of the I–V curve, the demodulation effect of the CNT can be maximized. The present invention is useful for possible nanoscale wireless communications systems, e.g., nanoscale radios.

A team of researchers with the U.S. Department of Energy's Lawrence Berkeley National Laboratory and the University of California in Berkeley have created the first fully functional radio from a single CNT, which makes it by several orders of magnitude the smallest radio ever made. Make way for the real nanopod and make room in the *Guinness Book of World Records*. Figure 4.31 presents a comparison between the different versions of radios and the nanotube radio. Meanwhile, Figures 4.31 and 4.32 show a microscopic image of the nanoradio.

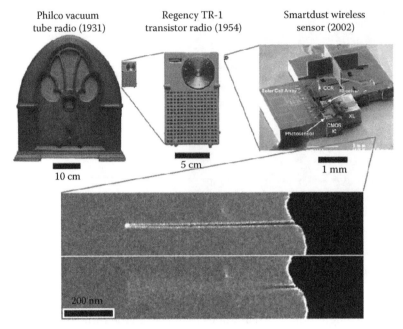

Nanotube radio (2007)

FIGURE 4.31
Nanotube radio. Over the past century, radio has shrunk dramatically from the wooden "cathedral" style radios of the 1930s to the pocket-sized transistor radios of the 1950s and more recently to the single-chip radios found in cell phones and wireless sensors. Continuing this trend, we have further miniaturized the radio by cleverly implementing multiple radio functions with a single component, the CNT. This nanotube radio is over 19 orders of magnitude smaller than the Philco vacuum tube radio from the 1930s. (Courtesy of Zettl Research Group, Lawrence Berkeley National Laboratory and University of California at Berkeley, Berkeley, CA.)

A single CNT molecule serves simultaneously as all essential components of a radio: antenna, tunable band-pass filter, amplifier, and demodulator. Using carrier waves in the commercially relevant 40–400 MHz range and both frequency and amplitude modulation (FM and AM), successful music and voice reception was demonstrated (Figure 4.33).

Given that the nanotube radio essentially assembles itself and can be easily tuned to a desired frequency band after fabrication, nanoradios will be relatively easy to mass-produce. Potential applications, in addition to incredibly tiny radio receivers, include a new generation of wireless communication devices and monitors. Nanotube radio technology could prove especially valuable for biological and medical applications.

It is also possible that the nanotube radio could be implanted in the inner ear as an entirely new and discrete way of transmitting information, or as a radically new method of correcting impaired hearing.

FIGURE 4.32

Images, taken by a TEM, show a single CNT protruding from an electrode. This nanotube is less than a micron long and only 10 nm wide, or 10,000 times thinner than the width of a single human hair. When a radio wave of a specific frequency impinges on the nanotube it begins to vibrate vigorously. An electric field applied to the nanotube forces electrons to be emitted from its tip. This electrical current may be used to detect the mechanical vibrations of the nanotube, and thus listen to the radio waves. (The waves shown in this image were added for visual effect, and are not part of the original microscope image. (Courtesy of Zettl Research Group, Lawrence Berkeley National Laboratory and University of California at Berkeley, Berkeley, CA.)

FIGURE 4.33

This simulation shows the electric field surrounding the nanotube radio during radio operation. Notice how the field is strongest at the tip of the nanotube and how the field varies as the nanotube vibrates. This effect allows the nanotube radio to demodulate radio signals. (Courtesy of Zettl Research Group, Lawrence Berkeley National Laboratory and University of California at Berkeley, Berkeley, CA.)

The CNT radio consists of an individual CNT mounted to an electrode in close proximity to a counter-electrode with a dc voltage source, such as from a battery or a solar cell array, connected to the electrodes for power. The applied dc bias creates a negative electrical charge on the tip of the nanotube, sensitizing it to oscillating electric fields. Both the electrodes and nanotube are contained in a vacuum in a geometrical configuration similar to that of a conventional vacuum tube.

Although it has the same essential components, the nanotube radio does not work like a conventional radio. Rather than the entirely electrical operation of a conventional radio, the nanotube radio is in part a mechanical operation, with the nanotube itself serving as both an antenna and a tuner.

Incoming radio waves interact with the nanotube's electrically charged tip, causing the nanotube to vibrate. These vibrations are only significant when the frequency of the incoming wave coincides with the nanotube's flexural resonance frequency, which, like a conventional radio, can be tuned during operation to receive only a preselected segment, or channel, of the electromagnetic spectrum.

The amplification and demodulation properties arise from the needle-point geometry of CNTs, which gives them unique field emission properties. By concentrating the electric field of the dc bias voltage applied across the electrodes, the nanotube radio produces a field-emission current that is sensitive to the nanotube's mechanical vibrations. Since the field-emission current is generated by the external power source, amplification of the radio signal is possible. Furthermore, since field emission is a nonlinear process, it also acts to demodulate an AM or FM radio signal, just like the diode used in traditional radios.

4.8 Conclusion

CNTs have recently received special attention in both academia as well as the industry because they exhibit extraordinary strength and unique mechanical and electrical properties that make them potentially useful in many applications in nanotechnology, electronics, telecommunications, and optics.

CNTs can be classified as SWNTs and MWNTs. They can be fabricated by four different methods: CVD, arc discharge, HiPco, and laser ablation or vaporization.

SWNTs can be either metallic or semiconducting depending upon the tube diameter and the chirality. Chirality is defined as the sheet direction in which the graphite sheet (graphene) is rolled to form a nanotube cylinder. CNTs can be categorized as armchair, zigzag, or chiral. The dielectric responses of the CNTs are found to be highly anisotropic.

The electronic transport in metallic SWNTs and MWNTs occurs ballistically over long lengths owing to the nearly one-dimensional electronic structure. This enables nanotubes to carry high currents with negligible heating. The electrical and electronic properties of nanotubes are affected by distortions like bending and twisting. Bending introduces a pentagon–heptagon pair in CNTs; which results in metal–metal and semiconductor–metal nanoscale junctions that can be used for nanoswitches. When bending angles are more than 45, the effect of bending becomes crucial. The ultimate effect of bending is a reduction in the conductivity of CNTs. Upon twisting, a bandgap opens up that turns metallic CNTs into semiconducting CNTs. Twisting above a certain angle results in the collapse of CNT structures. Conductivity in MWCNTs is quite complex. Some types of armchair-structured CNTs appear to conduct better than other metallic CNTs. It was found that CNT doping of MgB_2 enhanced the critical density and depressed the critical temperature.

Although today's state-of-the-art transistors are characterized by a physical gate length of approximately 50 nm, research groups in industry and academia have already demonstrated devices with gate lengths close to 30 nm. A 15 nm HEMT with a maximum frequency speed of 610 GHz was reported in 2008.

One primary method for the analysis of carbon nanostructures is Raman microspectroscopy. This approach has been proven to be effective since it is reliable, nondestructive, and requires little sample preparation. IEEE standard reporting practices are included in order to minimize confusion in analyzing reported data. A disclosure of environmental conditions and sample size are included so that the results can be appropriately assessed by the research community.

A Schrödinger dynamical analysis of a single particle along a single axis system behaves as a harmonic oscillator and the sinusoidal response. This simplified nanotube dynamics can be used in microtransistors, gates, or switches analysis.

Some of the promising applications of nanotubes in the telecommunications field are presented. They range from tunable resistors, neural commentators, spacecraft and satellite microwave diodes, and fiber-optics transistors to wireless and radio transmission as well as the IEEE fabrication standardization.

It was discovered that the electrical resistance between nanotubes and graphite surfaces changes according to the orientation of the nanotubes. Such results might have a fruitful impact on the future of telecommunications.

It has been successfully demonstrated that mats of SWCNTs can communicate electrical signals to neurons. Research results suggest that SWCNTs could be used as an electrical interface between neural prosthetics and the human body.

Using nanotube diode technology will help significantly in replacing conventional heavy, bulky, high temperature, microwave amplifiers. Typically, conventional satellites carry 50 microwave amplifiers on board, each weighing about 1 kg and measuring about 30 cm in length. CNT cathodes hold promise for a new generation of lightweight, compact, and efficient microwave devices for satellites or spacecraft telecommunications because of their tiny size and ability to generate and modulate the beam simultaneously without the need for high temperatures.

For computational and memory reasons, the transistors are getting smaller and smaller. One of the challenges that should be overcome is the speed limiting factor due to using metal wires to send signals through the systems.

A band of 40 nm was achievable with CNTs one order of magnitude higher than that observed in silicon semiconductor materials. Using CNT diodes would allow for wideband tunable laser telecommunications. Orientation manipulation of the nanotubes and eliminating impurities in the CNTs and matrix material could improve the telecommunication pulse frequency.

A single CNT molecule serves simultaneously as all essential components of a radio: antenna, tunable band-pass filter, amplifier, and demodulator. Using carrier waves in the commercially relevant 40 to 400 MHz range and both FM and AM, successful music and voice reception was demonstrated.

Finally, CNTs are proving their usefulness in cutting edge optoelectronic devices, where their unique combination of tunable electronic properties and nanoscale dimensions opens up a wide range of possibilities and applications in the telecommunication sector.

References

M. R. Abidian and D. C. Martin. February 24, 2009. Advances in advance: Multifunctional nanobiomaterials for neural interfaces. *Advanced Functional Materials*, 19(4): 503–664.

L. P. Biró, G. I. Márk, and P. Lambin. December 2003. Regularly coiled carbon nanotubes. *IEEE Transaction on Nanotechnology* 2(4): 362–367.

D. A. Buzatu, A. S. Biris, A. R. Biris, D. M. Lupu, J. A. Darsey, and M. K. Mazumder. September/October 2004. Electronic properties of single-wall carbon nanotubes and their dependence on synthetic method. *IEEE Transaction on Industry Applications* 40(5): 1215–1219.

J.-C. Charlier, M. Terrones, F. Banhart, N. Grobert, H. Terrones, and P. M. Ajayan. December 2003. Experimental observation and quantum modeling of electron irradiation on single-wall carbon nanotubes. *IEEE Transaction on Nanotechnology* 2(4): 349–354.

R. Chau,. B. Doyle, M. Doczy, S. Datta, S. Hareland, B. Jin, J. Kavalieros, and M. Metz. June 23–25, 2003. Silicon nano-transistors and breaking the 10 nm physical gate length barrier. *IEEE Device Research Conference*, Salt Lake City, UT, pp. 123–126.

D. Christensen, O. Englander, J. K. L. Lin. August 12–14, 2003. Room temperature local synthesis of carbon nanotubes. *Third IEEE Conference on Nanotechnology* (IEEE-NANO 2003), San Francisco, CA, vol. 2, pp. 581–584.

M. S. Dresselhaus, P. C. Eklund, and R. Saito. 1998. Carbon nanotubes. *Physical World*, 11: 33–38.

S. Iijima. 1991. Helical microtubules of graphite carbon. *Nature* 354: 56–58.

S. Iijima and P. M. Ajayan. 1992. Growth model for carbon nanotubes. *Physics Review Letter* 69(21): 3100–3103.

S. Iijima and T. Ichihashi. 1992. Single cell carbon nano tube of 1 nanometer diameter. *Nature* 363: 220–221.

K. Jensen, J. Weldon, H. Garcia, and A. Zettl. 2007. Nanotube radio: Supplementary materials, Department of Physics, University of California at Berkeley Center of Integrated Nanomechanical Systems, University of California at Berkeley, Materials Sciences Division, Lawrence Berkeley National Laboratory, Berkeley, CA.

C. Journnet, W. K. Master, P. Bernier, and A. Loiseque. 1997. Large scale production of single wall nanotubes by electric arc technique. *Nature* 388: 756–758.

S. Kar. October 4–6, 2004. Extraction of parameters of high permittivity ultrathin (0.5–2.0 nm) gate dielectrics. *Proceedings of the International Semiconductor Conference, 2004* (CAS 2004), Sinaia, Romania, vol. 2, pp. 341–344.

J. Karamdel, N. Talebi, M. Sattari, J. Derakhshandeh, F. L. Ayatollahi, A. Farrokhi, and A. R. Hadi. October 29, 2006. Fabrication of a single carbon nanotube for use in nanolithography of MOSFET gate. *IEEE International Conference on Semiconductor Electronics, ICSE 2006*, Kuala Lumpur, Malaysia pp. 122–126.

J. M. Ko and K. M. Kim. 2009. Electrochemical properties of MnO_2/activated carbon nanotube composite as an electrode material for supercapacitor. *Materials Chemistry and Physics* 114(2–3): 837–841.

F. Legorreta Garcia, C. Estourne`s, A. Peigney, A. Weibel, E. Flahaut, and Ch. Lauren. 2009. Spark-plasma-sintering of double-walled carbon nanotube–magnesia nanocomposites. *Scripta Materialia* 60(9): 741–744.

S. Li, Z. Yu, C. Rutherglen, and P. J. Burke. 2004. Electrical properties of 0.4 cm long single-walled carbon nanotube. *Nano Letters* 4(10): 2003–2007.

T. V. Lobanova, Y. I. Vorontsov, and S. V. Kalinin. July 1–5, 2006. The modern technology for silicon nanoFET. *Seventh Annual 2006 International Workshop and Tutorials Electron Devices and Materials*, Erlagol, Altai, pp. 27–29.

A. Naeemi and J. D. Meindl. July 2005. Impact of electron–phonon scattering on the performance of carbon nanotube interconnects for GSI. *IEEE Electron Device Letters* 26(7).

Z. Poh, B. S. Flavel, C. J. Shearer, J. G. Shapter, and A. V. Ellis. 2009. Fabrication and electrochemical behavior of vertically-aligned carbon nanotube electrodes covalently attached to p-type silicon via a thioester linkage. *Material Letters* 63(9–10): 757–760.

Z. Qi and W. Rui. June 20–25, 2004. Research on the possibility of nanotube antenna. *2004 IEEE Antennas and Propagation Society International Symposium*, vol. 2, pp. 1927–1930.

S. Ravindran, K. N. Bozhilov, C. S. Ozkan. March 16, 2004. Self-assembly of ordered artificial solids of semiconducting ZnS capped CdSe nanoparticles at carbon nanotubes ends. *Carbon* 42(8–9): 1537–1542.

W. H. Richardson. September 2003. Analysis of carbon nanotube intermolecular p–n tunnel junction transistors. *International Conference on Simulation of Semiconductor Processes and Devices*, Piscataway, NJ, pp. 753–755.

C. Rutherglen and P. Burke. April 11, 2008. Carbon nanotubes for wireless communication and radio transmission. US Patent: 703538.4158 (Provisional).

C. Rutherglen, D. Jain, and P. J. Burke. 2008. RF resistance and inductance of massively parallel single walled carbon nanotubes: Direct, broadband measurements and near perfect 50 ohm impedance matching. *Applied Physics Letters* 93: 083119.

N. Sinha and J. T.-W. Yeow. June 2005. Carbon nanotubes for biomedical applications. *IEEE Transaction on NanoBioSience* 4(2): 180–195.

M. Terrones. 2003. Science and technology of the twenty-first century: Synthesis, properties, and applications of carbon nanotubes. *Annual Review of Materials Research* 33: 419–501.

K. B. K. Teo, E. Minoux, L. Hudanski et al. October 12, 2005. Microwave devices: Carbon nanotubes as cold cathodes. *Nature* 437(7061): 968.

G. Timp, J. Bude, K. K. Bourdelle et al. 1999. The ballistic nano-transistors. *International Electron Devices Meeting* (IEDM Technical Digest), Washington, DC, pp. 55–58.

R. Vajtai, B. Wei, Y. J. Jung et al. December 2003. Building and testing organized architectures of carbon nanotubes. *IEEE Transaction on Nanotechnology* 2(4): 355–361.

R. L. WanderWal, G. M. Berger, and T. M. Ticich. 1993. Carbon nanotubes synthesis in a flame using laser ablation for insitu catalyst generation. *Applied Physics Letter C* 77(7): 885–889.

F. Wang, A. G. Rozhin, V. Scardaci, Z. Sun, F. Hennrich, I. H. White, W. I. Milne, and A. C. Ferrari. 2008. Wideband-tunable, nanotube mode-locked, fibre laser. *Nature Nanotechnology* 3: 738–742.

E. W. Wong, P. E. Sheehan, and C. M. Lieber. 1997. Nanobeam mechanics: Elasticity, strength, and toughness of nanotubes and nanorods. *Science* 277: 1971–1975.

P. Xu, X. J. Han, X. R. Liu, B. Zhang, C. Wang, and X. H. Wang. 2009. A study of the magnetic and electromagnetic properties of γ-Fe_2O_3–multiwalled carbon nanotubes (MWCNT) and Fe/Fe_3C–MWCNT composites. *Materials Chemistry and Physics* 114(2–3): 556–560.

W. K. Yeoh, J. Horvat, S. X. Dou, and P. Munroe. June 2005. Effect of carbon nanotube size on superconductivity properties of MgB_2. *IEEE Transaction on Applied Superconductivity* 15(2): 3284–3287.

A. Yin, H. Chik, and J. Xu. March 2004. Postgrowth processing of carbon nanotube arrays—Enabling new functionalities and applications. *IEEE Transaction on Nanotechnology* 3(1): 147–151.

G. Z. Yue, Q. Qiu, B. Gao, Y. Cheng, J. Zhang, H. Shimoda, S. Chang, J. P. Lu, and O. Zhou. 2009. Generation of continuous and pulsed diagnostic imaging x-ray radiation using a carbon-nanotube-based field-emission cathode. *Applied Physics Letters* 81(2): 355–357.

J. P. Zheng, C. M. Pettit, P. C. Goonetilleke, G. M. Zenger, and D. Roy. 2009. D.C. voltammetry of ionic liquid-based capacitors: Effects of faradaic reactions, electrolyte resistance and voltage scan speed investigated using an electrode of carbon nanotubes in EMIM–$EtSO_4$. *Talanta* 78(3): 1056–1062.

L. Zhu, Y. Sun, J. Xu, Z. Zhang, D. W. Hess, and C. P. Wong. May 31–June 3, 2005. Aligned carbon nanotubes for electrical interconnect and thermal management. *55th Electronic Components and Technology Conference*, Orlando, FL, vol. 1, pp. 44–50.

5

Silicon Nanostructures for Optical Communications

Adam A. Filios, Yeong S. Ryu, Kamal Shahrabi, and Raphael Tsu

CONTENTS

5.1 Introduction

Advances in the design, synthesis, and characterization of nanomaterials are expected to provide the unprecedented ability to manipulate matter at the most fundamental level, allowing the implementation of novel nanometer-scale devices and systems with unique properties and of utmost technological importance. When the physical dimensions of a device are reduced to the nanometer scale, quantum phenomena become prevalent, modifying the optical and electronic properties of the material. This may enable the design and manufacturing of materials with properties tailored for the particular application.

The tremendous growth in information technology and Internet applications over the past decade is due to a large extent to the advances in optical communications. Modern fiber-optic communication technologies—fueled in turn by advances in semiconductor laser diodes, optical filters, and other breakthroughs in materials and devices—have advanced to the point that data rates in excess of

40 gigabits per second per channel have already been achieved. Dense wavelength division multiplexing (DWDM) is a technique by which several optical signals of slightly different wavelengths are multiplexed and transmitted over the same optical fiber. Using DWDM, 20, 40, or even 100 channels can be transmitted simultaneously over the same optical fiber, increasing the system capacity accordingly. Optical technologies, however, are still considered to be high-cost solutions, mainly due to a lack of integration, a need for expensive materials, and a requirement for manual assembly and calibration of a large number of individual components. Furthermore, unless optical systems are effectively integrated with the silicon-based technologies that constitute the vast majority of computer microchips, the bottleneck posed by the traditional metallic interconnects on system speed will not be resolved.

For all the above-mentioned reasons, it is not surprising that the holy grail of the industry is the prospect of developing functional photonic-integrated circuits that are compatible and can be integrated with silicon-based technologies. Such advancement will lead to the significant cost reduction of optical technologies and will bring about the cost benefits of electronics to photonics. Even more importantly, a next generation photonic "superchip" will become feasible, enabled by optical interconnects among the various transistors, where both electrons and photons participate in the transmission and processing of information on the same chip, eliminating the bottlenecks and achieving operation at terahertz speeds.

Silicon is the cornerstone material in conventional VLSI systems clearly dominating the electronics industry [1]. However, traditional silicon devices suffer from some limitations inherent from its material properties. For example, silicon has a lower mobility than compound semiconductors such as GaAs. Even more importantly, there are no semiconductor materials that can form a sharp and high heterojunction barrier with silicon suitable for high-speed devices. Silicon devices today are all based on homojunction barriers using p–n junctions. Last but not least, conventional bulk silicon is very inefficient at emitting light. This is mainly due to its relatively small and indirect fundamental energy bandgap. Light emission from silicon is an important fundamental issue with enormous technological implications. If efficient silicon-based devices can be developed, the door will open for the integration of electronic and optical functionality on the same chip. This will have profound implications in the industry by enabling intra-chip optical interconnects, blending of photonics with electronics, and reducing the cost of optical technologies.

5.2 Nanotechnology Enables Unique Applications

Nanotechnology has emerged as one of the most rapidly developing fields with significant implications in engineering, technology, health, the environment, and society as a whole. Further advances in this interdisciplinary field will require the training of scientists and engineers able to work across traditional boundaries and utilize ideas and methods from several fields such as physics, chemistry, biology, and engineering in order to address challenges both in the basic understanding of the physical phenomena as well as in the engineering aspects involving fabrication, characterization, the effects of nanoparticles on humans and the environment, etc.

Although nanotechnology is considered to be a modern field in science and engineering, we should not forget that the underlying principles and the investigation of matter at the most fundamental level, i.e., at the atomic and molecular levels, have been carried out by scientists since the earliest days of scientific discovery. The difference is that until very recently, these studies were based either on theoretical analysis or indirect experimental observations. That is, when a scientist wanted to measure a property of matter at the atomic level, he or she would design an experiment to observe the consequences of that property at the macroscopic level and from those consequences he or she would obtain information indirectly about the atomic properties. Very rarely, an atomic level property could be measured directly. Some macroscale properties were measured and the atomic properties were inferred from those results. With the tremendous progress in the engineering and technology of electron microscopy over the last 10 to 15 years, now, for the first time in the history of science, we are able to directly observe the atomic and molecular arrangements of matter. Modern high-resolution transmission electron microscopy (TEM) provides the capability of directly seeing individual atomic layers in a material, achieving resolution of a few tenths of a nanometer, which would be unthinkable only a few years ago. In addition to allowing the observation of matter directly at the atomic level, these modern nanotechnology tools provide the capability to manipulate individual atoms as well. Although still at the experimental stage, scientists have succeeded in positioning individual atoms with atomic-scale precision by using the tip of a scanning tunneling microscope (STM) [2,3]. Other more established techniques such as molecular beam epitaxy (MBE) and atomic layer epitaxy (ALE) have been extensively used over the past 30 years to fabricate semiconductor devices atomic layer by atomic layer.

5.3 Superlattices and Quantum Wells

The majority of modern electronic and photonic devices that utilize nanotechnology tools for their fabrication and require quantum physics for a description of their operation are the outgrowth of the concept of superlattices and quantum wells originally proposed by Leo Esaki and Ray Tsu in 1969 and published in a milestone paper in 1970 [4]. These superlattices, considered to be only a scientific curiosity not too long ago, are at the heart of the most advanced semiconductor laser diodes, heterojunction transistors, and other high-speed electronic and photonic devices used commercially today.

Esaki and Tsu fathered the field of superlattices and quantum wells in III–V semiconductors in the early 1970s [4–6]. In 1969, in their proposal to the Army Research Office, they proposed the interruption of the periodicity of the lattice in a crystal structure by varying either the alloy compositions or the doping profiles during the epitaxial growth of a semiconductor device. They indicated that these superlattices constitute a new class of human-made semiconductor materials, which exhibit unique properties different than those of the host material. Forty years later, the impact of their work has been profound. Since then, there has been a myriad of publications and patents involving superlattices. A plethora of quantum functional devices have been proposed and fabricated. And over the past 20 years, many of these devices have already found many commercial applications.

For example, modern telecommunication lasers utilize superlattices of InGaAsP/InP or GaAs/AlGaAs in the active region resulting in much improved line width characteristics. Blue light-emitting gallium nitride lasers that were discovered in the mid-1990s use superlattice construction. Vertical cavity surface emitting lasers (VCSELs), which are considered to be an emerging technology offering many benefits in terms of cost and performance for metropolitan area optical networks, consist of superlattices and quantum wells typically grown by MBE. High mobility transistors based on GaAs/AlGaAs superlattices are been considered for high-speed electronics. In addition to the superlattices based on the III–V materials system, strain layer superlattices of Si/SiGe are being considered for use in high-speed electronic devices.

In the traditional electronic devices such as metal oxide semiconductor field effect transistors (MOSFET) and bipolar junction transistors (BJT), the electrons are treated as classical particles. For these large-scale devices, it is not necessary to consider the wave nature of the electrons and other quantum mechanical properties because only the statistical behavior of a very large number of electrons plays a role. However, for nanoscale devices consisting of very thin superlattice layers or comprising quantum dots (QDs) only a few nanometers in diameter, the wave nature of electrons plays a significant role in their electronic and optical properties. Therefore, as commercial electronic devices move into the nanoscale regime, it will become necessary to investigate and understand the quantum properties of these devices. In addition, the understanding of those characteristics may lead to the ability of scientists and engineers to design and manufacture materials and devices tailored for the particular application, circumventing restrictions imposed by the natural properties of the bulk material. Their physical operation based on quantum transport, as well as their characteristic dimensions, make the first superlattices discovered by Tsu and Esaki the precursor of today's nanotechnology concepts that hold the promise to revolutionize science, technology, and medicine in the future.

5.4 The Importance of Silicon

Semiconductor devices are key components in modern electronic systems. Silicon and gallium arsenide with its related III–V compounds form the basis of the most commonly used semiconductor materials. However, silicon is by far the major player in today's electronics market, dominating the microelectronics industry with about 90% of all semiconductor devices sold worldwide being silicon based.

But why is silicon still the most important semiconductor material, especially if one considers that its own electronic properties are rather mediocre? Its dominance is mostly due to the following aspects, which provide benefits that other electronic materials cannot easily match:

- Silicon possesses two of the most outstanding natural dielectrics, silicon dioxide (SiO_2) and silicon nitride (Si_3N_4), which are essential for device formation. In particular, SiO_2, which is the basis for the metal-oxide-semiconductor devices (MOS), can be grown thermally on a silicon wafer, it is chemically very stable, and it can achieve a very high breakdown voltage. The interface defects of the thermally grown SiO_2 by reaction of oxygen with a silicon wafer are several orders of magnitude lower than those of any deposited film.

- Silicon is nontoxic, relatively inexpensive (silicon comprises about 26% of the earth's crust, which makes it second in abundance only to oxygen), easy to process (a very well-established industrial infrastructure in silicon processing exists around the world), and has quite good mechanical properties (strength, hardness, thermal conductivity, etc.).

For all the above reasons, silicon is the cornerstone material in electronic systems. However, one of the most important limitations of bulk silicon is in optoelectronic applications, because of its inefficiency at emitting light. This is due to its indirect fundamental energy bandgap, which essentially makes optical transitions in the bulk material at room temperature a very rare phenomenon.

In semiconductor materials, the periodic arrangement of atoms results in the formation of bands of allowed energy levels for the electrons, separated by gaps of forbidden energies. Typically, we work with the valence band, which is mostly filled with electrons, and the conduction band, which can be reached by some electrons under certain circumstances, for example, if they are given some energy in the form of heat, voltage, or absorption of a photon. When an electron is raised to the conduction band, it leaves behind in the valence band a particle called a hole, which represents the absence of an electron and is treated as a positive particle. Between those two bands is the bandgap, a region of forbidden energies for the electrons. Electrons that have been excited to the conduction band will eventually return to the valence band, recombining with a hole. This represents the natural tendency of a system to return into the lowest energy state possible. When an electron recombines with a hole, conservation of energy dictates that their difference in energy (which is the energy difference between the bottom of the conduction band and the top of the valence band, i.e., the bandgap energy) must be conserved. Depending on the type of material, this energy difference might result either in light (emission of a photon) or it is mostly transferred to the lattice vibrations and dissipated as heat (emission of a phonon).

In addition to the conservation of energy, momentum must also be conserved during the transition. In a semiconductor with a direct fundamental energy bandgap (such as in GaAs), the maximum of the valence band and the minimum of the conduction band are found at the same value of the momentum in the *k*-space. Since the momentum of the photon is negligible, only vertical, momentum-conserving transitions between the conduction and valence bands are allowed. Therefore, an electron raised to the conduction band, will recombine in a very short time ($\sim 10^{-9}$ s) with a valence band hole to emit a photon of energy equal to the bandgap of the material. Thus, the probability of radiative recombination is very high in direct bandgap semiconductors. The band structure and the radiative recombination process for a direct semiconductor are illustrated in Figure 5.1.

In a semiconductor with an indirect fundamental energy bandgap, such as silicon, the maximum of the valence band and the minimum of the conduction band are found at different locations in the *k*-space. Recombination by a

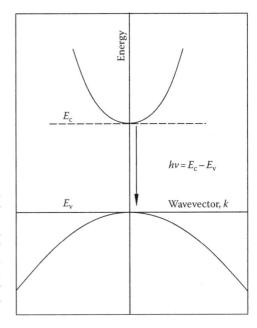

FIGURE 5.1
Band structure near the bandgap for a direct semiconductor. In an optical transition, crystal momentum $\hbar k$ is conserved. Since photons carry no momentum, direct optical transitions can occur only between conduction and valence band states that have the same value of *k*. (After Filios, A.A., PhD dissertation, University of North Carolina at Charlotte, Charlotte, NC.)

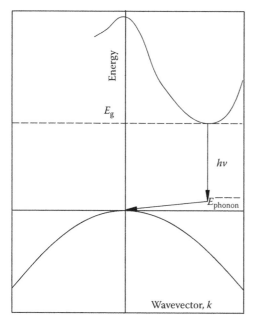

FIGURE 5.2
Band structure near the bandgap for an indirect semiconductor. Crystal momentum can be conserved only if an additional particle, such as a momentum-conserving phonon is emitted (or absorbed) along with the photon. Because this is a second-order process, the radiative efficiency for indirect materials such as silicon is quite low relative to that of direct-bandgap semiconductors. (After Filios, A.A., PhD dissertation, University of North Carolina at Charlotte, Charlotte, NC.)

single photon, which carries negligible momentum, is not allowed because of momentum conservation. Participation of a phonon with the right momentum is necessary to satisfy momentum conservation. Phonons are quantized modes of lattice vibrations that occur in a solid. In the bulk material, this phonon-assisted optical transition is very weak, allowing many other nonradiative processes to dominate resulting in a huge drop in the light emission efficiency. The band structure for an indirect semiconductor is illustrated in Figure 5.2.

Bulk silicon is therefore not suitable for the implementation of optoelectronic devices. To date, the semiconductor optoelectronics industry has been dominated by the III–V compound semiconductors because of their high efficiency in optical transitions primarily due to their direct fundamental energy bandgap.

5.5 Silicon Optoelectronics and Nanotechnology

The growth of the optical communications and related industries has generated a high demand for efficient and low-cost materials to be used for functions such as light emission, detection, and modulation. In addition, silicon-based materials with improved optical properties will find important applications in enhancing the efficiency of photovoltaic solar cells, which is a market also dominated by silicon, and which is expected to experience a tremendous growth in the near future. The importance of developing a technology that would allow optical and electronic devices to be easily integrated on a silicon wafer has long been recognized. Over the past 20 years, considerable efforts have been carried out within the research community to achieve this goal. Several materials and methods have emerged as possible contenders for silicon-based optoelectronics. These include silicon-based superlattices and QDs facilitating quantum confinement in silicon nanocrystals [7,8], SiGe and SiGeC devices [9], silicon devices doped with optically efficient rare earth impurities such as erbium [10,11], direct integration of III–V materials on silicon [12], porous silicon [13,14], silicon and carbon clusters embedded in oxide or nitride matrices [15], and superlattices of epitaxially grown silicon with adsorbed oxygen [16,17].

Most of the above-mentioned techniques involve devices that are based on nanoscale silicon. In a nanostructure, electrons in the conduction band and holes in the valence band are confined spatially by potential barriers. In the case of a QD, carriers are confined in all three dimensions (3D quantum confinement). In a nanowire, the carriers are confined in two dimensions and are free in only one dimension (2D quantum confinement). In a superlattice, carriers are confined in only one direction and free to move on the plane (1D quantum confinement). Such quantum confined superlattices based on gallium arsenide (GaAs) and indium phosphide (InP) have already found commercial applications in semiconductor distributed feedback lasers (DFB), semiconductor optical amplifiers, and VCSELs for optical communications [18,19]. Fundamentally, in all cases, quantum confinement pushes up the allowed energies effectively increasing the bandgap. The upshift of the quantum confined bandgap increases as the nanoparticle size becomes smaller. It also increases as the characteristic dimensionality of the quantum confinement increases (from 1D to 2D to 3D). Therefore, quantum confinement may be used to tune the energy of the emitted light in nanoscale optical devices based on the nanoparticle size and shape.

The other important fundamental issue is the requirement for momentum conservation in the optical transition. In silicon, even for the highest degree of confinement and nanoparticle size around 3 nm in diameter, the bandgap still remains indirect. Therefore, participation of a phonon with the right momentum is required to facilitate momentum conservation in the radiative transition. The light emission in indirect bandgap silicon nanocrystals can be explained in terms of phonon-assisted exciton recombination across the bandgap. An exciton is a pair of an electron and a hole bound to each other by Coulomb interaction. The exciton is in a way analogous to a hydrogen atom, but the binding energy of the exciton is much smaller than that of the hydrogen atom. In a defect-free crystal at room temperature, there are two competing processes involving phonons and excitons. One is the process of phonon-assisted radiative recombination and the other is the process of exciton break up due to the interaction with phonons. In order for a phonon to participate in phonon-assisted radiative recombination, it must have the right momentum to bridge the separation in momentum space between the top of the valence band and the bottom of the conduction band. However, any phonon can break up the exciton as long as it has enough energy. In bulk silicon, the exciton binding energy is small, about 15 meV, and thermal phonons with energy $kT \approx 26$ meV have enough energy to break up the exciton to a free electron and a free hole, which move away from each other through the continuum of states in the conduction and valence bands. Therefore, exciton break up dominates and radiative recombination becomes very unlikely. In a nanoparticle, the continuum of the valence band and conduction band states is modified into a discrete set of energy levels due to quantum confinement. Furthermore, in a nanoparticle, the exciton binding energy increases due to the confinement-induced overlap of the electron and hole wavefunctions. In a silicon QD of about 3 nm in diameter, the exciton binding energy has been calculated to be larger than 160 meV [20], much larger than the binding energy of the excitons in the bulk as well as the energy of the thermal phonons ($kT \approx 26$ meV). Therefore, in a nanoparticle, excitons cannot be broken up by thermal phonons, thus allowing the exciton enough time to wait for the phonon with the right momentum to participate in the phonon-assisted radiative recombination, producing an efficient light emission at room temperature.

5.6 Light Emission from Nanoscale Silicon

Several techniques have been explored toward developing silicon-based nanostructures utilizing quantum confinement for devices with engineered bandgap, increased functionality, and enhanced optical transitions. Porous silicon is a material produced by electrochemically etching silicon in aqueous hydrofluoric acid solutions, consisting of a network of nanometer-sized silicon crystallites in the form of nanowires and nanodots. Porous silicon exhibits bright room temperature photoluminescence (PL) in the visible region of the spectrum [13,14]. Several models have been proposed to explain the observed luminescence, including quantum confinement in silicon nanocrystals [21], luminescence from siloxene ($Si_6O_3H_6$) and other Si-O-H compounds [22], luminescence from silicon hydride complexes (SiH) [23], and several combinations of the above models [24].

Room temperature PL spectra [25] of a typical porous silicon sample excited by the 457.9 nm line of an Ar ion laser are shown in Figure 5.3. The vertical axis shows the relative intensity of the PL in arbitrary units (a.u.). In general, the term "arbitrary units" refers to the number of counts registered by the photodetector. The peak energy of the PL is at 1.85 eV.

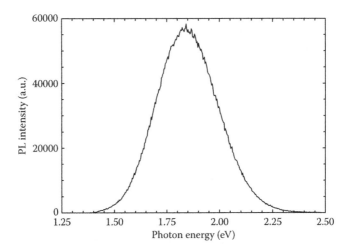

FIGURE 5.3
PL spectra from porous silicon at room temperature. (After Filios, A.A., PhD dissertation, University of North Carolina at Charlotte, Charlotte, NC.)

The full width at half maximum (FWHM) is about 350 meV. The large width of the spectrum is attributed to luminescence from nanocrystallites with a distribution of sizes. The peak energy and the intensity of the PL are strongly correlated with preparation conditions [10]. In addition, it has been shown that the luminescence can also be tuned by various post anodization treatments such as thermal oxidation [14] or rapid thermal oxidation [26]. The limitations of porous silicon are that the material is not very stable chemically and that the fabrication parameters and interface chemistry cannot be fully controlled.

In order to overcome these issues that plaque porous silicon, research efforts have been concentrated on manufacturing silicon nanoparticles utilizing standard semiconductor-processing techniques such as vacuum deposition by MBE or chemical vapor deposition (CVD), which offer much better control in the layer dimensions and interface quality. Devices consisting of thin silicon layers sandwiched between layers of oxide, as well as silicon QDs embedded in an oxide matrix have been extensively investigated [27–30].

In 1993, Tsu proposed that it would be possible to continue the growth of an epitaxial silicon layer on top of an oxygen-exposed silicon surface, as long as the oxygen adsorption remained within a couple of monolayers [31]. This idea led to the fabrication of a superlattice consisting of epitaxial layers of silicon sandwiched between monolayers of adsorbed oxygen. These silicon superlattices prepared epitaxially under ultrahigh vacuum deposition by MBE with silicon layers only 1–2 nm thick separated by adsorbed monolayers of oxygen exhibit strong PL and electroluminescence with a peak photon energy in the visible region, have very low interface defects, and show quantum confinement characteristics [16,30].

In Ref. [31], Tsu proposed using this silicon with adsorbed oxygen super-lattices as a means to implement an effective barrier for silicon as described in [31,32]. This would bring the benefits of the well-known III–V heterojunction devices to silicon, a major technology breakthrough with enormous economic implications. Figure 5.4 shows the scheme of an epitaxial Si/O effective barrier. Because the width b is much smaller than the width w, E_{b1} and E_{b2} are much higher than the quantum well states E_1 and E_2. The key for the fabrication of this kind of superlattice is that the epitaxial growth of silicon may be continued after the interruption with oxygen exposure. The method used for the experimental verification of the above proposal involves the adsorption of oxygen onto a clean Si surface followed by the deposition of 1–2 nm thick silicon layers using *in situ* reflection high-energy electron diffraction (RHEED) for monitoring the epitaxy, as described in detail in Ref. [27].

Although epitaxy is essential for heterojunction devices that operate at high speeds and for high-quality materials for lasers, several other strategies have also been explored if one is only interested in light emission applications, for example, for light-emitting diodes (LEDs) or displays. Still, most of these strategies make use, in one form or another, of quantum size effects in silicon nanostructures.

For example, one technique incorporates rare earth ions such as erbium into the silicon nanostructure. Since these rare earth ions are good light emitters, very efficient silicon LEDs have been reported in the literature utilizing this technique [10,11,33–37]. Silicon nanocrystals and rare earth ions are both incorporated within a silicon-rich oxide layer. In this approach, the main

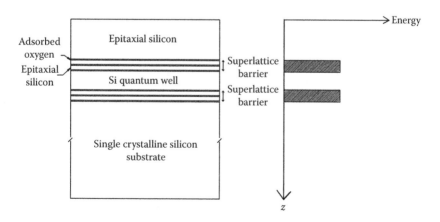

FIGURE 5.4
Schematic drawing of silicon-based quantum well, proposed by Tsu, using an epitaxial Si/O superlattice as effective barrier. (Redrawn and adapted from Tsu, R., *Nature*, 364, 19, 1993.)

role of the silicon nanocrystals is to transfer their energy to the rare earth ions that are responsible for the luminescence and to improve the conductivity of the silicon-rich oxide layer. In addition, bandgap-widening, resulting from the nanocrystals and the presence of oxygen, helps to mitigate the temperature quenching of the luminescence and low optical activity that has been observed in erbium-doped silicon [10,38]. Very good efficiency, approaching that of III–V light emitters, has been reported with this approach [10,39]. One of the advantages of this technique is that one can easily obtain various wavelengths of the emitted light, spanning the visible and near infrared regions of the spectrum, by simply incorporating different rare earth elements. Most of the reports use erbium, since erbium-doped silicon devices exhibit a luminescence peak at 1540 nm, which is the wavelength of interest in

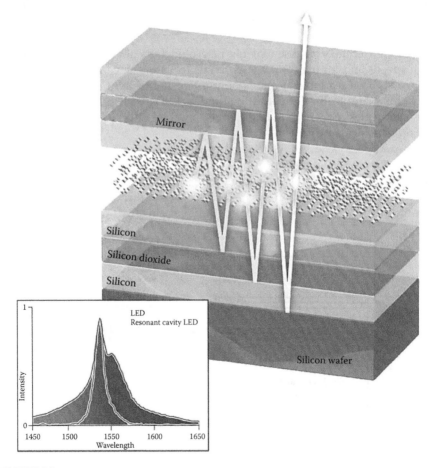

FIGURE 5.5
A silicon-based LED. Light emission is achieved by using silicon nanocrystals and rare earth ions. (From Coffa, S., *IEEE Spectrum*, 42(10), 44, Oct. 2005. With permission.)

optical communications. Light-emitting devices in the visible have also been reported in the literature by doping the silicon layers with samarium for red, terbium for green, and cerium for blue. A resonant cavity LED has been reported using this approach. The schematic of the structure is shown in Figure 5.5. It is fabricated by embedding a layer of silicon nanocrystals and rare earth ions between mirrors consisting of silicon with silicon dioxide multilayers.

One of the main disadvantages of this approach is the fact that there is an upper limit on how heavily one can dope the silicon with those rare earth ions. For example, the solid solubility limit for erbium in silicon is about $10^{18}/cm^3$, and at higher concentrations than this limit, erbium precipitates are formed. This situation can be somewhat improved if silicon nanocrystals and silicon-rich oxide are used. Since the light output depends on how densely the light emitting ions can be incorporated in the device, in general, these LEDs fabricated by doping with rare earth ions suffer from low output light power. There have been several attempts to demonstrate laser action in erbium-doped silicon [35]. However, the major problems are the difficulty in doping with the required high concentration of erbium for laser action, as well as the difficulties for effectively pumping the erbium ions and for maintaining low losses in the erbium-doped waveguides [35]. Figure 5.6 shows room temperature PL and electroluminescence

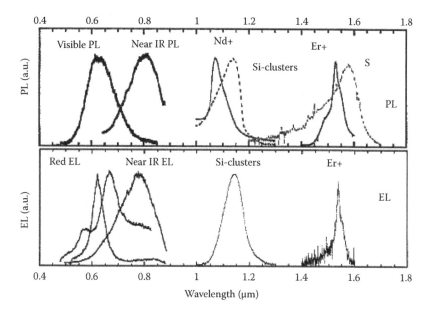

FIGURE 5.6

Room temperature PL and electroluminescence spectra from silicon-based nanocrystalline devices including devices doped with rare earth ions. (Reprinted from Fauchet, P.M., *IEEE J. Select. Topics Quantum Electron.*, 4(6), 1020, 1998. With permission.)

spectra from silicon-based nanocrystalline devices including devices doped with rare earth ions.

A different approach to achieving silicon-based active photonic devices has been extensively reported over the past few years. Some researchers have attempted to make use of the Raman effect to implement optical amplifiers and even lasers in silicon [40–45]. Fiber-based Raman amplifiers and lasers are extensively used in commercial optical fiber communication systems. However, one of the drawbacks of all Raman-based devices is that they require optical pumping. Therefore, this approach does not completely eliminate the requirement for traditional nonsilicon electrically pumped lasers used to optically pump the silicon Raman lasers.

5.7 Characterization of Silicon Nanoparticles by Raman Spectroscopy

From the above discussion, it becomes clear that one of the most important issues in the implementation of nanoscale devices is the ability to control the nanoparticle size and shape during the manufacturing process. One technique that can be used to indirectly measure the nanoparticle dimensions and crystalline structure in silicon is based on Raman spectroscopy [46–50].

Raman spectroscopy involves the inelastic scattering of light from a material. In Raman scattering experiments, monochromatic light, typically from a laser, interacts with the lattice vibrations (i.e., the phonons) in a semiconductor resulting in the energy of the scattered light being shifted up or down relative to the original laser line. This shift in energy provides information about the properties of the material. The three phases of silicon (i.e., amorphous silicon, crystalline bulk silicon, and nanoscale silicon) exhibit distinct signatures in their Raman spectra. For crystalline silicon, a sharp line is seen in the Raman spectra with a natural line-width of about 3 cm^{-1}. This sharp line for crystalline silicon is shown in Figure 5.7 and is shifted by 522 cm^{-1} relative to the laser line. Amorphous silicon exhibits a very broad peak at 480 cm^{-1} relative to the laser line. The typical Raman spectra of amorphous silicon are shown in Figure 5.8.

Raman spectra from nanocrystalline silicon are redshifted and broadened relative to the 522 cm^{-1} sharp Raman line for bulk crystalline silicon [51–55]. This Raman shift is attributed to the quantum confinement of the electronic wavefunction in silicon nanocrystals and can provide an estimate of the characteristic dimensions of the nanocrystalline structures [14]. A correlation of the PL and Raman spectra for several types of nanocrystals has been observed [14]. As the PL peak increases in energy, the Raman peak shifts to

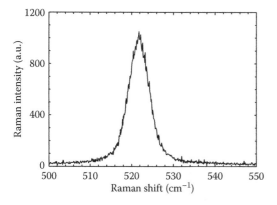

FIGURE 5.7
Raman spectra from crystalline silicon. (After Filios, A.A., PhD dissertation, University of North Carolina at Charlotte, Charlotte, NC.)

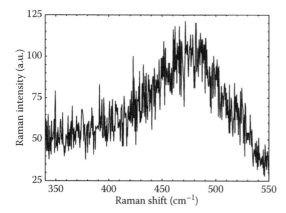

FIGURE 5.8
Amorphous silicon Raman spectra. (After Filios, A.A., PhD dissertation, University of North Carolina at Charlotte, Charlotte, NC.)

a lower energy indicating a smaller size of the nanoparticle. Typical Raman spectra from nanocrystalline silicon are shown in Figure 5.9.

The progress of the transformation of the amorphous silicon Raman peak to the crystalline silicon peak at 522 cm^{-1} for various annealing temperatures, namely, 600°C, 700°C, and 800°C, is shown in Figure 5.10. Note that since the as-deposited silicon layer is thick, there is no evidence of any nanocrystalline phase during the crystallization process. Figure 5.11 shows a comparison between porous silicon and bulk crystalline silicon Raman spectra. The red shift and broadening observed in the porous silicon Raman spectra is attributed to a nanocrystalline phase in porous silicon.

Raman spectroscopy provides indirect information about the crystal structure. It is a rather simple and relatively inexpensive technique that can be used to obtain qualitative information. However, a more direct method is now available in the field of nanotechnology by utilizing high-resolution TEM. Although TEM is a more expensive experimental technique, it provides direct information invaluable for device design and optimization.

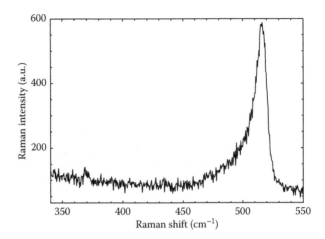

FIGURE 5.9
Nanocrystalline silicon Raman spectra. The redshift from the 522 cm^{-1} crystalline silicon line increases with decreasing particle size. (After Filios A.A. PhD dissertation, University of North Carolina at Charlotte, Charlotte, NC.)

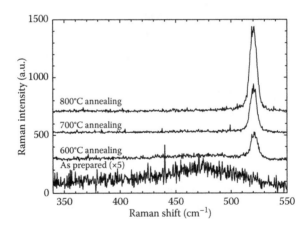

FIGURE 5.10
Raman spectra of a thick silicon layer deposited at room temperature on quartz as prepared and after annealing. (After Filios, A.A., PhD dissertation, University of North Carolina at Charlotte, Charlotte, NC.)

A cross-sectional high-resolution transmission electron microscope (XTEM) image of a multilayer structure of silicon nanocrystals in alumina is shown in Figure 5.12. The TEM image clearly shows the silicon nanocrystalline layers separated by layers of alumina (Al_2O_3). An XTEM is a very powerful characterization tool for obtaining information about the atomic structure of the material. Top of the line TEMs are capable of achieving sub-angstrom resolution [57] opening the door to new possibilities in scientific discovery.

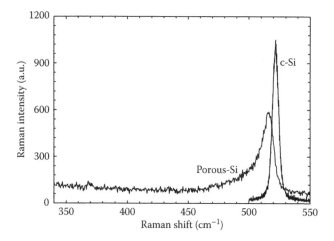

FIGURE 5.11
Raman spectra of porous silicon and bulk crystalline silicon (c-Si). (After Filios, A.A., PhD dissertation, University of North Carolina at Charlotte, Charlotte, NC.)

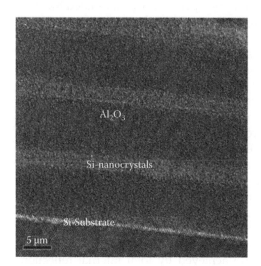

FIGURE 5.12
A high-resolution XTEM image showing the multilayer structure of silicon nanocrystals and alumina. (From Walters, R.J. et al., Luminescence properties of silicon nanocrystals in Al_2O_3 fabricated at low temperature, in *2008 5th IEEE International Conference on Group IV Photonics*, Sept. 17–19, 2008, Sorrento, Italy, paper WC3, pp. 41–42. With permission.)

5.8 Photonic Bandgap Crystals

Silicon is a suitable material for passive photonic devices such as waveguides, since it is transparent at 1.55 μm, which is the most relevant wavelength in optical communications. Indeed, silicon/silicon dioxide–based planar components constitute a large portion of the passive devices used in wavelength division multiplexed systems, namely, waveguides, optical filters, arrayed waveguide gratings, splitters, etc. [58–61]. However, these components are quite large in size with dimensions of several centimeters. In these conventional waveguides, light is guided based on the principle of total internal reflection. Therefore, the radius of curvature should be large enough to satisfy the critical angle requirements for total internal reflection. Very sharp bends are not possible with these conventional waveguides, as light will leak out at the corners. However, sharp bends are necessary in order to optically interconnect devices within a very small region in space.

Over the past few years, a new class of materials, photonic crystals, has emerged as a promising candidate for photonic devices. The potential and promise of photonic bandgap structures was first recognized in 1987, following the work of Yablonovitch [62] and John [63]. Since then, a plethora of research work has been done around the world, which led to the ability to fabricate photonic crystals with the appropriate size for the periodic structure so that they can operate at the communications wavelength, a task that poses many challenges in fabrication technology [64–68].

Photonic crystals consist of a periodic arrangement of regions with a high and low refractive index. This structure of a periodically modulated index of refraction leads to a photonic bandgap, in a way analogous to the electronic bandgap in a semiconductor, which results from the periodicity of the crystal lattice. To the photons, this contrast in the refractive index results in the formation of allowed energy regions separated by a forbidden region. The feature size must be in the order of the wavelength of the light to be guided. Therefore, in order to have relevance in the wavelength used in optical communications, the feature size of the periodic structures in the photonic crystal lies in the order of hundreds of nanometers. Therefore, advanced manufacturing techniques from the semiconductor industry are required to achieve these feature sizes. In addition, for a true photonic bandgap crystal, the periodic arrangement should extend to all three dimensions, which makes the fabrication very difficult. However, successful fabrication for some of those 3D photonic crystals has been reported in the literature. Most of the photonic crystals proposed and fabricated to date consist of a periodic structure that extends only to two dimensions (on a plane), while confinement in the z-direction is still achieved by total internal reflection. Those are the so-called 2D photonic crystals. Photonic bandgap crystals may find numerous important technological applications such as in integrated waveguides incorporating sharp bends, control of spontaneous emission, trapping of photons, etc. [69–71].

If the periodicity of the photonic crystal is broken by introducing a structural defect in the crystal, waveguide modes can be introduced within the photonic bandgap [66,67]. This is somewhat analogous to the energy levels within the electronic bandgap resulting from doping in semiconductors. Therefore, by introducing line defects, we can achieve strongly confined waveguides in the photonic crystal.

Silicon is a promising material for the fabrication of photonic bandgap waveguides. Typically, the silicon/silicon dioxide system is used (Si/SiO_2) due to the high refractive index contrast. An even higher index contrast results by using silicon with air holes for the periodic arrangement. The silicon acts as the high-index region and the SiO_2 or the air hole as the low-index region. The silicon-based line defect waveguide has received much attention as it can allow for very sharp bends [68]. A typical example of a 2D photonic crystal line defect waveguide is shown in Figure 5.13. This photonic crystal consists of silicon with air holes to take advantage of the high refractive index contrast. The photonic crystal is built on a commercial silicon-on-insulator (SOI) wafer. In the vertical direction, confinement is achieved by total internal reflection: The silicon layer acts as the core and the SiO_2 underneath the silicon on the SOI wafer acts as the cladding. The periodic structure is fabricated within the top silicon layer. A line defect waveguide with a very sharp bend can be seen in this figure and successful light propagation through this bend has been reported [66].

The fabrication of the air-hole pattern on the top silicon layer is done by using standard semiconductor-processing techniques involving steps such as thermal oxidation, photoresist patterning, and dry etching. The patterned silicon dioxide layer acts as an etching mask to define the location of the holes in the silicon layer. Several variations of the above described fabrication process may be employed. For example, the technique described in Ref. [68] uses evaporated Ni and Ti as etching masks and then the photonic crystal pattern is drawn by electron beam lithography [68]. A final chemical etching step in hydrofluoric acid solution may be used to remove the underlying SiO_2 layer, resulting in an airbridge-type waveguide. Silicon-based photonic bandgap waveguides are expected to play an important role in next generation photonic platforms, especially when matched with silicon-based light-emitting devices that may be available in the future [72–76].

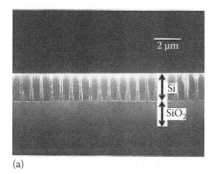
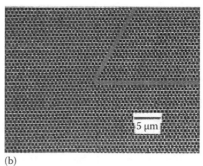

(a) (b)

FIGURE 5.13

TEM micrographs of a photonic bandgap slab that contains a single-line defect waveguide with a sharp bend. (a) Cross-sectional view and (b) top view. (Reprinted from Tokushima, M. and Yamada, H., *IEEE J. Quantum Electron.*, 38(7), 753, July 2002. With permission.)

5.9 Applications of Nanoscale Silicon in Photovoltaic Solar Cells

Nanocrystalline silicon films have recently attracted attention for use in photovoltaic solar cells since they show promise for providing an approach that results in lower cost and higher efficiency than conventional solar cells [77–79]. Traditionally, most of the commercially available solar cells are made out of various forms of silicon wafers.

Single crystalline silicon solar cells provide relatively good efficiencies; however, they suffer from high cost. In addition, crystalline silicon has a low absorption coefficient and narrower bandgap than that desired for solar cell applications. Solar light with energy much larger than the bandgap when absorbed by the material is converted into heat rather than electricity thus reducing the efficiency. Polycrystalline silicon or amorphous silicon solar cells are available at a lower cost; however, their conversion efficiency is lower than their single crystalline counterparts. In addition, amorphous silicon shows photodegradation and reduced stability under long-term illumination.

The use of nanocrystalline silicon holds the promise to realize both high performance and low cost in photovoltaic solar cells. Nanoscale silicon provides much better chemical stability than thin film polycrystalline or amorphous silicon under long-term operation. Its wider bandgap is much more suitable for solar cell applications than the narrow indirect bandgap of single crystalline silicon. Furthermore, silicon nanostructuring may provide the means for achieving a bandgap-engineered material to be used in tandem solar cells. Silicon nanoparticles of appropriate sizes or superlattices of Si/SiO_2, $Si/SiGe$, and/or Si/C can be integrated with other materials with appropriate bandgap, such as amorphous silicon, polycrystalline silicon, or single crystalline silicon to manufacture stacked solar cells. These types of solar cells can provide a variation of the effective bandgap across the material, allowing a significant portion of the solar spectrum to be efficiently coupled into the device and produce electricity.

5.10 Ideas May Work

The following is a summary of what has been discussed:

- Silicon is remarkable, constituting 95% of the electronic industry, but it cannot serve the optical needs because it is simply not an emitter and absorber of photons.
- Very small QDs continue to be indirect down to 10 Å. Because of the large surface to volume ratio, small QDs are terribly affected by

surface, or interface "crude," causing telegraph-like oscillation, switching in *I–V* [80], and blinking in PL. Unless means are found to passivate these defects, it is highly unlikely that making small QDs is the answer. However, it is possible that an appropriate matrix may be found, which will provide sufficiently good passivation of the interface defects. Then quantum composites of Si QDs embedded in such an appropriate matrix may serve to represent silicon quantum composites for optoelectronic applications. Many research groups are still working toward achieving this endeavour.

- Small QDs of c-Si embedded in an a-Si matrix seemed like a good idea almost 20 years ago [80], but the problem is mainly because the number of electrons participating in the effects are only a handful. Of course defects play a role, but certainly the small number of states involved may be the major culprit. Even in a 35 nm channel length MOSFET, there are more than 1000 electrons involved in the process of switching. So the problem is really not because of the degradation of materials; rather, the small number of electrons is the main culprit. Therefore, we must work hard to find the matrix for embedding silicon. The scheme of using excess silicon in SiC by Green and coworkers [77] seems to be on the right track.

- In essence, we propose an undertaking of a program of looking for the right kind of matrix for the QDs. We think that starting from putting Si into SiC is something in which we would like to get involved. Particularly, the polymorphs of SiC, e.g., 4H and 6H together with the cubic SiC, constitute the ideal matrix for the study of embedding QDs.

5.11 Conclusion

Optoelectronic properties and characteristics of nanoscale silicon that are fundamentally different from those of bulk crystalline silicon were presented. Several techniques toward developing silicon-based nanostructures, including devices composed of silicon nanocrystallites embedded in an amorphous matrix as well as multilayer structures such as superlattices and quantum wells, erbium-doped silicon LEDs, and Raman-based silicon lasers were reviewed. Potential applications of nanoscale silicon-based materials include silicon-based light-emitting devices for use in optical communications, including chip-to-chip and intra-chip optical interconnects, fully integrated silicon-based optoelectronic circuits, photovoltaic solar cells, heterojunction devices, and optical sensors. Lastly, looking for the ideal matrix to embed QDs of silicon seems to be the thing to do along the direction of the silicon-based optoelectronic system, not only for nonlinear optics, but it may also find its niche in photovoltaic applications.

Acknowledgments

Two of the authors (Adam A. Filios and Yeong S. Ryu) are grateful to the Centre for Functional Nanomaterials, Brookhaven National Laboratory, for providing access to their facilities under the facility users program, to carry out part of their research during the years 2008–2009.

References

1. J.C. Bean, in *High-Speed Semiconductor Devices*, S.M. Sze (Ed.), John Wiley & Sons, Inc., New York, 1990.
2. D.M. Eigler and E.K. Schweizer, *Nature*, 344(6266), 524 (1990).
3. M. Ternes, C.P. Lutz, C.F. Hirjibehedin, F.J. Giessibl, and A.J. Heinrich, *Science*, 319(5866), 1066–1069 (Feb. 22, 2008).
4. L. Esaki and R. Tsu, *IBM J. Res. Dev.*, 14, 61–65 (1970).
5. R. Tsu and L. Esaki, *Appl. Phys. Lett.*, 22, 562–564 (1973).
6. L. Chang, L. Esaki, and R. Tsu, *Appl. Phys. Lett.*, 24, 593–595 (1974).
7. G.Y. Sung, N.-M. Park, J.-H. Shin, K.-H. Kim, T.-Y. Kim, and C. Huh, *IEEE J. Select. Topics Quantum Electron.*, 12(6), Part 2, 1545–1555 (Nov.–Dec. 2006).
8. R. Tsu, Q. Zhang, and A. Filios, *Proc. SPIE*, 3290, 246–256, Optoelectronic Integrated Circuits II, (invited talk), San Jose, CA (1998).
9. A.K. Okyay, D. Kuzum, S. Latif, D.A.B. Miller, and K.C. Saraswat, *IEEE Trans. Electron Dev.*, 54(12), 3252 (Dec. 2007).
10. S. Coffa, *IEEE Spectrum*, 42(10), 44–49 (Oct. 2005).
11. S. Coffa, G. Franzo, and F. Priolo, *Appl. Phys. Lett.*, 69(14), 2077 (1996).
12. A.W. Fang, H. Park, R. Jones, O. Cohen, M.J. Paniccia, and J.E. Bowers, *IEEE Photon. Technol. Lett.*, 18(10), 1143 (May 2006).
13. L.T. Canham, *Appl. Phys. Lett.*, 57(10), 1046 (1990).
14. A.A. Filios, S.S. Hefner, and R. Tsu, *J. Vac. Sci. Technol. B*, 14(6), 3431–3435 (December 1996).
15. R. Tsu, A. Filios, and Q. Zhang, *Advances in Science and Technology*, vol. 27, pp. 55–66, in *9th Cimtec-World Forum on New Materials, Symposium X—Innovative Light Emitting Materials (Invited Lecture)*, P. Vincenzini and G.C. Righini (Eds.), Techna Srl, Faenza, Italy (1999).
16. R. Tsu, A. Filios, C. Lofgren, K. Dovidenko, and C.G. Wang, *Electrochem. Solid-State Lett.*, 1(2), 80–82 (1998).
17. Q. Zhang, A.A. Filios, C. Lofgren, and R. Tsu, *Physica E*, 8, 365–368 (2000).
18. R.G. Castrejon and A. Filios, *J. Lightwave Technol.*, 24(12), 4912–4917 (December 2006).
19. A. Filios, R. Gutiérrez-Castrejón, I. Tomkos, B. Hallock, R. Vodhanel, A. Coombe, W. Yuen, R. Moreland, B. Garrett, C. Duvall, and C. Chang-Hasnain, *IEEE Photon. Technol. Lett.*, 15(4), 599–601 (April 2003).
20. D. Babic and R. Tsu, *Superlattices Microstruct.*, 22(4), 581–588 (1997).
21. V. Lehmann and U. Gosele, *Appl. Phys. Lett.*, 58(8), 856 (1991).

22. J. Sarathy, S. Shih, K. Jung, C. Tsai, K. Li, D. Kwong, J.C. Campbell, S. Yau, and A. Bard, *Appl. Phys. Lett.*, 60(13), 1532 (1992).
23. S. Prokes, J. Freitas, and P. Searson, *Appl. Phys. Lett.*, 60, 3295 (1992).
24. S.M. Prokes, *Appl. Phys. Lett.*, 62(25), 3244 (1993).
25. A.A. Filios, PhD dissertation, University of North Carolina at Charlotte, Charlotte, NC.
26. P.M. Fauchet, C. Peng, L. Tsybeskov, V. Vandyshev, A. Dubois, L. McLoud, and S. Duttagupta, *Advanced Photonics Materials for Information Technology, SPIE Proc.*, Vol. 2144, Bellingham, WA (1994).
27. R. Tsu, A.A. Filios, C. Lofgren, D. Cahill, J. Van Nostrand, and C.G. Wang, *Solid-State Electron.*, 40(1–8), 221–223 (1996).
28. B.T. Sullivan, D.J. Lockwood, H.J. Labbe, and Z.-H. Lu, *Appl. Phys. Lett.*, 69(21), 3149 (1996).
29. M. Zacharias et al., *Appl. Phys. Lett.* 69(21), 3149 (1996).
30. R. Tsu, A. Filios, C. Lofgren, J. Ding, Q. Zhang, J. Morais, and C.G. Wang, *Proceedings of the Electrochemical Society*, Vol. 97–11, Quantum Confinement: Nanoscale Materials, Devices and Systems, Montreal, Quebec, Canada (May 4–7, 1997).
31. R. Tsu, *Nature*, 364, 19 (1993).
32. R. Tsu, U.S. Patent No. 5216262 (1993).
33. F. Priolo et al., *J. Appl. Phys.*, 74(8), 4936 (1993).
34. P.K. Giri, S. Coffa, and E. Rimini, *Appl. Phys. Lett.*, 78(3), 291 (2001).
35. S. Coffa, S. Libertino, G. Coppola, and A. Cutolo, *IEEE J. Quantum Electron.*, 36(10), 1206 (2000).
36. H. Ennen, J. Schneider, G. Pomrenke, and A. Axmann, *Appl. Phys. Lett.*, 43, 943 (1983).
37. J. Michel, J.L. Benton, R.F. Ferrante, D.C. Jacobson, D.J. Eaglesham, E.A. Fitzgerald, Y.H. Xie, J.M. Poate, and L.C. Kimerling, *J. Appl. Phys.*, 70, 2627 (1991).
38. J.H. Shin, J. Lee, H. Han, J.-H. Jhe, J.S. Chang, S.-Y. Seo, H. Lee, and N. Park, *IEEE J. Sel. Topics Quantum Electron.*, 12(4), 783 (2006).
39. P.M. Fauchet, *IEEE J. Select. Topics Quantum Electron.*, 4(6), 1020 (1998).
40. A. Liu, H. Rong, M. Paniccia, O. Cohen, and D. Hak, *Opt. Express*, 12, 4261–4267 (2004).
41. O. Boyraz and B. Jalali, *Opt. Express*, 12, 5269–5273 (2004).
42. H. Rong et al., *Nature*, 433, 292–294 (2005).
43. H. Rong, R. Jones, A. Liu, O. Cohen, D. Hak, A. Fang, and M. Paniccia, *Nature*, 433, 725–728 (2005).
44. H. Rong, Y.-H. Kuo, S. Xu, A. Liu, R. Jones, M. Paniccia, O. Cohen, and O. Raday, *Opt. Express*, 14, 6705–6712 (2006).
45. B. Jalali, V. Raghunathan, D. Dimitropoulos, and O. Boyraz, *J. Select. Topics Quantum Electron.*, 12(3), 412 (2006).
46. J.J. Laserna (Ed.), *Modern Techniques in Raman Spectroscopy*, John Wiley & Sons, New York, 1996.
47. R. Tsu, *Disordered Semiconductors*, M.A. Kastner and G.A. Thomas (Eds.), Ovshinsky, Plenum Publishing Corp., New York, p. 479, 1987.
48. R. Tsu, *SPIE*, 276, 78 (1981).
49. W. Hayes and R. Loudon, *Scattering of Light by Crystals*, John Wiley & Sons, New York, 1978.

50. R. Tsu, *Solar Cells*, 21, 19 (1987).
51. F.H. Pollak and R. Tsu, *SPIE Proc.*, 452, 26 (1983).
52. R. Tsu, S.S. Chao, M. Izu, S.R. Ovshinsky, G.J. Jan, and F.H. Pollak, *J. Phys. Coll.* 42(10), C4 (1981).
53. I.H. Campbell and P.M. Fauchet, *Solid State Commun.*, 58(10), 739 (1986).
54. H. Richter, Z.P. Wang, and L. Ley, *Solid State Commun.*, 39, 625 (1981).
55. R. Tsu, *Mater. Res. Soc. Symp. Proc.*, 38, 383 (1985).
56. R.J. Walters, R. van Loon, A. Polman, I. Brunets, and J. Schmitz, Luminescence properties of silicon nanocrystals in Al_2O_3 fabricated at low temperature, in *2008 5th IEEE International Conference on Group IV Photonics*, Sept. 17–19, 2008, Sorrento, Italy, paper WC3, pp. 41–42.
57. FEI Company web site: www.fei.com. Titan family of Transmission Electron Microscopes, Hillsboro, OR.
58. N. Izhaky, M. Morse, S. Koehl, O. Cohen, D. Rubin, A. Barkai, G. Sarid, R. Cohen, and M. Paniccia, *IEEE J. Select. Topics Quantum Electron.*, 12(6), 1688 (2006).
59. M. Paniccia and S. Koehl, *IEEE Spectrum*, 42(10), 38–43 (October 2005).
60. R. Soref, *IEEE J. Select. Topics Quantum Electron.*, 12(6), 1678 (2006).
61. B. Jalali and S. Fathpour, *J. Lightwave Technol.*, 24(12), 4600 (2006).
62. E. Yablonovitch, *Phys. Rev. Lett.*, 58, 2059–2062 (1987).
63. S. John, *Phys. Rev. Lett.*, 58(23), 2486–2489 (June 1987).
64. J.D. Joannopoulos, R.D. Meade, and J.N. Winn, *Photonic Crystals*, Princeton Press, Princeton, NJ, 1995.
65. J.D. Joannopoulos, P.R. Villeneuve, and S. Fan, *Nature*, 386, 143–149 (1997).
66. M. Tokushima and H. Yamada, *IEEE J. Quantum Electron.*, 38(7), 753–759 (July 2002).
67. M. Notomi, A. Shinya, K. Yamada, J. Takahashi, C. Takahashi, and I. Yokohama, *IEEE J. Quantum Electron.*, 38(7), 736–742 (July 2002).
68. S. Assefa et al. *Appl. Phys. Lett.*, 85, 25, 6110 (2004).
69. T. Baba, A. Motegi, T. Iwai, N. Fukaya, Y. Watanabe, and A. Sakai, *IEEE J. Quantum Electron.*, 38(2), 743–752 (July 2002).
70. M. Qi et al., *Nature*, 429, 538–542 (June 2004).
71. S. Noda et al., *IEEE J. Quantum Electron.*, 38(7), 726 (July 2002).
72. S. Weis and P. Fauchet, *IEEE J. Select. Topics Quantum Electron.*, 12(6), 1514 (2006).
73. M. Fujita, Y. Tanaka, and S. Noda, *IEEE J. Select. Topics Quantum Electron.*, 14(4), 1090 (2008).
74. D. Prather et al., *IEEE J. Select. Topics Quantum Electron.*, 12(6), 1416 (2006).
75. T. Tsuchizawa et al., *IEEE J. Select. Topics Quantum Electron.*, 11(1), 232 (2005).
76. C.W. Wong et al. *Appl. Phys. Lett.*, 84(8), 1242 (2004).
77. D. Song, E. Cho, G. Conibeer, Y. Huang, and M.A. Green, *Appl. Phys. Lett.*, 91(12), 123510 (2007).
78. H. Kawauchi, M. Isomura, T. Matsui, and M. Kondo, *J. Non-Cryst. Solids*, 354, 2109–2112 (2008).
79. D. Song, E. Cho, G. Conibeer, C. Flynn, Y. Huang, and M.A. Green, *Solar Energy Mater. Solar Cells*, (92), 474–481 (2008).
80. R. Tsu, *Microelectron. J.*, 39, 335–343 (2008).

6

Nanostructured Optoelectronic Materials for Short-Wavelength Devices

M. Khizar and M. Yasin Akhtar Raja

CONTENTS

6.1 Introduction

Over the last decade, there has been an explosion of immense interest and activity in the field of nanophotonics. This is an emerging area of research, which would potentially revolutionize the fields of nanomanufacturing, nanobiotechnology, nanoemitters, digital electronics, medicine, optical computing, sensing, and short-wavelength optical communication [1].

Nanostructures serve a critical need using nanofabrication techniques for the pursuit of ever-decreasing feature sizes in integrated photonic and optoelectronic devices and chips. One of the important thrust in nanotechnology research is the discovery and development of new materials that lend themselves to novel device fabrication. The main objective in this chapter is to document and explore most of such avenues, which may effectively be used for nanoengineered materials and devices pertaining to the nanophotonics arena. Necessary details on the design, fabrication, and characterization of new materials systems at the nanoscale level, and the fabrication of devices from such or other materials are covered within the scope of the chapter.

At its roots, it is a science of light–matter interactions, which take place, on the one hand, within the light wavelength and subwavelength scales and, on the other hand, are determined by the physical, chemical, and structural nature of artificially or naturally synthesized nanostructures. It is envisaged that nanophotonics has the potential to provide ultrasmall optoelectronic devices and components with high speed and greater bandwidth [2]. One of the driving forces behind nanophotonics is its great promise to act as the key to accessing the atomic scale since it would dispense with the need to make electrical contacts. While the field is still in its infancy, with much research in fundamental aspects to be done, research into alternative ways of fabricating nanometer-scale devices would open the way for mass-producing future light handling devices that may be only tens or few hundreds of nanometers in size—a strong motivation behind the fabrication of efficient short-wavelength photonic devices for communications [3], sensing, lighting, and numerous other applications.

Presently, at the advent of nanophotonics, we need to understand how electromagnetic waves behave in the presence of periodic or nearly periodic arrays of nanoparticles, clusters, or even the voids (hole) in the dielectrics, semiconductors, and metallic materials, and how subwavelength-size material particle and features give rise to phenomena that classically would not be allowed. Scientists and engineers are bringing their resources together to design and optimize the tools, which may effectively be used to map topology, size, and shapes with its optical properties of smart structures at nano- and submicron scales [4]. In the past, these tools were not available and still need further improvements, which are the subject of active research area. It is the locally modified electromagnetic field, which allows us to make use of molecular photonics, taking nanophotonics to the molecular scale, i.e., of a few nanometers [5]. This, in turn, opens up possibilities not just for further miniaturization but also for radically enhanced light–matter interactions by invoking quantum size effects. At present, one may have some very genuine engineering concerns when producing ultracompact, high-power, and high-sensitivity short-wavelength optical devices toward the level of single photon detection and emission, and onward manipulations.

Researchers in the field of nanophotonics are just beginning to understand the behavior of light in periodic media [6], subwavelength periodic and nanostructured surfaces. We have an incipient understanding of what happens in the subwavelength periodic and aperiodic structures and, at the other end, optical processes of single nanoparticles and clusters. We need to understand, what would happen, if we have nearly-periodic or quasiperiodic nanoparticles or cluster arrays? The description of their optical properties is very much in progress as is the behavior of light in complex media.

The tentative contents of this chapter will be spread across a range of topics covering

- III–V semiconductor micro/nanostructures and targeted outcomes such as better visible wavelength materials and highly efficient light-emitting devices

- Optoelectronic materials and devices, with goals such as a better understanding of the pathways to radiative and nonradiative interactions within the host material often used in fabrication of new types of near–ultraviolet (UV) and UV light emitters, detectors, and sensors

- Development of optoelectronic devices that could be used as "light engines" for the improvement of flat-panel display technologies

- Design and development and fabrication of high efficiency near-UV light-emitting devices

- Design and development and fabrication of ultrabright UV and deep-UV (DUV) light-emitting devices

- Design and development of white light-emitting diodes (LEDs) for solid state light applications

- Development of biologically inspired materials and devices with targeted outcomes being the development of carbon-based surface self-assembly techniques, and the establishment of new techniques for manipulating miniaturized cellular diagnostics for improved biomedical devices

In many cases, these subtopics will help the scientists and engineers' community to gain a better understanding of materials and devices at the nanoscale. There will also be interesting information on the new technologies development that will be incorporated into the products and processes. This diversity of research results are from an interdisciplinary research program being carried out in the micro- and nanophotonic research labs at the Center for Optoelectronics and Optical Communication at UNC Charlotte, Charlotte, North Carolina.

In the scope of this chapter, our main focus is to explore the avenues related with the design, growth, and fabrication of III-nitrides wide-bandgap materials nanostructures, which may effectively be used for the fabrication of short-wavelength light-emitting devices [7].

6.2 Main Focus of the Chapter

In this chapter, the applications of nanostructures for the next generation short-wavelength devices are discussed in detail. Studies show that significant improvement in the performance of various devices can be achieved by decreasing dimensionality from 3D crystalline solids to 2D quantum wells (QWs), to 1D quantum wires, and finally to 0D quantum dots (QDs). In order to address some of the problem associated with 0D space, many new physical techniques with interesting phenomena have been introduced.

This should not be surprising because these phenomena may create new opportunities to vary the Seebeck coefficient and electrical and thermal conductivities independently. Previous studies on the quantum size effects regarding the transport properties of electrons and holes has predicted an improved electric conductivity, along with a reduction in the thermal conductivity. All these concepts have been theoretically studied over the last decade, and reports on experimental verifications of pioneer work have been published [8]. III-nitrides-based photonic devices are at the heart of nanophotonics to unleash the demand of emerging nanotechnology. However, it is indeed quite challenging for scientists and the engineers for making a successive and productive transition from micro- to nanoscale devices. There are several unknowns when incorporating the lower dimensionality within the epi-structures of such devices. Recently, several breakthroughs have been made to uncover such mysteries with the integration of QDs, nanowires, and nanopatterns during the in situ and lateral regrowth of photonic structures. A comprehensive review of all such approaches is summarized here that utilizes the principle of nanotechnology through novel methods such as nanostructures, bandgap engineering, and novel nanodevice architectures.

6.3 III-Nitrides-Based Nanostructures for Photonic Devices

During the past few years, tremendous progress has been made in the area of photonics and it has shown an immense potential for almost all disciplines in realizing the number of the unexplored applications [9]. In its broader sense, this is one of the most vibrant and growing research areas in pure and applied sciences for next generation optoelectronic and communication applications as well as the biomedical devices and materials. Although nanostructured materials have a long list of interesting candidates, some of the most important ones include various thin films and bulk nanostructured materials for short-wavelength photonic devices. One of the most common features of these materials is the nanoscale dimensionality, i.e., at least one dimension less than 100 nm, and typically less than 20 nm [10].

Recently, the world's energy crisis has attracted more attention toward green energy resources and energy conversion systems. The awareness to boost research efforts to discover and exploit such avenues, which could help resolve our requirements have been on the rise [11]. Therefore, it is mandatory to look into all possibilities for novel devices. At present, we need a reliable technology to develop a foundation for contemporary and next generation photonic resources. Nanostructured technology blended with wide-bandgap III-nitride semiconductors (GaN, AlN, InN) based on a variety of substrates are currently the materials of choice for the latest optoelectronic technologies [12]. In fact, worldwide efforts for the development of nanoengineered nitride optoelectronics for short-wavelength communication applications are already underway [13]. The outstanding physical and chemical stability of nanoengineered nitride semiconductors has made it possible to use this technology in harsh environments. Furthermore, their biocompatibility and huge piezoelectric coefficients render them suitable for the fabrication of efficient sensors, emitters, detectors, and biochips [14].

Quite recently, nanostructure-based photonic devices have been recognized as one of the most important future technologies involving several disciplines of science including physics, chemistry, solid-state ionics, materials engineering, medical science, and biotechnology [15]. Nanotechnology is primarily the manipulation of matter at the nanometer scale, using building blocks with dimensions in the nano-size range, makes it possible to design and create new materials with unprecedented functionality and novel properties [16]. Nanostructures made of nano-sized grains or nanoparticles, and even voids (holes, lines) in crystalline or amorphous materials always play a role of building blocks for photonics.

These structures contain a significant fraction of grain boundaries with a high degree of disorder of atoms along the grain boundaries (or particle surfaces), and a large ratio of interface (or surface) area to volume [17]. However, it is essential that the chemical composition of the phases and the interfaces, between nano-grains, must be controlled efficiently in order to exploit this technology. Among the key characteristics of nano-engineered and nanostructured materials is their dependence of certain properties upon the sizes in the nanoscale region [18]. For example, electronic properties, with quantum size effects, caused by spatial confinement of delocalized valence electrons, are directly dependent on the particle size [19].

Usually, small particle sizes permit conventional restrictions of phase equilibriums and kinetics to be overcome during the synthesis and processing with the combination of short diffusion distances and high driving forces of available large surfaces and interfaces. There is a long list of materials, which are being used as nanostructures for the fabrication of photonic devices for numerous applications. It has been found that the large surface area always gives higher reactivity; therefore, its properties may result in somewhat relatively high surface "defects." Some of the high-resolution images of nanostructures used for the fabrication of short-wavelength range devices are shown in Figure 6.1 [20].

It has been found that there are several features which primarily depend on the synthesis and processing techniques, such as surface pores, grain boundary junctions, and related crystal lattice defects. The observable changes in their lattice parameters are sometimes attributed to surface stress, whereas the decline in melting temperature comes due to an increase in surface free energy [21]. This opens the doors for devising given properties by careful synthesis of the building blocks and their assembly to fabricate functional short-wavelength photonic devices with improved features [22].

FIGURE 6.1
High-resolution images of nanostructures. (Courtesy of Mervyn Miles, Nanophysics and Soft
Matter Research Group, University of Bristol, Bristol, England.)

6.4 Nanostructure-Based Photonic Materials and Devices

Low-dimensional nanostructures, have recently been integrated by various
researchers with the photonic devices for their high conversion and extraction
efficiencies, for example, QDs-based UV and DUV LEDs. Based on fundamentals, the formation of semiconductor QDs on the nanometer scale are usually
controlled by the Stranski–Krastanow growth mode [24], where the self-organ-
ized formation of nano-islands is driven and delimited by the strain energy in
large lattice-mismatch systems. Alternatively, QDs are also formed by droplet
epitaxy, which relies on the self-assembly of liquid metal nanoparticles of
group III atoms and their crystallization into II–V semiconductor QDs by a
subsequent supply of group V atoms [25,26]. Interestingly, the random distribution of the QDs sizes and locations usually occurs, that limits the benefits
from low-dimensional structures.

It has also been established that the dynamics of excitons decay are mono-
exponential and that the emission from a single dot displays a very narrow
spectrum. However, the disorder observed in QDs system controls the
recombination dynamics to give a non-exponential photoluminescence (PL)

decay of the entire QDs ensemble. One of the serious problems linked with the quantum- and nano-dots is the control on their homogeneity in size and position/location. Consequently, several techniques are being used to achieve the objective, e.g., nanoscale selective area epitaxy on a pre-patterned substrate or growth template for high-efficiency UV photonic devices.

On the other hand, the nano-patterning of substrates/templates can also be realized by a variety of lithographic or non-lithographic techniques, such as interference lithography [27–29], anodic oxidation of aluminum to form nano-porous alumina [30–33], nano-sphere lithography [34], focused ion beam sputtering [35], and electron-beam lithography (EBL) [36–38]. At present, state-of-the-art EBL allows the creation of patterns with versatile physical geometry and localization, with homogeneity and tunable sizes. Major challenge in the growth of III-nitride materials is the difficulty to realize nano-dot formation due to the high growth temperatures. In order to overcome such barriers, Si is used as an antisurfactant to modify GaN or AlGaN surface states to realize nitride nano-dots formation, which has been reported to occur by means of metal-organic chemical vapor deposition (MOCVD) [39].

The first evaluation in the fabrication of a UV LED was the use of GaN-based QDs as the active layer [40] over two decades ago. The presence of Si atoms as an anti-surfactant on (Al)GaN surfaces modified the nitride expitaxial growth kinetics. The anti-surfactant is achieved by depositing SiO_2 on the native substrate [41]. However, those were distributed randomly. Further, it had been reported that InGaN nano-dots and nano-rings grown on patterned GaN surfaces by nanoscale selective area epitaxiay (NSAE), in which the growth template patterned by nano-patterns in the porous aluminum oxide are achievable [42]. However, it is very hard to obtain long-range-ordered nanostructure arrays because of the inherent feature of self-assembled porous aluminum oxide.

The fabrication of long-range-ordered AlInGaN nanostructures grown on GaN/sapphire substrates by using NSAE is among the few latest techniques for efficient short-wavelength light-emitting devices. As a key step in the fabrication of such nanostructures, dense periodic nano-hole patterns with different hole diameters were defined and transferred to an oxide layer, serving as the growth template. AlInGaN nano-dots and nano-rings were grown through the nanoporous oxide template by MOCVD. From the surface topographic results, it has been found that periodic AlInGaN nano-dot arrays with a narrow size distribution were formed during the growth. A CCD image of AlInGaN-based short-wavelength LED is depicted in Figure 6.2. Optical properties of the integrated periodic nano-dots were also studied by power-dependent PL spectroscopy at room temperature as well as low temperatures.

The periodic AlInGaN nano-dots showed improved optical performance, which was attributed to the strong localization of photogenerated carriers giving rise to high-efficiency photonic devices due to good control over nano-dot size and their controlled positioning and distributions.

FIGURE 6.2
An AlInGaN nano-dots array-based short-wavelength (200×200 μm^2) LED.

6.5 Photonic Devices Fabrication Using Nanostructures

For the realization of high-efficiency short-wavelength light-emitting devices, extensive research efforts have shown enhancement in their performances [43]. The very nature of micro- and nano-photonic devices allows the development of cost-effective and mass-reproducible photonic integrated circuits (PICs). These in turn allow denser, faster, cheaper, and more efficient signal processing in the optical domain. Although many of the ideas and potentials of these devices have been identified long ago, only recently a transition from basic research to engineering application has been made for micro- and nanostructured UV and DUV photonic devices due to various technological advances pioneered almost two decades ago [44].

III-nitride wide-bandgap semiconductors (WBGS), with energy bandgap varying from 0.8 eV (InN) to 3.4 eV (GaN) to about 6.2 eV (AlN), have been recognized as technologically important materials [45] as they all are direct bandgap. III-nitride optoelectronic devices offer special benefits including UV/blue emission, allowing higher optical storage density and resolution as well as the ability for chemical- and biohazard-substance detection. Much larger band offsets of 2.8 or 4.3 eV for GaN/AlGaN or InGaN/AlGaN heterostructures allow readily QW-based device designs. DUV photonic devices and their material properties allow them to withstand extreme conditions of temperature, chemical, and radiation exposures. The progresses in high-brightness LEDs based on III-nitride WBGS place them as a strong candidate for solid-state lighting technology. Due to their low power consumption [46], smaller size, and long lifetime (compared to conventional lamps), these LEDs would advantageously replace conventional technology being used for lighting, sensing, and strategic optical communications. In addition, visible LEDs with high external efficiency are currently in high demand for a variety of applications including flat panel displays, printers, and optical interconnects in computers [47].

High-efficiency UV/DUV emitters are particularly sought for applications including chemical and biological agent detection and short-wavelength optical communications. In addition to that high-intensity, UV/DUV-LEDs are also used as transceivers for covert non-line-of-sight (NLOS) optical communications [48]. The internal quantum efficiency (QE) of InGaN/GaN QW blue LEDs is close to 100%, but most of the light is lost due to the parasitic absorption of lateral guided modes arising due to total internal reflection (TIR) in the semiconductor materials. The need for improvement of extraction efficiency in LEDs is exceptionally great, especially for DUV LEDs ($\lambda < 300$ nm) with AlGaN as an active layer. In addition to improving the material quality, novel device architectures are required too.

Reports on unique polarization property of optical emission from AlGaN alloys suggests that the nano-application engineering of AlGaN is also responsible for the low QE of the AlGaN-based DUV LEDs [49]. Numerous process routes have been investigated that allow very close control of the composition, bandgap energy, and lattice constant(s) of ternary III–V semiconductor alloys to control the properties of the associated devices. Wurtzite structure GaN and AlN have lattice mismatches of only 2.4% and 4% along the *a*- and *c*-axes, respectively. The direct bandgap energies (UV wavelengths), of $Al_xGa_{1-x}N$ alloys ($0 \leq x \leq 1$) range from 3.4 eV (360 nm in GaN) to 6.2 eV (200 nm in AlN). Thus, the $Al_xGa_{1-x}N$ alloys are attractive for optoelectronic devices such as UV/DUV LEDs and LDs (laser diodes). The large bandgap and high-field electron mobility also make these materials candidates for high-power and high-frequency microelectronic applications.

A schematic of DUV LEDs is given in Figure 6.3. In this diagram, the junction layer of the order of few nanometers only is closer to the *p*-metallization contact pads compared to that of the *n*-contacts; because hole mobility is much lower than electrons.

Therefore, extra care will be introduced to improve the quality of the junction layer because *p*-AlGaN/GaN is already suffering from several issues by its growth quality standpoint [50]. There have been great efforts to develop III-nitrides-based DUV emission with $\lambda \leq 300$ nm using the QW epi-structures. Inevitably, the Al-fraction in alloy composition for the *p*- and *n*-cladding should be more than those of the active layer to achieve the transparency at the emission wavelengths.

Based on advanced bandgap engineering, different material structures have been utilized to improve the material quality and the UV LED power level. Some difficulties have been encountered because of current crowding in the standard 300 µm × 300 µm geometry lateral injection LEDs. The lateral structures, nonuniformity, and the poor conductivity of the *n*- and *p*-cladding layers lead to current crowding and the inhomogeneous effects. This in turn can significantly limit the achievable output optical power density. Commercial success of such materials requires the epitaxial growth of films with low

FIGURE 6.3
A schematic of a DUV LED epi-structure grown by
PE-MOCVD.

densities of structural defects and controlled *n*- and *p*-type doping. Studies
have been performed to enhance the extraction efficiency of such devices
based on their nanolayers and bandgap engineering. Further, a brief overview of the growth of GaN/Al$_x$Ga$_{1-x}$N alloys, introduction of donor and
acceptor elements to achieve *n*- and *p*-type doping via pulsed-enhanced
metal organic vapor phase epitaxy (PE-MOVPE) and related challenges are
discussed in detail.

During the past few years, there has been a significant improvement in the
growth quality of III-nitride materials. In brief, Hagan et al. [51] synthesized
Al$_x$Ga$_{1-x}$N films by chemical vapor deposition (CVD) and showed the exist-
ence of an alloy system throughout the range from GaN to AlN. Branov
et al. [52], grew Al$_x$Ga$_{1-x}$N films up to $x = 0.45$ by chloride vapor phase
epitaxy (VPE) at 1050°C and measured the electrical and optical properties
whereas Yosida et al. [53] deposited Al$_x$Ga$_{1-x}$N films that contained the
entire composition range on sapphire and Si substrates using reactive
molecular beam epitaxy (MBE). The elimination of gas-phase pre- (or parasitic) reactions between NH$_3$ and metalorganics is very important for achieving the growth of high Al-content Al$_x$Ga$_{1-x}$N films. Gas-phase adduct
formation in the trimethylgallium/ammonia (TMG)/NH$_3$ system has been
suggested by Mazzarese et al. [54]. However, despite all such efforts, there
have been several issues related to the poor quality of the grown material
such as the high density of defects and localized strain effects. Quite recently,
Khan et al. [55] were able to grow Al$_x$Ga$_{1-x}$N solid solutions over the entire
binary system using MOVPE at low pressure (5–100 torr), a relatively good
quality material for practical devices. However, at atmospheric pressure,
Koide et al. [56], found that white, Al-rich deposits were observed on
AlGaN layers and that control over the solid composition was impossible

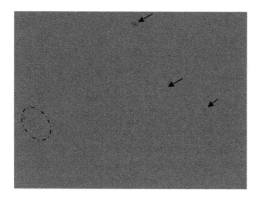

FIGURE 6.4
Al-rich micrograph of a DUV LED epi-structure grown by PE-MOCVD.

at low gas velocities (0.7–2 cm/s). They attributed these deposits to in volatile adduct formation, which depleted the Al precursor in the gas phase. An AFM micrograph of Al-rich epi-structure is shown in Figure 6.4. Here, the density of defects can be seen, which are randomly distributed on the sample. These defects can further propagate into the epi-structure if the concentration of Al increases from certain thresholds. In this situation, it was essential to have control on the doping and the associated conductivity of AlGaN solid solutions for the success of this epi-structure for optoelectronic and high-power and high-frequency electronic devices. However, one of the major challenges has been to achieve highly conductive $Al_xGa_{1-x}N$ solid solutions with higher Al content. Several controlled experiments have been designed to address this issue and it was found that a steady decrease in the electron concentration and the associated n-type conductivity of these alloys is linked with the increase in Al fraction [57–60].

In the study, DUV LEDs epi-structure was grown on a basal-plane sapphire substrates using a custom-designed vertical pulsed-enhanced metal organic chemical vapor deposition (PE-MOCVD) system [61]. Tri-methyl aluminum (TMA), trimethyl gallium (TMG), trimethyl indium (TMI), and NH_3 are used as precursors. A typical structure features a high-temperature AlN buffer layer and an AlN/AlGaN superlattice grown by the PE-MOCVD. The main structure consists of 2–3 μm Si-doped AlGaN, a multiple quantum-well (MQW) active region, a Mg-doped p-AlGaN blocking layer, and a graded p-AlInGaN p-contact layer.

The growth material was characterized by Hall measurements, x-ray, PL, and atomic force microscopy to study the quality of AlN buffer and AlN/AlGaN superlattices on the n-AlGaN. The AlNAlGaN superlattices helped in reducing the tensile strain resulting from the fine grain structure. It was found that the improved growth techniques based on nanolayers and bandgap engineering could help reduce the dislocation density to increase the UV/DUV emitter's efficiency [61–63]. The epi-structures grown on sapphire substrate using AlN epilayer template have been used in fabrication of high-efficiency UV/DUV lighting devices. The benefits of inserting an AlN epitaxial layer as a template for the growth of subsequent III-nitride device structures have been exploited [64–68] to reduce defect densities.

Based on improved nanolayers and bandgap engineering, AlN epilayer was used as a template for III-nitride devices and it helped improve the long existing lattice mismatch issues. Further, with its excellent UV transparency down to 200 nm, the insertion of AlN epilayer as a template for UV LED structures can be revolutionaries to filter defects in the epi-structure for high power, long lifetime light-emitting devices. In brief, a high-quality AlN epilayer with a thickness of about 1 μm was grown as the epitaxial template for the subsequent epilayers within micro- and nano-ranges. On the AlN epilayer, a 1.8 μm Si-doped n-$Al_{0.6}Ga_{0.4}N$ was grown as the n-type contact layer followed by the active region consisting of a three-period $Al_{0.35}Ga_{0.65}N$ multiple QWs for these LEDs. Since it is difficult to achieve a desirable hole density in an AlGaN alloy with high Al composition, a p-$Al_{0.7}Ga_{0.3}N$ layer was employed as an electron blocking layer to effectively block the electron overflow, thereby enhancing the electron–hole radiative combination in the QWs. The structure was then completed with a 70 Å Mg-doped p-$Al_{0.6}Ga_{0.4}N$ and 180 nm heavily doped p-$Al_{0.1}Ga_{0.9}N$ as the p contact layer [69–71].

The device fabrication starts with the magnesium activation of the epi-wafers using rapid thermal annealing. The mesa etching to expose the n-$Al_{0.6}Ga_{0.4}N$ is performed, followed by Ti/Al metal deposition for the n contacts and Ni/Au for p contacts with rapid thermal annealing at 650°C for 2 min. The sapphire side of the process devices was patterned using AZ-4620 photoresist and dry-etched with a series of etching cycles in 790-series ICP-System. An SEM micrograph of a patterned substrate is shown in Figure 6.5. As a next step, the processed wafers were lifted off for the photoresist and diced into chip-scale devices, flip-chip bonded with Au–Sn bumps onto ceramic AlN submount, and, finally, mounted on TO headers. A schematic of a flip-chip package of the nano-patterned DUV devices on TO header is shown in Figure 6.5.

A calibrated UV integrating optical sphere was used to measure the light output power from the sapphire side since the submount side is

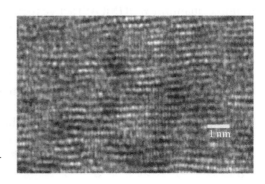

FIGURE 6.5
An SEM micrograph of a nano-patterned substrate.

nontransparent. Compared with the usual square geometry, nano-patterned geometry with n contact encircling the p-type mesa disk showed a significant improvement in their output power.

From the spectral response, it is found that the long wavelength emission is caused mainly by electron recombination in the p-cladding layer with deep-level impurities associated most likely with cation vacancies. By incorporating an electron blocking layer, the injected electrons can be more effectively confined in the active region, diminishing the long-wavelength emission. Another observed contribution to the pure emission spectrum comes from the high-quality AlN template and the subsequent low-defect structure, especially, reduction of other recombination channels in the n cladding layer. Further, based on advanced device processing techniques, it was discovered that the mesa size can influence these LEDs' performance, including the turn-on voltage, differential resistance, output power, and power density. As noted above, before, the characterization of these devices, the chip-scale LEDs were flip-chip packaged (Figure 6.6).

From the data shown in Figure 6.7, it is evident that LEDs from the same batch with and without nano-patterning show a factor three enhancement. The higher extraction efficiency of patterned LEDs was attributed to the reduction of TIR within the epi-structure, whereas, the better saturation current limits is mainly because of an excellent control on the threading dislocation in the epilayers. However, a relatively, higher value of the V_F and the turn-on voltages are the result of the use of relatively high Al concentration (70%) in the n- and p-cladding layers, and consequently lower quality n- and p-ohmic contacts. Furthermore, the inset shows the electroluminance (EL) spectra of a 290 nm DUV LED. A full width half maximum (FWHM) of ~16 nm was measured using EL micrograph, as shown in Figure 6.7. It can be seen that the rise in the profile of nano-patterned

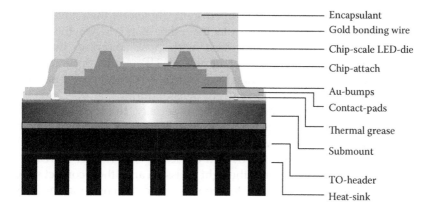

FIGURE 6.6
A schematic of a flip-chip packaging of the nano-patterned DUV devices.

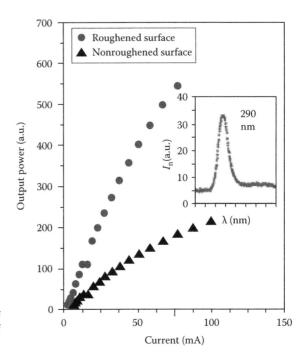

FIGURE 6.7

EL spectra and comparison of *I–V*, *L–I* characterization of nano-patterned DUV LEDs.

devices is much better compared to that of the conventional device. Furthermore, the conventional device had started saturating at $\sim$75 mA whereas the patterned device was still rising even at the double injected current of 150 mA.

In brief, it has been established that at a low driving current $\sim$20 mA, the light output is similar, with no significant dependence on the nano-patterning. With increased input current, the thermal effect becomes important and the light output gradually saturates. Under the same driving current, the non-patterned flip-chip packaged devices have larger thermal losses and a larger thermal intensity due to trapped light; therefore, they have a lower saturation optical power output. On the other hand, when the device's sapphire side (substrate) is processed with nano-texturing, the output power enhances mainly because of improved TIR effects. Specifically, we found that for a 300 μm^2 disk, LEDs have saturation optical power density of $\sim$8 W/cm^2 at 2 kA/cm^2 injection current density during CW operation tests. As the material quality further improves, especially, the conductivity of *n*- and *p*-cladding layers, we believe that the optimized device power can further be improved along with the lifetime. Since the current spreading problem puts limitations on the overall device size, to further increase the total output power by increasing the light emission area, the *n* contact should be introduced into the mesa area to keep the mesa dimension in the limitation of current spreading distance.

Further, we describe MQW-based quaternary InAlGaN-based DUV LEDs that are often fabricated using the advanced processing techniques.

As usual, micro/nano epi-structures are designed using quaternary AlInGaN and AlGaN as the active region. The comparative study of the grown epi-structures for InAlGaN, AlGaN, and GaN is usually performed by fabricating the LEDs and thereafter, characterizing by employing PL study.

In short, the LED structures are grown on SiC and sapphire substrates. The structure consists of a 50 nm-thick Si-doped $Al_{0.25}Ga_{0.75}N$ buffer layer, a 400 nm-thick Si-doped $Al_{0.17}Ga_{0.83}N$ layer, a 60 nm-thick undoped quaternary $In_{0.05}Al_{0.34}Ga_{0.61}N$ active layer, a 200 nm-thick Mg-doped $Al_{0.18}Ga_{0.82}N$ layer, and a 30 nm-thick Mg-doped GaN capping layer, as illustrated in Figure 6.8.

The samples were annealed at 800°C in a N_2 ambient in order to activate the Mg acceptors for enhancing *p*-conductivity. The Mg incorporation in the Mg-doped $Al_{0.18}Ga_{0.82}N$ layer was analyzed and determined by secondary-ion mass spectroscopy measurements. The hole and electron concentrations in the Mg-doped $Al_{0.18}Ga_{0.82}N$ layer and in both the Si-doped $Al_{0.25}Ga_{0.75}N$ and the $Al_{0.15}Ga_{0.85}N$ layers were also optimized and confirmed using Hall effect measurements. As noted earlier, schematic of AlInGaN MQWs-based LED epi-structure is depicted in Figure 6.8. From this schematic, one can see that along with the incorporation of a single-crystal AlN template layer, multilayers of superlattices are also introduced to avoid further propagation of the defects toward the junction layer, which is in the nanometric order. A viable metallization process based on Ni/Au electrodes are used for the *p*-type surface and Ti/Al for the *n*-type contacts. The diameter of the *p*-side electrode is ∼0.6 mm. It was found that the injection current density changes within the range of 0–300 A/cm^2. Single-peaked bright emission with a

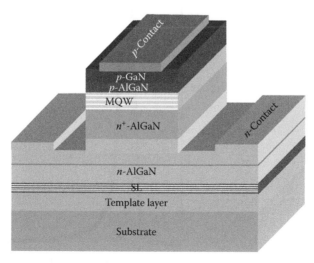

FIGURE 6.8
Schematic of AlInGaN-based MQW DUV LEDs.

wavelength of 290 nm from the ternary AlGaN LEDs has been observed. No noticeable deep-level emissions were observed such as emission from Mg-acceptor levels or 540 nm-band yellow emission. In addition, there was neither a significant wavelength shift nor any output power saturation with increasing injection current in the current-density range between 0 and 300 A/cm^2.

It was observed that in some cases the DUV output power of the LED was not high because of absorption losses through the p-GaN cap layer and the Ni/Au p electrode. The spectral properties of the fabricated DUV LEDs on sapphire substrates using AlGaN-based active region have been studied and an emission at $\lambda \sim 290$ nm under RT CW operation have been achieved, as illustrated in Figure 6.9. This shows that a single, well-defined peak exists at ~290 nm and there are no signatures of any long wavelength tail in the emission spectrum. The DUV intensity from quaternary III-N (AlInGaN) LEDs was more than one order of magnitude higher than that obtained from AlGaN or GaN LEDs. From these results, we revealed that the use of quaternary AlInGaN as the emitting regions of DUV emitters is advantageous in comparison with AlGaN or GaN, particularly, when they are fabricated on wafers with high-TDD nanometric buffer layers.

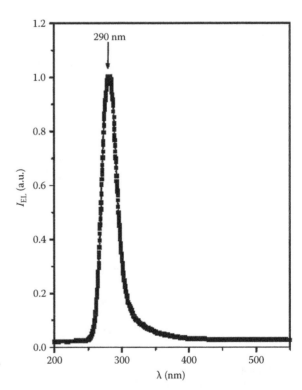

FIGURE 6.9
Emission spectra of AlInGaN-based MQW DUV LEDs.

6.6 High-Efficiency Multiquantum-Wells Deep-UV LEDs with High Al Content

For the realization of nitride-based short-wavelength light-emitting devices, the achievement of high-quality *p*-type AlGaN with high Al-content is one of the key requirements. In particular, for DUV LEDs operating in the wavelength range below 300 nm, the requirement for a sufficiently high hole concentration in *p*-type AlGaN is important for suppressing electron overflow and for obtaining high injection efficiency. The highest optimized limit of the Al-content for Mg-doped $Al_xGa_{1-x}N$ is approximately 40% in order to obtain the desired hole conductivity [60]. A single-crystal, high-quality AlN template layer is also introduced in the epi-structures in order to improve the quality of the grown epi-structures. Further, single-crystal AlN has a transparency down to 200 nm, which is ideal for short-wavelength emission devices. A PL spectra of the incorporated AlN as template layer within the epi-structure is shown in Figure 6.10. From this micrograph, well-defined peaks at 3.44 and 5.96 eV can clearly be seen, as an evidence of the best growth quality of a single-crystal GaN and AlN epi-structures.

In the past few years, significant research efforts have been involved to realize high hole conductivity in group-III-nitride materials, by incorporating the high-quality superlattices within the epi-structures [72–75]. It is difficult to

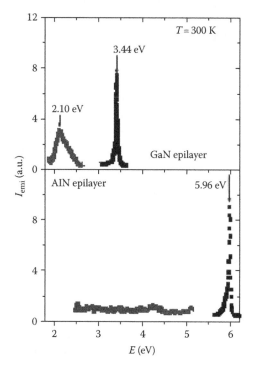

FIGURE 6.10
PL spectra of GaN and AlN epilayers for DUV LEDs.

obtain a sufficiently high hole concentration in p-type AlGaN with a high Al content. The main reason for this is that the energy level of the Mg acceptor is deep for high-Al-content AlGaN, and thus the activated hole concentration is quite low. In order to obtain hole conductivity, heavy Mg-doping level is required. Such a heavily Mg-doped AlGaN easily becomes n type due to an increase in the number of vacancies or defects in the crystal. Therefore, in order to obtain optimal hole conductivity for high-Al-content p-type AlGaN, the identification of growth conditions for obtaining high-crystalline quality AlGaN for even heavy Mg doping is necessary. As a step forward toward improving the overall quality of the epi-structures, an alternating gas flow growth technique can also be incorporated for the growth of high-Al-content p-type AlGaN [62]. The crystalline quality of high-Al-content p-type AlGaN is considered to be remarkably improved due to enhanced migration of the precursors when using the alternating gas flow sequence. Moreover, by using alternating gas supply, the vapor reactions between ammonia and TMA or ammonia and the Mg source are considered to be significantly suppressed in front of the graphite susceptor. The crystalline quality of high-Al-content AlGaN is considered to be improved significantly by the suppression of these gas-phase reactions.

By growing with alternating gas flow, we achieved hole conductivity in Mg-doped AlGaN with Al contents between 40% and 60% [63]. On the other hand, we could not obtain p-type character for high-Al-content Mg-doped AlGaN grown with a conventional continuous gas flow regime. Based on the obtained results, it was found that the alternating gas flow growth method is advantageous for obtaining high-Al-content p-AlGaN. In brief, the AlGaN-based DUV LEDs with high Al contents using p-type AlGaN on sapphire substrates has been described. We employed a high Al content greater than 40% for the p-type AlGaN in order to suppress electron overflow adequately. If a lower Al-content p-AlGaN layer is used, the LED intensity remains weak since the electron injection efficiency into the QW is low due to electron overflow. Due to the incorporation of AlN as a template layer (transparent down to 200 nm), the emitted power was detected from the substrate side. The output power radiated into the backside of the LED was measured using a Si photodetector located behind the LED sample. L–I and I–V characterizations of the processed example devices are presented in Figure 6.11. The maximum relative output power was observed ~1700 μW at an injection current of 160 mA. It is worth mentioning here that local thermal management has also been introduced to improve the overall performance of these devices. Compared to conventional devices, which usually saturate at ~50 mA of the injected current, the presented example devices have higher saturation limits with a better value of the forward voltage. The measured external QE was approximately 0.8% for short-wavelength LEDs at around peak maximum output power.

It has been established now that the efficiency of these LEDs can further be improved by controlling the density of threading dislocation densities within

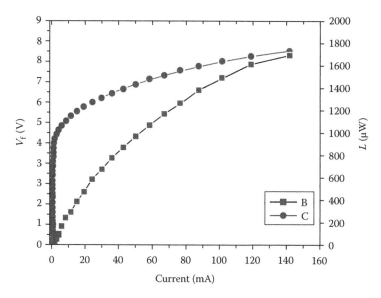

FIGURE 6.11
L–I–V characteristics of high Al-content, AlGaN-based DUV LEDs.

the grown nanometric epi-structures. Further, much higher efficiency could be obtained by controlling the generation of strong polarization and piezo-electric current within the junction layer of the epi-structure for short wavelength. It is believed that the efficiencies of such devices can also be improved with the incorporation of nanostructures such as colloidal nanoparticles within the active layers of their epi-structures.

Integration of 2D, 3D photonic crystal [64] within the nanoscales could be another effective technique to improve the output power of DUV light-emitting devices. By incorporating AlN as a template layer and good quality conductive *n*-AlGaN into the actual device structure, significantly improved performance of DUV LEDs ($\lambda < 300$ nm) can effectively be achieved. It was also found that III-nitride PCs provide three times enhancement in output power and four times enhancement in modulation speed in DUV LEDs [65]. The demonstration of photonic bandgap (PBG) materials active in the UV and visible wavelengths opens many new applications [66–68].

6.7 Conclusion

An overview of the research and development, and future prospects of III–V WBGS nanostructures and related photonic devices (specifically LEDs) for short-wavelength applications are presented in this chapter. Some of the recent developments in nano-epi-structures based on SiC and Al_2O_3 substrates have been discussed. After a careful review of the fundamentals of nanostructures and their related physical properties, this chapter addresses the need of nanotechnology based on smart nanostructures such as incorporation of nanoparticles within the junction layers, improved bandgap engineering of the MQWs in order to overcome the inherent polarization effects.

Nanowires and nanocrystals can also help improve the performance of such devices. A comprehensive note on the quaternary and ternary InAlGaN and AlGaN has been considered within the scope of this chapter. Furthermore, some of the practical issues and problems such as the challenges in the epi-structure growth and related effect of defects on device performance are also highlighted. A good portion of the chapter also reviews the critical issues related with the advance growth processing for better bandgap engineering of short-wavelength photonic devices. Such WBGS materials have shown exciting applications in optoelectronics and show a great potential for the next generation communication technology.

In brief, significant efforts are still underway to fabricate efficient short-wavelength photonic devices at the scale of mesoscopic widths of mere hundreds of nanometers and nanoscopic heights of several tens of nanometers. An unprecedented novelty in these new nanostructures is that they are quasi-single crystalline, having extraordinary potential that may effectively be tailored via functionalization of their extreme optical and electro-optical properties. While III-N materials are gaining acceptance, another group of nanomaterials based on II–VI binary and ternary semiconductors with direct wide bandgap, e.g., ZnO [69] is also gaining acceptance in the research community.

References

1. Agranovich, V. M., G. C. LaRocca, and F. Bassani. *Chem. Phys. Lett.* **247**: 355 (1995).
2. Burroughes, J. H., D. D. C. Bradley, A. R. Brown, R. N. Marks, K. Mackay, R. H. Friend, P. L. Burns, and A. B. Holmes. *Nature* **347**: 539 (1999).
3. Burrows, P. E. and S. R. Forrest. *Appl. Phys. Lett.* **64**: 2285 (1994).
4. Burrows, P. E., S. R. Forrest, S. P. Sibley, and M. E. Thompson. *Appl. Phys. Lett.* **69**: 2959 (1996).
5. De Leeuw, D. M., A. R. Brown, M. Matters, K. Chmil, and C. M. Hart. *Mat. Res. Soc. Spring Mtg.*, San Francisco, (1997).
6. Yee, C. K. et al. *Adv. Matter.* **16**: 1184 (2004).
7. Dodabalapur, A., H. E. Katz, L. Torsi, and R. C. Haddon. *Science* **269**: 1560 (1995).
8. Dodabalapur, A., L. J. Rothberg, and T. M. Miller. *Appl. Phys. Lett.* **65**: 2308 (1994).

9. Dodabalapur, A., L. Torsi, and H. E. Katz. *Science* **268**: 270 (1995).
10. Dubovsky, O. and S. Mukamel. *J. Chem. Phys.* **15**: 417 (1992).
11. Fang, S., K. Kohama, H. Hoshi, and Y. Maruyama. *Jpn. J. Appl. Phys.* **32**: L1418 (1993).
12. Adivarahan, V., J. P. Zhang, A. Chitnis, W. Shuai, J. Sun, R. Pachipulusu, M. Shatalov, and A. Khan. *Jpn. J. Appl. Phys.* **41**: L435 (2002).
13. Adivarahan, V., S. Wu, A. Chitnis, R. Pachipulusu, V. Mandavilli, M. Shatalov, J. P. Zhang, M. A. Khan, G. Tamulaitis, A. Sereika, I. Yilmaz, M. S. Shur, and R. Gaska. *Appl. Phys. Lett.* **81**: 3666 (2002).
14. Zhang, J. P., X. Hu, Y. Bilenko, J. Deng, A. Lunev, R. Gaska, M. Shatalov, J. W. Yang, and M. A. Khan. *Appl. Phys. Lett.* **85**: 5532 (2004).
15. Sun, W. H., V. Adivarahan, M. Shatalov, Y. Lee, S. Wu, J. W. Yang, J. P. Zhang, and M. A. Khan. *Jpn. J. Appl. Phys.* **43**: L1419 (2004).
16. Zhang, J. P., A. Chitnis, V. Adivarahan, S. Wu, V. Mandavilli, R. Pachipulusu, M. Shatalov, G. Simin, J. W. Yang, and M. A. Khan. *Appl. Phys. Lett.* **81**: 4910 (2002).
17. Zhang, J. P., S. Wu, S. Rai, V. Mandavilli, V. Adivarahan, A. Chitnis, M. Shatalov, and M. A. Khan. *Appl. Phys. Lett.* **83**: 3456 (2003).
18. Adivarahan, V., S. Wu, J. P. Zhang, A. Chitnis, M. Shatalov, V. Mandavilli, R. Gaska, and M. A. Khan. *Appl. Phys. Lett.* **84**: 4762 (2004).
19. Sun, W. H., J. P. Zhang, V. Adivarahan, A. Chitnis, M. Shatalov, S. Wu, V. Mandavilli, J. W. Yang, and M. A. Khan. *Appl. Phys. Lett.* **85**: 531 (2004).
20. http://cms.iopscience.iop.org
21. Yasan, A., R. McClintock, K. Mayes, D. Shiell, L. Gautero, S. R. Darvish, P. Kung, and M. Razeghi. *Appl. Phys. Lett.* **83**: 4701 (2003).
22. Mayes, K., A. Yasan, R. McClintock, D. Shiell, S. R. Darvish, P. Kung, and M. Razeghi. *Appl. Phys. Lett.* **84**: 1046 (2004).
23. Fischer, A. J., A. A. Allerman, M. H. Crawford, K. H. A. Bogart, S. R. Lee, R. J. Kaplar, W. W. Chow, S. R. Kurtz, K. W. Fullmer, and J. J. Figiel. *Appl. Phys. Lett.* **84**: 3394 (2004).
24. Hanlon, A., P. M. Pattison, J. F. Kaeding, R. Sharma, P. Fini, and S. Nakamura. *Jpn. J. Appl. Phys.* **42**: L628 (2003).
25. Zhang, J. P., H. M. Wang, M. E. Gaevski, C. Q. Chen, Q. Fareed, J. W. Yang, G. Simin, and M. A. Khan. *Appl. Phys. Lett.* **80**: 3542 (2002).
26. Wang, H. M., J. P. Zhang, C. Q. Chen, Q. Fareed, J. W. Yang, and M. A. Khan. *Appl. Phys. Lett.* **81**: 604 (2002).
27. Nikishin, S. A., V. V. Kuryatkov, A. Chandolu, B. A. Borisov, Gela D. Kipshidze, I. Ahmadl, M. Holtzl, and H. Temkin. *Jpn. J. Appl. Phys.* **42**: L1362 (2003).
28. Kim, K. H., Z. Y. Fan, M. Khizar, M. L. Nakarmi, J. Y. Lin, and H. X. Jiang. *Appl. Phys. Lett.* **85**: 4777 (2004).
29. Khizar, M., Z. Y. Fan, K. H. Kim, J. Y. Lin, and H. X. Jiang. *Appl. Phys. Lett.* **86**: 173504 (2005).
30. Forrest, S. R., M. L. Kaplan, P. H. Schmidt, W. L. Feldmann, and E. Yanowski. *Appl. Phys. Lett.* **41**: 90 (1982).
31. Forrest, S. R. and F. F. So. *J. Appl. Phys.* **64**: 399 (1988).
32. Garnier, F., G. Horowitz, D. Fichou, and A. Yassar. *Supramolec. Sci.* **4**: 155 (1997).
33. Garnier, F., G. Horowitz, X. Peng, and D. Fichou. *Adv. Mater.* **2**: 592 (1990).
34. Lam, J. F., S. R. Forrest, and G. L. Tangonan. *Phys. Rev. Lett.* **66**: 1614 (1991).

35. Leegwater, J. A. and S. Mukamel. *Phys. Rev. A* **46**: 452 (1992).
36. Lin, Y.-Y., D. J. Gundlach, and T. N. Jackson. *MRS Spring Mtg.*, San Francisco, CA (1997).
37. Maruyama, Y., H. Hoshi, S. L. Fang, and K. Kohama. *Synth. Met.* **71**: 1653 (1995).
38. Mukamel, S., A. Takahashi, H. X. Huang, and G. Chen. *Science* **266**: 250 (1994).
39. So, F. F. and S. R. Forrest. *IEEE Trans. Electron. Dev.* **36**: 66 (1989).
40. Tang, C. W. *Appl. Phys. Lett.* **48**: 183 (1986).
41. Tang, C. W. and S. A. VanSlyke. *Appl. Phys. Lett.* **51**: 913 (1987).
42. Tang, C. W., S. A. VanSlyke, and C. H. Chen. *J. Appl. Phys.* **65**: 3610 (1989).
43. Taylor, R. B., P. E. Burrows, and S. R. Forrest. *IEEE Photon. Technol. Lett.* **9**: 365 (1997).
44. Wang, N. and S. Mukamel. *Chem. Phys. Lett.* **231**: 373 (1994).
45. Wohrle, D. and D. Meissner. *Adv. Mater.* **3**: 129 (1991).
46. Zang, D. Y. and S. R. Forrest. *IEEE Photon. Technol. Lett.* **4**: 365 (1992).
47. Zang, D. Y., Y. Q. Shi, F. F. So, S. R. Forrest, and W. H. Steier. *Appl. Phys. Lett.* **58**: 562 (1991).
48. Madec, R., B. Devincre, L. Kubin, T. Hoc, and D. Rodney. *Science* **301**, 1879 (2003). [Abstract/Free Full Text].
49. Meyers, M. A. and K. K. Chawla. *Mechanical Behavior of Materials* (Prentice-Hall, Upper Saddle River, NJ, 1999).
50. Weertman, J. R. In *Nanostructured Materials: Processing, Properties and Potential Applications*, C. C. Koch, Ed. (William Andrew, Norwich, NY, 2002), pp. 397–421.
51. Hagan, J., R. D. Metcalfe, D. Wickenden, and W. Clark. *J. Phys.* **C 11**: L143 (1978).
52. Branov, B., L. Daweritz, V. B. Gutan, G. Jungk, H. Neumann, and H. Raidt. *Phys. Stat. Solidi (a)* **49**, 629 (1978).
53. Yosida, S., S. Misawa, and S. Gonda. *J. Appl. Phys.* **53**: 6844 (1982).
54. Mazzarese, D., A. Tripathi, W. Conner, K. Jones, L. Caleron, and D. Eckart. *J. Electron. Mater.* **18**: 369 (1989).
55. Khan, M. A., R. A. Skogman, and R. G. Schulze. *Appl. Phys. Lett.* **43**: 492 (1983).
56. Koide, Y., H. Itoh, N. Sawaki, I. Akasaki, and M. Hashimoto. *J. Electrochem. Soc.* **133**: 1956 (1986).
57. Meyers M. A. et al. *Acta Mater.* **51**: 1211 (2003).
58. McNaney, J. M., J. Edwards, R. Becker, T. Lorenz, and B. A. Remington. *Metall. Trans. A* **35**: 2625 (2004).
59. Schuh, C. A. and A. C. Lund. *Nat. Mater.* **2**: 449 (2003).
60. Lund, A. C. and C. A. Schuh. *Acta Mater.* **53**: 3193 (2005).
61. Jiang, B., G. J. Weng. *J. Mech. Phys. Solids* **52**: 1125 (2004).
62. www.sciencemag.org, *Science*, **320**: 1768 (2008).
63. Van Swygenhoven, H. and A. Caro. *Phys. Rev. B* **58**: 11246 (1998).
64. Lin, J. Y. and Jiang, H. X. *Lasers and Electro-Optics, 2005.* (CLEO), pp. 1441–1443, vol. 2 (2005).
65. Fan Z. Y. et al. *J. Phys. D Appl. Phys.* **41**: 094001 (12pp) (2008).
66. Khizar, M. and Y. A. Raja. *Proc. SPIE* **6473**: 64730V (2007).
67. Khizar, M. and Y. A. Raja. *5th Annual Symposium at the Center for Optoelectronics and Optical Communications UNCC*, October 27, Charlotte, NC (2006).
68. Khizar, M. and Y. A. Raja. *Symposium on Photonics at the Frontiers of Science and Technology*, September 28–29, Durham, NC (2006).
69. Mosbacker, Y. M. et al. *Appl. Phys. Lett.* **91**: 072102 (2007).

7

Nanophotonics for Emerging Optical Networks

Kavan Acharya and M. Yasin Akhtar Raja

CONTENTS

7.1 Introduction

Multi-core microprocessors are becoming a norm in computing systems to increase the multi-threading and parallel processing as well as reducing the power dissipation. But the bandwidth limitation of the electrical interconnections is increasingly becoming the restricting factor in limiting the chip performance. To overcome this bottleneck, fiber is being considered to replace copper on chip level in the near future, as it did in the case of long-distance communications. The optical domain provides a theoretically unlimited bandwidth of fiber or the free-space option, with low power-dissipation and latency. Fibers have already started replacing copper in data centers to perform various rack-to-rack and board-to-board communications. A pioneering work on optical interconnecting both free-space [Morris 92, Feldman 94, Raja 95, Cheng 95] and waveguide (fiber) [Chen 00, Cheng 95] was demonstrated decades ago [Feldman 94, Morris 94] but the essential tools, perfection, and need has taken a long time to create justification. This will soon be extended to on-board and later on-chip communications. Micro-photonics has played a very important role in the case of board-to-board communications where, in major scenarios, a combination of vertical-cavity surface-emitting lasers (VCSELs) and photodiodes are used for transferring data signals between boards. VCSELs' underlying active structure and the epitaxially integrated distributed Bragg reflectors in the microcavity are nanometric layers [Raja 88, 89, Jewell 89, Corzine 89] and nanostructures. Closely following the footsteps of this big brother (VCSELs) is the nanophotonics. In reality, the VCSEL itself is an outcome of nanotechnology where 10 nm thick quantum wells and 50–60 nm thick epitaxial reflectors are the building blocks of such devices [Raja 88, 89]. The technology evolution as well as size is poised to replace copper on-board and, in the very near future, on-chip. This chapter concentrates on the role of nanophotonics in on-board and on-chip communications. Also, the role of this emerging technology in current access-, metro-, and long-haul networks will be highlighted briefly in the context of self-healing and optical-packet-based networks. We shall discuss the background of the state of the art in on-board communications and later will also focus on the building blocks of inter-chip and intra-chip high-capacity communications.

7.2 Chip-to-Chip Communication

Optoelectronics has moved from its fiber-based long-distance communications in the 1980s and early 1990s [Mukherjee 94] to short-distance communications such as optical access [Acharya 07, Acharya 08], system-to-system, and board-to-board communications [Sakano 95, Rode 97, Mederer 01]. The next step is the efficient signal transmission between chips on board and intra-chip signal transport. A number of researchers have tried to address the challenges of chip-to-chip communication over the last decade to realize system-on-package (SoP)-based systems [Chang 04] and applications from the perspective of power dissipation and distribution on board, interconnects, and system integration. The broad consensus among them was that (1) the main problem with large-scale optical interconnects on board is related to the low yield and long-term reliability of VCSEL arrays in areas of low threshold current and power supply [Krishnamoorthy 96, Miller 00]; (2) the detector is a major source of power dissipation in on-chip and chip-to-chip communications. This is because the bias current is always on in the case of optical communication, unlike CMOS circuits where power is drawn only during switching functions. Using a common bias circuit for the array of transceivers may help address this problem [Kibar 99]; (3) the major impact of using optics on-chip is that optical links have much lower latency as compared with their copper counterparts [Kyriakis-Bitzaros 01]; and (4) optical interconnect density is comparable with the off-chip interconnect density as the optical waveguides are of the order of wavelength and can be close together due to low crosstalk [Miller 00, Kibar 99].

Figure 7.1 illustrates the concept of SoP that enables digital-optical integration. The basic technologies required for such complete integration are

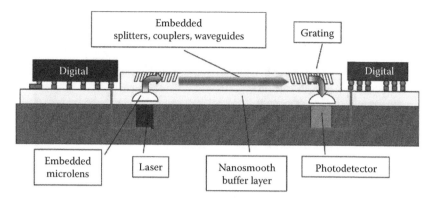

FIGURE 7.1
An SOP concept for low-cost opto/digital integration on FR-4 board. (Reprinted from Chang, G. et al., *IEEE Trans. Adv. Packag.*, 27(2), 386, May 2004. With permission.)

shown in Figure 7.1. A laser array is directly modulated by the high-frequency output port from the processor. This optical signal is further coupled to an on-board network via a microlens array. This network comprises various optical components like splitters, waveguides, gratings, etc. The output signal at the other end of the package is detected by a photodiode array. The signal received by the receivers is converted to an electrical signal that is fed into a specific port of the receiving processor. The entire microsystem is built directly on a buffer layer that is fabricated on flame retardant 4 (FR-4) or APPE boards [Chang 04]. The work that has been done and is being done toward building various building blocks to realize a fully integrated SoP is as follows.

Nan Jokerst [Jokerst 04] and her colleagues at Duke University studied planar optical interconnections integrated onto boards to realize embedded optical interconnections from a design, fabrication, test, and theoretical perspective. They demonstrated both beam-turning and embedded interconnections to the active optoelectronic devices that utilize active thin-film optoelectronic components embedded in the board using silicon, ceramic, and high temperature FR-4 substrates [Jokerst 04]. Figure 7.2 shows photomicrographs of a thin-film inverted metal–semiconductor–metal photodiode (I-MSM PD) embedded in a benzocyclobutane (BCB)-based polymer waveguide, with the PD/BCB structure integrated onto a SiO/Si substrate.

Researchers at the Fraunhoffer Institute IZM in Germany have developed a hybrid carrier [Krabe 00] that provides complete compatibility between electrical and optical surface-mounted components. The key approach is the formation of a separate optical-layer consisting of multimode waveguides. These waveguides are incorporated by hot embossing [Heckele 04] and standard printed wiring board technology [Whitaker 05]. They have also developed hybrid polymer–silica vertical coupler switches (Figure 7.3). Here, the silica provides a low-loss optical signal transmission while the polymer

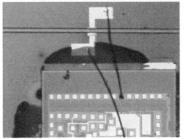

FIGURE 7.2
Photomicrograph of I-MSM PD embedded in polymer waveguide. Optical signal is coupled to the photodetectors via the waveguides. (Reprinted from Jokerst, N. et al., *IEEE Trans. Adv. Packag.*, 27(2), 376, May 2004. With permission.)

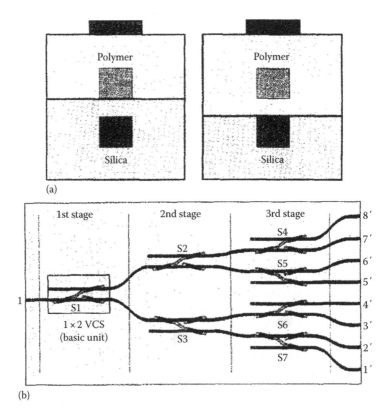

FIGURE 7.3
(a) Cross-section of vertical-coupler waveguide structure and (b) schematic of a 1 × 8 switch. (Reprinted from Keil, N. et al., *Electron. Lett.*, 37, 89, 2001. With permission.)

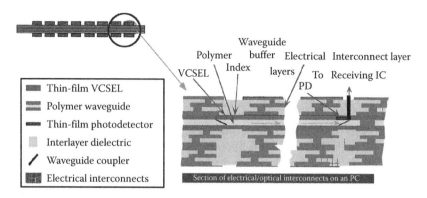

FIGURE 7.4
Fully embedded guided-wave optical interconnect system architecture. (Reprinted from Chen, R.T. et al., *IEEE Proc.*, 88, 780, 2000. With permission.)

provides coupling and thermal switching as it has a much higher thermo-optic coefficient [Keil 01].

A collaborative effort led by the University of Texas at Austin developed optical-clock synchronization architecture for a Cray multilayer mother-board. The distribution network is shown in Figure 7.4. It consists of VCSEL transmitters, grating couplers, metal–semiconductor–metal (MSM) detectors and polyimide waveguides [Chen 00].

7.3 Intra-Chip Communication

In the drive for higher speed and better performances of multiprocessors, power efficiency is one of the major design consideration factors. The scaling of transistor speeds and integration densities already approaching against the wall can no longer drive the processor performances. The local processing frequencies have clearly reached a tipping point leading to an exponential increase in power dissipation [Horowitz 05, Brodersen 02, Mudge 01]. To cope with the power dissipation problems, manufacturers use multi-core processors operating multiple parallel processors at lower clock frequencies [Horowitz 05, Mudge 01]. Over the next few years, this paradigm shift in processor chip design would lead to another serious bottleneck of the intra-chip communication infrastructure [Shacham 07]. Hence, the main challenge for future processor systems is to offer very high-bandwidth capacity and low-latency in connecting a large number of processor cores with low power-dissipation.

Therefore, this intra-chip communication has been the focus of significant research efforts to realize system-on-a-chip [Benini 02, Dally 01, Hemani 00].

These "micro-networks," also termed as networks-on-chip (NoC) [Benini 02], are highly scalable and provide huge bandwidth in order to replace the conventional bus-based electronic links on chips. Electronic NoCs will suffer saturation because of a fixed upper limit to the total high power dissipation [Shacham 07]. Therefore, photonic interconnection networks or photonic NoCs offer a potential solution to this problem as power dissipation is greatly reduced in the case of photonic intra-chip communications. This is due to the fact that unlike electronic NoCs where messages get buffered, regenerated, and transmitted on the intra-chip links several times during the transmission of the whole message, in a photonic NoC, the data is transmitted without any regeneration or buffering after the route is established. This leads to huge power savings [Shacham 06]. A power comparison analysis was carried out in the article by Shacham et al. [Shacham 06] between electronic and photonic NoCs. The results were astounding: Power expended in intra-chip communication can be reduced by two orders of magnitude by employing photonic NoCs when high bandwidth communications are required among the cores.

7.3.1 Architecture Overview

A photonic NoC comprises a hybrid design, combining optical- and electronic-networks. The optical network that comprises photonic integrated circuits (PICs) and silicon photonic switches that are connected via waveguides is used to transmit all data messages [Shacham 07]. Many of the building blocks in PICs and waveguides rely on the nanophotonic technologies. While an electronic network, which is identical in topology to the optical network, is used to send control messages and thus controls the optical network. Also, photonic technology cannot provide, currently, two important functionalities in a communication network, i.e., buffering and processing. That is, there are neither optical memories available nor a practical photonic logic for signal processing. Hence, these are implemented in the electronic network.

An electronic control packet precedes every photonic data packet. It helps in routing and path setup; this is the same paradigm used in long-haul optical burst switching using a protocol known as "just in time" (JIT) [Wei 99]. Any buffering that may be needed is done during this route setup process in the electronic medium. Once the route is set, the optical data packets are transmitted at one go, without any kind of buffering. As noted above, this is quite similar to optical circuit (burst) switching.

There are two main advantages of using photonics as light-paths for the transfer of data packets. One is the bit-rate transparency of optical medium. Because the transfer of the optical packet is done all at once, the optical switching elements switch on and off only once per message. While in the case of electronic packets, electronic switches must switch on and off for every bit transmitted. It scales as the "higher the bit-rate, the higher the number of switches." This ultimately results in high power dissipation. Therefore, optical bit-rate transparency results in low power dissipation. The second advantage is the low-loss of optical-waveguides compared with copper transmission lines. At the chip scale, the attenuation (power loss) in the optical link is independent from the distance. So, practically, if a data is transmitted between two cores that are 3 mm apart or 3 cm apart, the power dissipation is the same. Low-loss optical fibers further add to the advantage when this data is transmitted between two chips.

The main component of a photonic NoC is a broadband 2×2 photonic switching element (PSE) that is capable of switching parallel data signals on several wavelengths with a sub-μs time response. These switches are arranged in a two-dimensional matrix in a group of four. Each group is controlled by an electronic circuit to construct a 4×4 switch [Shacham 07]. The optical–electrical–optical (O–E–O) conversions are necessary for the end-to-end transmission of signals on the on-chip network. Hence, each node includes a so-called network gateway to serve this purpose. The component that makes up this gateway is a micro-ring-based or Mach–Zehnder-based silicon modulator [Xu 07, Gunn 06]. Clock synchronization and the recovery circuit as well as the serializer/deserializer are also included in the gateway. Laser sources acting as a transmitter can be located off-chip and coupled into the chip using fibers [Gunn 06, Johansson 05] or they can be bonded on-chip [Fang 06]. A SiGe photodetector [Gupta 04] can be used as a receiver (Figure 7.5).

7.3.2 Building Blocks

The two main building blocks for an on-chip optical network are the PSEs and silicon optical modulators (SOM). One of the key components needed for any such optical network is an SOM, which has the job of transferring high-speed electrical signals traveling on wires into laser-light

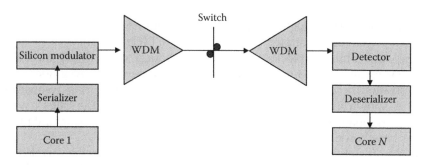

FIGURE 7.5
Block diagram outlining the building blocks of core-to-core on-chip communication. (From Vlasov, Y., Silicon photonics for next generation computing systems, tutorial in *European Conference on Optical Communications 2008*, Brussels, Belgium, Sept. 2008. With permission.)

pulses, traveling along a silicon waveguide. For on-chip interconnect applications, these SOMs need to have three very important characteristics. First, the SOMs should be of a very compact and small size. This is because many such modulators will be needed on a single-chip. Second, the SOMs should require minimal power. The third and final requirement is that the SOMs must be temperature insensitive. That is because the temperature within a chip can change dramatically depending on the operating conditions based on data rates, and these changes in temperature should not adversely affect the SOMs.

Among the variety of modulators, two different types of SOMs have been reported in literature. The first is the ring-resonator-based SOM [Xu 05, Xu 07] and the second is the 10 Gbps Mach–Zehnder type SOM [Gunn 06, Green 07]. Ring resonators enable high modulation in silicon using compact devices, again based on nanofabrication techniques and photonic processes [Xu 05, Xu 06, Lipson 05]. The micro-ring resonators have the ability to circulate light within the cavity at the resonance wavelengths. Hence, the optical path length can be increased by multiple circulations without the need for increasing the size of the device. But, in the case of the Mach–Zehnder interferometer, the physical length of the device needs to be increased in order to increase the optical path length. Hence, the micro-ring resonator seems to be a better choice of the two mentioned here.

Micro-ring resonators, with their physical features ranging from a few to hundreds of nm have been described in literature in detail. Typically, those consist of a ridge-waveguide with a $p+$ region and an $n+$ region defined in the slab at each side of the ridge [Lipson 05]. Such waveguides are fabricated on silicon on insulator (SoI) wafers. These SoI wafers are made up of three layers: a single-crystal layer of silicon that acts as the device layer, a base silicon substrate, and a thin insulator that acts as a bottom cladding layer and hence prevents leakage of light into the substrate. The silicon layer has an n-type background doping concentration of 10^{15} cm^{-3}, whereas a uniform doping concentration of $\sim 10^{19}$ cm^{-3} for both $p+$ and $n+$ regions are considered. The whole device structure is covered by a cladding layer (SiO$_2$) [Lipson 05]. A high modulation depth with modulation speeds up to 12.5 Gbps was demonstrated by Xu et al. [Xu 07] (Figure 7.6).

The other building blocks for an on-chip optical network are the PSEs. These switching elements are essentially two intersecting waveguides with the position of intersection located between sets of two ring resonators. This is shown in Figure 7.7 [Shacham 07, Shacham 08]. The rings have a specific resonance frequency depending on the material and structural properties based on nanometric features.

In the off-state, when the resonance frequency differs from the wavelength of the data signal being transmitted, the signal travels straight through the intersection without any interruption. But, when the switch is in the *on*-state and the resonance frequency difference matches the wavelength of the data signal, the ring switches and the light travels at right angles (left or right) on

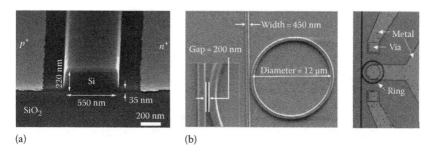

FIGURE 7.6
(a) 10 Gbps Mach–Zehnder-based silicon modulator. (From Green, W.M.J. et al., *Opt. Express*, 15(25), 17106, Dec. 2007. With permission.) (b) Top view SEM image of the micro-ring resonator–based SOM coupled with the waveguide. (Reprinted from Xu, Q. et al., *Nature*, 435, 325, May 19, 2005; Lipson, M., Manipulating light on chip, *The 18th Annual Meeting of the IEEE LEOS 2005*, Piscataway, NJ, Oct. 22–28, 2005, pp. 278–279. With permission.)

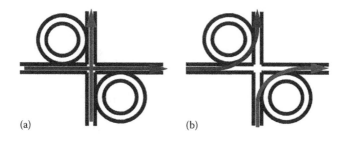

FIGURE 7.7
PSE: (a) *OFF*-state and (b) *ON*-state. (Reprinted from Shacham, A. et al., *IEEE Trans. Comput.*, 57(9), 1246, Sept. 2008. With permission.)

the intersection and gets coupled to the micro-ring. When the switch is in the off-state, it acts as a passive device and consumes negligible power. When it is in the on-state, it only consumes ~0.5 mW power [Shacham 07]. Hence, the main advantages of these PSEs are their small footprint (~12 μm) and their low power consumption. They also have low insertion loss (~1.5 dB) and good crosstalk isolation properties (>20 dB) [Shacham 07].

These basic PSEs are interconnected by using silicon waveguides and are organized in groups of four. Each group is controlled by an electronic circuit also called an "electronic router" and the whole system forms a 4 × 4 switch as shown in Figure 7.8. Control packets are received in the electronic router, processed, and sent to their next destination, while the PSEs are turned *ON* and *OFF*. Once a packet reaches the final destination, a series of such PSEs are now ready to transmit the optical signal. The estimated size of such a 4 × 4 switching fabric is 70 μm [Shacham 07].

Typically, in mesh-networks, 5 × 5 switches are used with one port dedicated to the local injection and ejection of packets. Such switches are easier

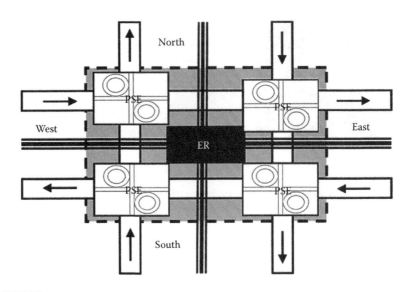

FIGURE 7.8
Four PSEs controlled by an electronic network—a 4 × 4 switch. (Reprinted from Shacham, A. et al., On the design of a photonic network on chip, *Proceedings of the First IEEE International Symposium Networks-on-Chips*, Washington, DC, pp. 53–64, May 2007. With permission.)

to implement in electronic transistor–based circuits. But those are difficult to construct using the 2 × 2 PSEs. Therefore, the injection and ejection of packets is done through one of the four existing ports of the 4 × 4 switching fabric, thus blocking through traffic. Such designs would lead to some constraints in designing network topology. Typically, these switches are narrowband. Research efforts are ongoing to fabricate wideband switching structures. One such structure is shown in Figure 7.9. Researchers at

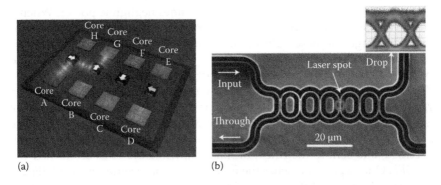

(a)   (b)

FIGURE 7.9
(a) Illustration of traffic routing between cores and (b) silicon-coupled micro-ring resonator-based nanophotonic switches. (From Vlasov, Y.A. et al., *Nat. Photon.*, 2(4), 242, Apr. 2008. With permission.) The inset shows eye diagram of drop port at 40 Gbps.

IBM [Vlasov 08] first demonstrated a broadband optical switch as shown by a black-box in Figure 7.9.

The device illustrated in Figure 7.9 is a fifth order ring resonator optical filter based on submicron silicon photonic wires [Vlasov 08]. It has been fabricated on a SoI wafer. Such filters are characterized by a ~1 dB flat-top passband of 310 GHz with ripples smaller than 0.4 dB [Xia 07]. The insertion loss of the device is only 1.8 dB with an out-of-band rejection ratio better than 40 dB. The major factor giving it a small form factor is the bending radii of just 4 μm. This allows reducing the size of the whole device chip as well as obtaining a large free spectral range of around 18 nm [Vlasov 08, Vlasov2 08]. It has been shown that the switching times of this switch is only limited by the carrier recombination lifetime of <2 ns for up to nine 40 Gbps optical channels [Vlasov 08, Vlasov2 08].

Another fundamental issue facing the researchers and engineers is the interface between the optical fiber and the die. The difficulty in designing this interface is due to the extremely small SoI waveguides, with a core cross-sectional area of 0.1 μm^2. By comparison, an optical fiber core has a cross-sectional area of at least 50 μm^2. Hence, it is very difficult to couple photonic signals from optical fibers into the silicon waveguides. Conventional approaches to solve these problems involve the use of lenses. But accurate alignment using lenses is extremely difficult and also an expensive option. These waveguides are to be routed to the edge of the die that must be diced, polished, and coated with antireflection coating [Gunn 06]. Besides being expensive, this process does not support wafer-scale testing. Luxtera™ Inc. recently tried to address many such problems with the holographic lenses. The holographic lens is designed to accept light at normal incidence. It is etched into silicon on the surface of the die. Thus, when a fiber is brought close enough to the surface, the light gets focused into the lens, turns 90°, and is simultaneously coupled into the waveguide [Gunn 06]. Figure 7.10 shows the image of a holographic lens connected to the waveguides for off-chip light coupling.

7.3.3 The Big Picture

After talking about all the building blocks for an NoC, now seems the right time to present the complete picture. It is very likely that in order to implement a photonic NoC, future CMOS ICs will implement 3D integration (3DI) [Topol 06, Haensch 07]. An artist's rendition of a 3D-integrated chip with an on-chip photonic network is shown in Figure 7.11 [Vlasov3 08]. This futuristic 3D-integrated chip will consist of several layers interconnected with each other. The lower layer is a processor layer that consists of ~200 CMOS microprocessor cores, grouped into ~20 "supercores." On top of it is the memory layer to provide fast access to local caches. On top of the stack is the photonic NoC that handles communication between the supercores. It consists of many individual optical devices including modulators, detectors,

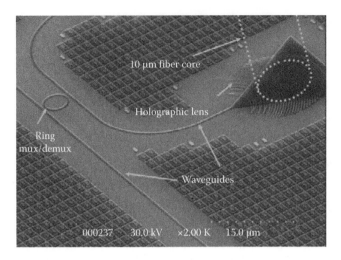

FIGURE 7.10
CMOS compatible silicon waveguides and holographic lens connected to them for off-chip coupling. (Reprinted from Gunn, C., *IEEE Micro.*, 26(2), 58, Mar./Apr. 2006. With permission.)

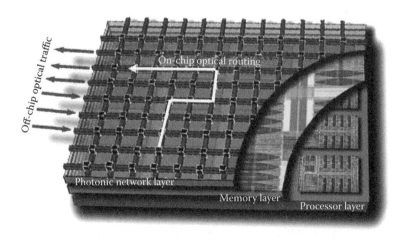

FIGURE 7.11
3D silicon processor chip with optical IO layer featuring on-chip nanophotonic network (an artist's rendition). (From Kash, J.A., Leveraging optical interconnects in future supercomputers and servers, in *16th IEEE Symposium on High Performance Interconnects*, August 26–28, 2008, HOTI. IEEE Computer Society, Washington, DC, 190–194, 2008; doi = http://dx.doi.org/10.1109/HOTI.2008.29; Vlasov, Y. Silicon photonics for next generation computing systems, in *European Conference on Optical Communications* 2008, Brussels, Belgium, Sept. 2008; Courtesy of Dr. Yurii Vlasov, IBM Research, Yorktown Heights, NY.)

waveguides, and photonic switches. The primary purpose of this layer is not only to provide a point-to-point high-speed optical link between different cores, but also to route this traffic on- as well as off-chip with an array of nanophotonic switches. Vertical interconnections between the planes are accomplished with electrical vias [Kash 08].

These NoCs enabled by the dense chip-scale integration of optical components can reduce the power consumption while delivering low latencies and high throughputs. To meet the stringent requirements for these NoCs, a novel engineering approach from devices to system architecture will be necessary. Hence, we can safely say that there is still more room for innovations.

7.4 Conclusion

Currently, nanophotonics is used only to replace copper in order to obtain high-bandwidth high-speed transport. In the next decade or two, optical interconnects will move inwards following the push to integrate photonics with electronic-chip packaging. On-card and on-chip waveguides will replace the fiber-based modules and enable higher densities of multiprocessors. Electrical NoCs will be replaced by photonic NoCs and associated switching fabrics. These NoCs, available commercially in the future, will have the energy efficiency and bandwidth capacity to tackle the zetabyte data transport and exaflop performance of the computers in the future.

References

[Acharya 07] K. Acharya and M. Y. A. Raja, Stimulated Raman scattering impairments in WDM-PON, *Proceedings of FTTH Conference 2007*, Orlando, FL, Oct. 2008.

[Acharya 08] K. Acharya and M. Y. A. Raja, Raman amplified PON (RA-PON): Mitigating SRS using SRS, *HONET 2008 Proc IEEE ComSoc Xplore*, 2009.

[Benini 02] L. Benini and G. D. Micheli, Networks on chip: A new SoC paradigm, *IEEE Comput.*, 49(2/3):70–71, Jan. 2002.

[Brodersen 02] R. W. Brodersen, M. A. Horowitz, D. Marković, B. Nikolić, and V. Stojanović, Methods for true power minimization, *International Conference on Computer Aided Design*, San Jose, CA, pp. 35–42, Nov. 2002.

[Chang 04] G. Chang, D. Guidotti, F. Liu et al, Chip-to-chip optoelectronics SOP on organic boards or packages, *IEEE Trans. Adv. Packag.*, 27(2):386–397, May 2004.

[Chen 00] R. T. Chen, L. Lin, C. Choi et al., Fully embedded board-level guided-wave optoelectronic interconnects, *IEEE Proc.*, 88:780, 2000.

[Cheng 95] J. Cheng, Y.-C. Lu, B. Lu, A. Alduino, and G. Ortiz, High speed optical interconnect technology for a reconfigurable multi processor network, *Proc. SPIE*, 2400:160 (1995); doi:10.1117/12.206327.

[Corzine 89] S. W. Corzine, R. S. Geels, R. H. Yan, J. W. Scott, L. A. Coldren, and P. L. Gourley, Efficient, narrow-linewidth distributed-Bragg-reflector surface-emitting laser with periodic gain, *IEEE Photon. Technol. Lett.*, 1(3):52–54, 1989.

[Dally 01] W. J. Dally and B. Towles, Route packets, not wires: Onchip interconnection networks, *Proceedings of the Design Automation Conference*, Las Vegas, NV, pp. 684–689, June 2001.

[Fang 06] A. W. Fang, H. Park, O. Cohen, R. Jones, M. J. Paniccia, and J. E. Bowers, Electrically pumped hybrid AlGaInAs-silicon evanescent laser, *Opt. Express*, 14(20):9203–9210, Oct. 2006.

[Feldman 94] M. R. Feldman, J. E. Morris, I. Turlik, G. Adema, and M. Y. A. Raja, Holographic optical interconnect for VLSI multichip modules, *IEEE Trans. Compon. Packag. Manuf. Technol. B*, 17:223, 1994.

[Green 07] W. M. J. Green, M. J. Rooks, L. Sekaric, and Y. A. Vlasov, Ultra-compact, low RF power, 10 Gb/s silicon Mach–Zehnder modulator, *Opt. Express*, 15(25):17106, Dec. 2007.

[Gunn 06] C. Gunn, CMOS photonics for high-speed interconnects, *IEEE Micro.*, 26(2):58–66, Mar./Apr. 2006.

[Gupta 04] A. Gupta, S. P. Levitan, L. Selavo, and D. M. Chiarulli, High-speed optoelectronics receivers in SiGe, *17th International Conference on VLSI Design*, Mumbai, India, pp. 957–960, Jan. 2004.

[Haensch 07] W. Haensch, Is 3D the next big thing in microprocessors?, *Proceedings of International Solid State Circuits Conference* (ISSCC), San Francisco, CA, 2007.

[Heckele 04] M. Heckele, Hot embossing: A flexible and successful replication technology for polymer MEMS, *Proc. SPIE*, 5345:108, 2004; doi:10.1117/12.537197.

[Hemani 00] A. Hemani, A. Jantsch, S. Kumar et al., Network on chip: An architecture for billion transistor era, *18th IEEE NorChip Conference*, Turku, Finland, Nov. 2000.

[Horowitz 05] M. A. Horowitz, E. Alon, D. Patil, S. Naffziger, R. Kumar, and K. Bernstein, Scaling, power, and the future of CMOS, *IEEE International Electron Devices Meeting*, Washington, DC, Dec. 2005.

[Jewell 89] J. L. Jewell, A. Scherer, S. L. Mccall et al., Low-threshold electrically pumped vertical-cavity surface-emitting microlasers, *Electron. Lett.*, 25(17):1123–1124, 1989.

[Johansson 05] L. A. Johansson, Z. Hu, D. J. Blumenthal, L. A. Coldren, Y. A. Akulova, and G. A. Fish, 40-GHz dual-mode-locked widely tunable sampled-grating DBR laser, *IEEE Photon. Technol. Lett.*, 17(2):285–287, Feb. 2005.

[Jokerst 04] N. Jokerst, T. Gaylord, E. Glytsis et al., Planar lightwave integrated circuits with embedded actives for board and substrate level optical signal distribution (invited paper), *IEEE Trans. Adv. Packag.*, 27(2):376–385, May 2004.

[Kash 08] J. A. Kash, Leveraging optical interconnects in future supercomputers and servers, *Proceedings of the 2008 16th IEEE Symposium on High Performance Interconnects*, August 26–28, 2008, HOTI. IEEE Computer Society, Washington, DC, pp. 190–194; doi = http://dx.doi.org/10.1109/HOTI.2008.29.

[Keil 01] N. Keil, H. H. Yao, C. Zawadzki et al., Hybrid polymer/silica vertical coupler switch with <-32 dB polarization-independent crosstalk, *Electron. Lett.*, 37:89, 2001.

[Kibar 99] O. Kibar, D. A. Van Blerkom, C. Fan, and S. C. Esener, Power minimization and technology comparisons for digital free-space optoelectronic interconnections, *J. Lightwave Technol.*, 17:546, 1999.

[Krabe 00] D. Krabe, F. Ebling, N. Arndt-Staufenbiel, G. Lang, and W. Scheel, New technology for electrical/optical systems on module and board level: the EOCB approach, in *Proceedings of the 50th Electronic Components and Technology Conference*, Las Vegas, NV, May 21–24, 2000, p. 970.

[Krishnamoorthy 96] A. V. Krishnamoorthy and D. A. B. Miller, Scaling optoelectronic-VLSI circuits into the 21st century: A technology roadmap, *IEEE J. Select. Topics Quantum Electron.*, 2:55, 1996.

[Kyriakis-Bitzaros 01] E. D. Kyriakis-Bitzaros, N. Haralabidis, Y. Moisiadis, M. Lagadas, A. Georgakilas, and G. Halkias, Comparison of the signal latency in optical and electrical interconnections for interchip links, *Optic. Eng.*, 40:144, 2001.

[Lipson 05] M. Lipson, Manipulating light on chip, *The 18th Annual Meeting of the IEEE LEOS 2005*, Piscataway, NJ, Oct. 22–28, 2005, pp. 278–279.

[Mederer 01] F. Mederer, R. Jäger, H. J. Unold et al., 3-Gb/s data transmission with GaAs VCSEL's over PCB integrated polymer waveguides, *IEEE Photon. Technol. Lett.*, 13(9):1032–1034, Sept. 2001.

[Miller 00] D. A. B. Miller, Rationale and challenges for optical interconnects to electronic chips, *Proc. IEEE*, 88:728, 2000.

[Morris 92] J. E. Morris, M. R. Feldman, W. H. Welsh et al., Prototype optically interconnected multichip module based on computer generated hologram technology, *SPIE Proc. OE'LASE*, 1849-11, 1992.

[Morris 94] J. E. Morris, W. Heyward, H. Yang, F. Kiamilev, M. Y. A. Raja, and M. R. Feldman, Experimental demonstration of free-space optical interconnects for multichip modules, *Optical Society of America and ILS-X Annual Meeting*, Dallas, TX, 1994, paper MJJ5.

[Mudge 01] T. Mudge, Power: A first-class architectural design constraint, *IEEE Comput.*, 34(4):52–58, 2001.

[Mukherjee 94] B. Mukherjee, *Optical Communication Networks*, McGraw-Hill, New York, July 1997.

[Raja 88] M. Y. A. Raja, S. R. J. Brueck, M. Osinski et al., Novel wavelength-resonant optoelectronic structure and its application to surface-emitting semiconductor laser, *Electron. Lett.*, 24:1140, 1988.

[Raja 89] M. Y. A. Raja, S. R. J. Brueck, M. Osinski et al., Resonant periodic gain surface-emitting semiconductor laser, *IEEE J. Quantum Electron.*, QE-25:1500, 1989.

[Raja 95] M. Y. A. Raja, S. T. Srinivasan, R. R. Gauthier, S. D. Hersee, and S. Z. Sun, Flip-chip bondable low-threshold edge-emitting laser arrays, *Micro-Phase (μφ) News*, III(1): Jan. 1–4, 1995.

[Rode 97] M. Rode, J. Moisel, O. Krumpholz, and O. Schickl, Novel optical board-to-board interconnection, *Proceedings of ECOC'97*, Edinburgh, U.K., 1997, pp. 2.228–2.231.

[Sakano 95] T. Sakano, T. Matsumoto, and K. Noguchi, Three-dimensional board-to-board free-space optical interconnects and their application to the prototype multiprocessor system: COSINE-III, *Appl. Opt.*, 34:1815–1822, 1995.

[Shacham 06] A. Shacham, K. Bergman, and L. P. Carloni, Maximizing GFLOPS-per-Watt: High-bandwidth, low power photonic on-chip networks, *P = ac2 Conference*, New York, Oct. 2006, pp. 12–21.

[Shacham 07] A. Shacham, K. Bergman, and L. P. Carloni, On the design of a photonic network on chip, *Proceedings of the First IEEE International Symposium on Networks-on-Chips*, Washington, DC, May 2007, pp. 53–64.

[Shacham 08] A. Shacham, K. Bergman, and L. P. Carloni, Photonic networks-on-chip for future generations of chip multiprocessors, *IEEE Trans. Comput.*, 57(9):1246–1260, Sept. 2008.

[Topol 06] A. W. Topol, J. D .C. La Tulipe, L. Shi et al., Three-dimensional integrated circuits, *IBM J. Res. Dev.*, 50:491–506, 2006.

[Vlasov 08] Y. A. Vlasov, W. M. J. Green, and F. Xia, High-throughput silicon nanophotonic wavelength-insensitive switch for on-chip optical networks, *Nat. Photon.*, 2(4):242–246, Apr. 2008.

[Vlasov2 08] Y. Vlasov, W. M. Green, and F. Xia, High-throughput silicon nanophotonic deflection switch for on-chip optical networks, *Optical Fiber Communication Conference and Exposition and The National Fiber Optic Engineers Conference, OSA Technical Digest (CD)*, Optical Society of America, Washington, DC, 2008, paper OTuF5.

[Vlasov3 08] Y. Vlasov, Silicon photonics for next generation computing systems, tutorial in *European Conference on Optical Communications 2008*, Brussels, Belgium, Sept. 2008.

[Wei 99] J. Y. Wei, J. L. Pastor, R. S. Ramamurthy, and Y. Tsai, Just-in-time optical burst switching for multiwavelength networks, *IFIP Broadband Commun.*, Hong Kong, Nov. 1999, pp. 339–352.

[Whitaker 05] J. C. Whitaker, *The Electronics Handbook*, 2nd edn., illustrated, CRC Press, Boca Raton, FL, 2005.

[Xia 07] F. Xia, M. J. Rooks, L. Sekaric, and Y. A. Vlasov, Ultra-compact high order ring resonator filters using submicron silicon photonic wires for on-chip optical interconnects, *Opt. Express*, 15(19):11934–11941, Sept. 2007.

[Xu 05] Q. Xu, B. Schmidt, S. Pradhan, and M. Lipson, Micrometrescale silicon electro-optic modulator, *Nature*, 435:325–327, May 19, 2005.

[Xu 06] Q. Xu, B. Schmidt, J. Shakya, and M. Lipson, Cascaded silicon micro-ring modulators for WDM optical interconnection, *Opt. Express*, 14(20):9430–9435, Oct. 02, 2006.

[Xu 07] Q. Xu, S. Manipatruni, B. Schmidt, J. Shakya, and M. Lipson, 12.5 Gbit/s carrier-injection-based silicon microring silicon modulators, *Opt. Express*, 15(2):430–436, Jan. 2007.

8

Crystalline Colloidal Array Photonic Crystal Optical Switching

Marta K. Maurer

CONTENTS

8.1 Introduction

The field of photonic crystals, also called photonic bandgap materials, is a very exciting research area that incorporates aspects of physics, chemistry, and engineering. Photonic crystals are three-dimensional structures in which the dielectric constant differences between the particles and surrounding medium are great enough to create a bandgap that does not allow the passage of photons of a specific wavelength (Joannopoulos et al. 1995). Photonic crystals play a unique role in controlling the propagation of electromagnetic waves, and the development of novel ways to manipulate such waves is expected to have an important influence on future technology and science.

Photonic crystal materials are finding increasing applications in optics, optical computing, sharp bending light guides, very low threshold lasers, and sensors (Joannopoulos et al. 1995, Rotello 2004, Sakoda 2004). There is an increased understanding of light propagation within materials with periodic optical dielectric constant modulations, and a corresponding growth of fabrication methods for development of these materials (Jiang et al. 2001, Ozin and Yang 2001, Soukoulis 2002, Wong et al. 2003, 2006). Photonic crystal materials with a three-dimensional photonic bandgap in the ultraviolet (UV), visible, and near-infrared (IR) spectral region have been developed; however, some fabrication processes require the use of complicated and expensive nanofabrication techniques (Noda et al. 2000, Norris and Vlasov 2001, Braun et al. 2006, Wong et al. 2008).

In response, a number of groups have recently developed methods to grow highly ordered, close-packed arrays from dispersions of monodisperse particles, relying on the controlled growth of particles in layers on a crystal lattice surface (van Blaaderen et al. 1997, Park and Xia 1998, Wijnhoven and Vos 1998, Gates et al. 1999, Jiang et al. 1999). The growth of colloidal crystals proceeds via heterogeneous nucleation mechanisms on the container wall, and via homogeneous nucleation mechanisms inside the suspension (Okubo 1988). The development of these methods to synthesize sufficiently monodisperse colloids, i.e., those with particles having size distributions with a relative standard deviation of about 8% or less, has led to the intense study of colloidal crystals and their properties, particularly, diffraction (Hachisu and Kobayashi 1974, Pusey van Megen 1986).

Opal is a naturally occurring colloidal crystal, composed of a three-dimensional array of 150–400 nm silica spheres that form and bind together naturally in sediment. Visible light diffracts from the lattice planes, resulting in an iridescence that earns opal a place among the world's most-prized gemstones. Synthetic opals can be made by sedimenting or centrifuging synthesized monodisperse silica spheres in the laboratory for several months to a year or more, after which they are dried in an autoclave and sintered in a furnace at temperatures of up to 1000°C (Filin et al. 2002, van der Beek et al. 2007).

An inverse opal is another type of synthetic photonic crystal in which a high refractive index material, such as titania, is infiltrated into the interstices between close-packed silica spheres. The silica spheres are then removed, leaving the material in the interstices between close-packed spherical voids. These materials can be expected to produce highly efficient filters and low-threshold lasers. However, high-quality crystalline opals are required to make the inverse opals, including the use of silica spheres with a narrow size distribution (which determines the wavelength of the photonic bandgap) and absence of particle aggregation and adhesions during the formation process (Holland et al. 1998, 1999, Schroden et al. 2001, Stein and Schroden 2001).

The fabrication of three-dimensional photonic crystals in a controllable and inexpensive manner is one of the key challenges in this field. Microlithographic techniques have been developed, but are complex and expensive (Joannopoulos et al. 1995, Lin et al. 1998, Fleming and Lin 1999). An alternative approach to fabricating these crystals is based on the self-assembly of colloidal particles by electrostatic repulsive interactions. Self-assembly methods that have been explored include phase separation of block copolymers, template-directed synthesis, and crystallization of monodisperse colloidal spheres (Wijnhoven and Vas 1998, Yoshino et al. 1998, Subramania et al. 1999). This chapter focuses on the third approach.

In the self-assembly of colloidal particles into photonic crystals, highly charged monodisperse spherical colloidal nanoparticles repel each other over macroscopic distances. This system finds a well-defined minimum energy state where the particles self-assemble into ordered lattice (Hiltner et al. 1971, Asher et al. 1986, Vlasov et al. 2001). These materials are interesting for many technological applications since the electromagnetic radiation that enters the three-dimensional bandgap can travel through the material without losing its intensity. Applications for these crystals have been found as narrowband filters that efficiently reject Rayleigh scattered incident light in Raman spectroscopy (Flaugh et al. 1984), in nanosecond optical switches (Pan et al. 1997), and in chemical sensors (Holtz and Asher 1997).

The chemical approach to the development of photonic crystals based on crystalline colloidal self-assembly and their applications is explained in this chapter. The objective of this chapter is to describe the fabrication of a photonic crystal material which changes its volume with temperature and light, and can therefore be used for optical switching materials, tunable filters, sensors, and display devices.

8.2 Background

8.2.1 Crystalline Colloidal Array Photonic Crystal

Monodisperse polystyrene colloidal particles were first synthesized in the 1950s by a group from Dow Chemical (Alfrey et al. 1950). These monodisperse particles form highly ordered, close-packed arrays when their aqueous dispersions are dried on transmission electron microscopy grids. When dispersions of these particles are cleaned from ionic impurities, they diffract visible light according to Bragg's law (Krieger and O'Neill 1968,

Hiltner et al. 1971). In addition, they form a large three-dimensional crystal by self-assembly into a crystalline colloidal array (CCA) (Asher et al. 1986), which efficiently diffracts light by orienting face-centered cubic (111) planes parallel to the nearby surface. These periodic arrays of mesoscopic colloidal particles can be used to fabricate photonic crystals that filter UV, visible, and IR spectral regions. The CCA Bragg diffraction results in unique optical phenomena, which give these materials applications in optics and spectroscopic instrumentation.

Monodisperse colloidal particles can be developed by emulsion polymerization (Reese et al. 2000), seeded emulsion polymerization (Seung-Man et al. 2008), emulsifier-free emulsion polymerization (Reese and Asher 2002), precipitation polymerization (Li and Stöver 2000), and dispersion polymerization (Klein et al. 2003, Schmid et al. 2007). CCAs fabricated by emulsion polymerization of polystyrene is discussed below; however, other polymers such as polymethylmethacrylate (Ye et al. 2002 and Qiang et al. 2002) and heptafluorobutyl methacrylate (Pan et al. 1998a,b) have also been used successfully. In addition, photonic crystals based on inorganic materials such as silica (Esquena et al. 1997), titania (Olson and Liss 1988), and zirconia (Widoniak et al. 2005) have also been developed. The first photonic crystal patents from the University of Pittsburgh (Asher 1986) led to commercialization of CCA-based photonic crystals for laser light rejection filters by EG&G Princeton Applied Research Corp.

8.2.1.1 Synthesis and Characterization

A jacketed cylindrical reaction vessel equipped with reflux condenser, a Teflon stirrer, and a temperature sensor are used for emulsion polymerization of highly charged, monodisperse polystyrene spheres. Nanopure water and sodium bicarbonate (buffer) are added to the vessel, and the mixture is stirred and deoxygenated by bubbling nitrogen for 40 min. Aerosol MA-80-I (surfactant) is added to the vessel and the temperature is increased to 50°C. Finally, styrene (monomer), divinyl benzene (cross-linker), and sodium 1-allyloxy-2-hydroxypropane sulfonate (ionic comonomer) are added and stirred under nitrogen atmosphere. The styrene and divinyl benzene are deinhibited prior to addition by passing though an aluminum oxide column. The temperature is then increased to 70°C, ammonium persulfate (thermal initiator) is added, and the reaction is left to proceed for 3 h (Reese et al. 2000).

The resulting colloidal suspension, which appears milky white, is allowed to cool and then filtered through previously boiled glass wool. This suspension is then dialyzed through a 10,000 molecular weight cut-off dialysis tube against nanopure water. The water is replaced daily until the suspension's conductivity decreases to the conductivity of nanopure water. This step removes nonionic and ionic impurities from the suspension and enables electrostatic repulsion between the highly negative charged monodisperse particles. Additional cleaning steps can include an ion-exchange resin to further decrease the ionic strength of the solution.

The temperature, stirring speed, amounts of monomer and cross-linker, and size of the micelles formed by the addition of surfactant determine the size and monodispersity of the resulting colloidal particles. The addition of the ionic comonomer adds surface sulfonate groups, which ionize in nanopure water. For example, 118 nm spheres contain ~9,000 charges/sphere and 322 nm spheres contain ~95,000 charges/sphere. Particle size can be measured using quasi-elastic light scattering and transmission electron microscopy, and the particle charge densities are measured by conductometric titration (Roberts et al. 1998).

8.2.2 Diffraction Properties of Crystalline Colloidal Array Photonic Crystals

CCAs are mesoscopically periodic fluid materials, which efficiently diffract UV, visible, or near-IR light meeting the Bragg condition (Carlson and Asher 1984, Thirumalai 1989) due to the repulsion of the highly charged spheres over macroscopic distances. The system finds a well-defined minimum energy state where particles self-assemble into a face-centered cubic array (however, a body-centered cubic array is also possible under certain conditions) (Carlson and Asher 1984). The interparticle Coulombic repulsive interaction between spheres is modeled by the Derjaguin, Landau, Verwey, and Overbeek (DLVO) potential. The potential energy of the interactions increases with the square of the colloidal particle charge (z), and inversely with the solution dielectric constant (ε). The potential energy decreases with the Debye layer length (κ^{-1}), which decreases with the solution ionic strength. This repulsion can be significant over macroscopic distances, ~700 nm in water due to H^+ and OH^- concentrations (Rotello 2004).

$$U(r) = \frac{z^2 e^2}{\varepsilon} \left[\frac{e^{\kappa a}}{1 + \kappa a} \right]^2 \frac{e^{-\kappa r}}{r} \qquad \kappa^2 = \frac{4\pi e^2}{\varepsilon k_B T}(n_p z + n_i)$$

where
  $U(r)$ is the interaction potential energy between the spheres
  ε is the medium dielectric constant
  a is the sphere radius
  κ is the Debye layer thickness
  n_p is the number of dissociated ions
  n_i is the number of ionic impurities (Crocker and Grier 1994, Vondermassen et al. 1994, Sader and Chan 1999)

In general, the optical dielectric constant of these particles differs from that of the surrounding medium, which results in a periodic variation in the material's refractive index. The modulation in the refractive index results in Bragg diffraction, similar to the diffraction of x-rays from atomic and molecular crystals. A major difference, however, is that the periodicity in the refractive index for the CCA is much larger (10 nm to 3 μm) than that found in atomic and molecular crystals. Thus, the CCA efficiently diffracts electromagnetic radiation in the UV, visible, and near-IR spectral regions (Flaugh et al. 1984).

CCA photonic crystals show brief instability under mechanical shocks and presence of ionic impurities. They show over 10 years of stability if special care is taken to insure purity of the compounds and sample container, and prevent water evaporation (Rotello 2004). The repulsive interactions between colloidal particles kinetically stabilize the CCA lattice and prevent it from reaching the primary energy minimum at which flocculation would occur. However, the CCA ordering will be disturbed and colloidal particles will aggregate when ionic impurities (>1 μm) are present due to screening of the electrostatic repulsive interaction.

A CCA with ~100 nm particle diameter in water very efficiently diffracts visible light (Flaugh et al. 1984). Diffracted wavelength from the system can be approximated (Rundquist et al. 1989) by Bragg's law:

$$m\lambda_0 \sim 2nd \sin \theta$$

where

  m is the order of diffraction
  λ_0 is the wavelength of light in vacuum
  n is the average refractive index of the system (solvent and colloidal particles)
  d is the spacing between diffracting lattice planes
  θ is the Bragg glancing angle (the glancing angle between the incident light propagation direction and the diffracting planes)

The interparticle spacing can vary from a close-packed system to a system where the interparticle spacings are many times the particle diameter (Rundquist et al. 1991a,b, Chang et al. 1994, Tse et al. 1995).

This model assumes a weak diffraction kinematic limit. Essentially all light that meets the diffraction condition is diffracted from a 40 μm thin CCA of polystyrene spheres in water. The diffraction bands are narrow (<10 nm), and light that is not diffracted transmits through the crystal. Optical rejection filters were developed based on these colloidal crystals (Kesavamorthy et al. 1991, 1992a,b, Kamenetzky et al. 1994, Asher and Pan 1996), and earlier work established that the diffraction of light from these colloidal crystals falls within the so-called dynamic diffraction limit (Rundquist et al. 1991a,b).

The finite bandwidth (5–10 nm) of the extinction peak from a CCA occurs because the polymer spheres in water show a large scattering cross section from visible light resulting in an efficient diffraction. Attenuation of the incident beam limits the number of layers involved in diffraction and determines the bandwidth of diffraction. The stronger the diffraction, the fewer

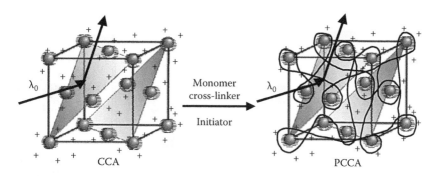

FIGURE 8.1
Synthesis of PCCA. (From Maurer, M.K. et al., *Adv. Funct. Mater.*, 15, 1401, 2005. With permission.)

layers involved in the scattering, and therefore, the broader the diffraction (Rundquist et al. 1989).

Dynamic diffraction theory predicts highly efficient diffraction and explains diffraction dependence on sphere diameter and angle of incident light. However, more sophisticated theoretical approaches are needed to model the photonic crystal diffraction of particles with higher refractive indices and for polystyrene spheres sizes above ~150 nm (Mittelman et al. 1999).

8.2.3 Polymerized Crystalline Colloidal Array Synthesis

A rugged photonic crystal can be developed by embedding the CCA within a hydrogel matrix (see Figure 8.1) (Haacke et al. 1993, Asher and Jagannathan 1994). A nonionic monomer (such as acrylamide) and cross-linker (*N,N'*-methylenebisacrylamide) is dissolved in aqueous solution of CCA (~10%). A photoinitiator or thermal initiator is added, and the mixture is injected into the container, which is then either sealed and heated, or exposed to UV light to initiate free radical polymerization. This procedure results in a polymerized crystalline colloidal array (PCCA) consisting of ~90% water. The photonic crystal in PCCA form can be further used to develop more sophisticated materials such as sensors and memory storage devices.

8.3 Polymerized Crystalline Colloidal Array Optical Switching

8.3.1 Thermally Switchable Photonic Crystals

8.3.1.1 *Poly(N-Isopropylacrylamide)-Based Photonic Crystal*

The well-known temperature-induced volume phase transition of poly(*N*-isopropylacrylamide) (PNIPAM) was utilized to create a CCA photonic crystal with variable sphere size and array periodicity. Below 30°C, PNIPAM

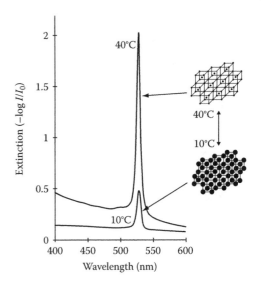

FIGURE 8.2
Diffraction from a CCA of PNIPAM spheres at 10°C, where PNIPAM spheres are swollen, and at 40°C, where the spheres have collapsed. (From Weissman, J.M. et al., *Science*, 274, 958, 1996. With permission.)

colloidal particles are hydrated and swollen; however, when heated above their lower critical solution temperature (~32°C), they undergo a reversible volume phase transition to a collapsed, dehydrated state (Figure 8.2). The temperature increase causes the PNIPAM polymer to expel water and contract into a more hydrophobic state (Weissman et al. 1996).

PNIPAM monodisperse, highly charged colloidal particles are synthesized by copolymerizing PNIPAM with an ionic comonomer, 2-acrylamido-2-methyl-1-propane sulfonic acid, to increase the colloid surface charge which facilitates CCA self-assembly (Weissman et al. 1996). The sphere diameter of the PNIPAM crystalline colloidal particles increases from ~100 nm at 40°C to ~300 nm at 10°C (Pelton and Chibante 1986, McPhee et al. 1993), which corresponds to a 27-fold increase in volume. At low temperatures, the CCA particles are highly swollen, almost touching, and diffract weakly; however, above the phase-transition temperature, the particles are compact and diffract nearly all incident light at the Bragg wavelength (Figure 8.2). The diffraction peak intensity increases as temperature increases from 10°C to 40°C, enabling this material to act as a thermally controllable optical switch and optical limiter.

A wavelength-tunable diffraction material can be developed by embedding a polystyrene CCA into the PNIPAM hydrogel to control the periodicity of the polystyrene particles. As the hydrogel shrinks and swells continuously and reversibly between 10°C and 35°C, the CCA follows, changing the lattice spacing and the diffracted wavelength from ~700 to ~460 nm (Figure 8.3). This material functions as an easily controlled tunable optical filter since the diffracted wavelength varies with temperature.

The above-mentioned materials have dimensions that are controllable by temperature and can be used for light modulation in tunable diffracting and

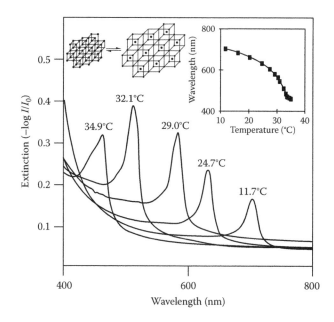

FIGURE 8.3
Shift in diffraction wavelength of polystyrene spheres embedded in a PNIPAM gel, as a result of a temperature-induced volume change. (From Weissman, J.M. et al., *Science*, 274, 958, 1996. With permission.)

transmitting optical devices. They have potential technological applications for display devices and for image processing. Diffraction from the materials can also be used to monitor the swelling properties of hydrogels. The diffracted wavelength gives information on the hydrogel volume, and an optical microscope can be used to examine the Bragg diffraction from small areas within the hydrogel to monitor the phase transition homogeneity of the hydrogel.

8.3.1.2 Oil Blue N-Based Photonic Crystal

Another way to control diffraction of the CCA is to prepare highly charged colloidal particles containing a low refractive index absorbing dye and embedding them in a polyacrylamide hydrogel. These colloidal particles can be matched to an aqueous medium of the same refractive index. Under low light intensities, because the refractive indices of the CCA and medium match, no optical dielectric constant modulation occurs and light is freely transmitted. However, high intensity illumination within the dye absorption band heats the particles within nanoseconds, decreasing their refractive index and resulting in a mesoscopically periodic refractive index modulation with the periodicity of the CCA lattice. This refractive index difference enables the array to diffract incident light meeting the Bragg condition.

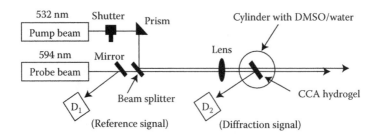

FIGURE 8.4

Experimental setup used to monitor diffraction switching. Probe beam diffraction intensities are normalized to the reference signal, and are measured in the presence and absence of the pump beam. (From Pan, G.S. et al., *J. Am. Chem. Soc.*, 120, 6525, 1998. With permission.)

These materials have applications in optical limiting, optical computing, and nanosecond fast optical switching devices (Kesavamoorthy et al. 1992a,b, Pan et al. 1997, 1998a,b).

Figure 8.4 depicts the experimental setup used to monitor the diffraction switching. The performance of the dyed and undyed PCCA is illustrated in Figure 8.5, comparing the ratios of the normalized diffracted probe beam intensities in the presence and absence of the pump beam as a function of the pump beam energy. When the colloidal particles have a refractive index less than that of the medium, the pump beam pulse energy decreases the particle refractive index, resulting in a greater refractive index mismatch, increasing the diffraction. Alternatively, if the colloidal particles have a refractive index greater than that of the medium, the pump pulse energy will decrease the refractive index mismatch, decreasing the diffraction efficiency. For undyed PCCA spheres, the pump pulse energy does not change the refractive index of the particles.

8.3.2 Photoswitchable Photonic Crystals

Photoresponsive materials can be manufactured by adding a photoresponsive monomer or cross-linker within the hydrogel matrix around the photonic crystal. This section describes the development of azobenzene and spiro-pyran-functionalized PCCAs as well as the development of photoresponsive spiropyran-based photonic crystal polystyrene particles. In addition, a photo-responsive interpenetrating network can be incorporated around the photonic crystal materials.

8.3.2.1 Photoresponsive Photonic Crystals Based on Pendant Azobenzene

A PCCA can be functionalized with epoxide groups by copolymerizing glycidyl methacrylate with acrylamide and *N,N'*-methylenebisacrylamide

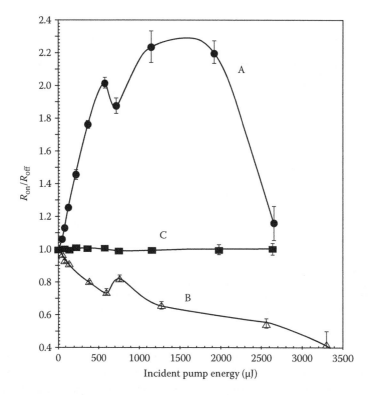

FIGURE 8.5
Diffraction ratio of the probe beam in the presence and absence of the pump beam. Curves A and B were measured for a dyed PCCA with the medium refractive index greater than and less than the particle refractive index, respectively. Curve C is for an undyed PCCA. (From Pan, G.S. et al., *J. Am. Chem. Soc.*, 120, 6525, 1998. With permission.)

by free radical polymerization. Glycidyl methacrylate has a good balance of hydrolytic stability and reactivity with amine groups, and does not destroy the CCA order. An epoxide-functionalized PCCA enables incorporation of photoresponsive units into the PCCA to produce light-sensitive photonic crystal materials without hydrolyzing the hydrogel network. Epoxide groups are available for further reaction, such as with amine groups, immediately after the PCCA is formed. Using this procedure, a photoresponsive PCCA was developed by covalently attaching a water-soluble azobenzene derivative containing primary amine (Kamenjicki et al. 2004).

Figure 8.6 shows the progress of the photoisomerization of the azobenzene derivative over time. In the dark equilibrium state, the azobenzene molecules are in their *trans*-form and absorb light strongly at 350 nm. The 365 nm excitation occurs within this strong 350 nm $\pi \rightarrow \pi^*$ transition, which photoisomerizes the *trans*-azobenzene to the *cis*-form. This absorption band decreases and the 440 nm *cis*-azobenzene $n \rightarrow \pi^*$ transition band

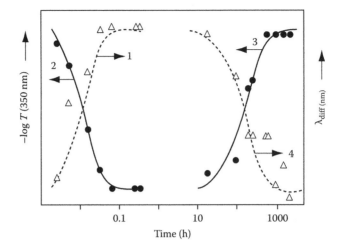

FIGURE 8.6
Time dependence of the absorbance (lines 2 and 3) and diffraction wavelength (lines 1 and 4) with moderate-intensity illumination (lines 1 and 2) and dark relaxation (lines 3 and 4). (From Kamenjicki, M. et al., *J. Phys. Chem. B*, 108, 12637, 2004. With permission.)

appears. This process can be reversed either by excitation within the $\sim$440 nm absorption band or by thermal relaxation in dark, which promotes $cis \rightarrow trans$ conversion, restoring the 350 nm *trans* absorption band and the original diffraction. The response time of the diffraction shift is controlled by actinic light.

The photochemical conversion from *trans* to *cis* is very efficient (quantum yield of $\sim$0.2) and the photoswitching effect can be actuated by a single nanosecond laser pulse (1.2 mJ/cm^2). The diffraction shifts are limited by the polymer and water time response, and occur $\sim$10 s after the laser pulse. The PCCA absorption and diffraction can be toggled back and forth indefinitely by illuminating alternatively with UV light followed by visible light. The 670 nm diffraction peak redshifts by 15 nm due to the increased dipole moment (Maack et al. 1995) of the *cis*-azobenzene derivative, resulting in a more favorable mixing with the water medium.

An alternate approach to functionalizing a photonic crystal hydrogel with azobenzene is by copolymerizing N,N'-cysteinebisacrylamide (additional cross-linker) within acrylamide and N,N'-methylenebisacrylamide. The disulfide bonds formed this way within the hydrogel are cleaved by exposing the PCCA to dithiolthreitol. The resulting sulfhydryl groups are reacted with a maleimide-functionalized azobenzene derivative.

Excitation with 365 nm UV light ($\sim$13 mW/cm^2) results in a decrease in the 322 nm *trans* absorption and an increase in the 430 nm *cis* absorption due to the photoconversion of the *trans*-form to the *cis*-form (Figure 8.7). The conversion of *trans*-azobenzene to the *cis*-form causes the (111) plane

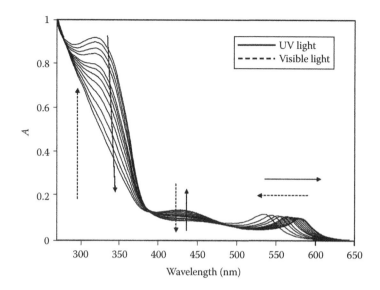

FIGURE 8.7
Absorption spectrum of an azobenzene-functionalized PCCA showing a 50 nm diffraction red-shift upon UV irradiation. (From Kamenjicki, M. et al., *Adv. Funct. Mater*, 13, 774, 2003. With permission.)

diffraction to shift from ~530 to ~580 nm. Excitation with visible light shifts the diffraction back. The reading of this device is limited only by the speed with which the material can be scanned by a laser beam.

The volume of the PCCA hydrogel is controlled by the balance between the free energy of mixing of the polymer hydrogel with water and the elastic restoring force of the hydrogel cross-links (Holtz and Asher 1997, Kamenjicki et al. 2003). The photoresponsive PCCA materials described above function as novel recordable and erasable memory devices. Light absorption actuates a redshift in diffraction, which is read out at wavelengths that are not absorbed and cannot induce photochemistry. This information is erased by exciting the photonic crystal with visible light. The pixel area can be smaller than 10 μm^2, an area which can easily result in a narrow diffraction band. The response of the material is limited by the collective diffusion time of the hydrogel polymer and the required flow of water into and out of the hydrogel.

Both of the systems explained above result in photoresponsive pendant groups freely exposed in the space between the CCA. The diffraction redshift is due to an increase in the free energy of mixing of the hydrogel polymer network with the medium upon UV light illumination, while the blueshift results from a decrease in the free energy of mixing upon return of the azobenzene to its *trans*-form. This photochemically driven diffraction change is reversible and can be cycled many times. Another way to introduce a photoresponsive molecule into the photonic crystal is to cross-link the hydrogel network, embedding the CCA with azobenzene molecules.

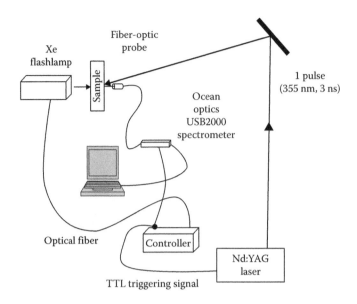

FIGURE 8.8
Experimental setup used to monitor the diffraction kinetics of the azobenzene-functionalized PCCA upon exposure to a single 355 nm pulse. (From Kamenjicki, M. et al., *J. Phys. Chem. B*, 108, 12637, 2004.)

Diffraction switching in a photoresponsive PCCA results from both azobenzene photoisomerization and laser heating. In the fast nanosecond time regime, heating increases the materials volume and redshifts the diffraction. This thermally induced volume change relaxes in the microsecond time domain, as determined using the experimental setup shown in Figure 8.8. At longer times, the volume of the PCCA is controlled by the balance between the free energy of mixing of the polymer with solvent and the elastic restoring force of the hydrogel cross-links.

8.3.2.2 Photochemically Controlled Cross-Linking in PCCA Photonic Crystals

Instead of being able to freely move inside the hydrogel matrix, azobenzene molecules can be tethered to a PCCA hydrogel matrix. This is accomplished by copolymerizing N,N'-cysteinebisacrylamide within an acrylamide and N, N'-methylenebisacrylamide network, breaking the disulfide bond with dithiolthreitol, and reacting the resulting sulfhydryl groups with azophenyl-p-N,N'-dimaleimide, an azobenzene derivative (Figure 8.9). UV irradiation by a 3 ns pulse of 355 nm light (1.2 mJ/cm^2) converts most of the *trans*-azobenzene to the *cis*-form (Kamenjicki and Asher 2004). This results in a 10 nm blueshift of the 475 nm diffraction peak. The blueshift observed in this system upon UV irradiation is exactly opposite to the redshift observed in the pendant azobenzene PCCA described above. The observed blueshift is

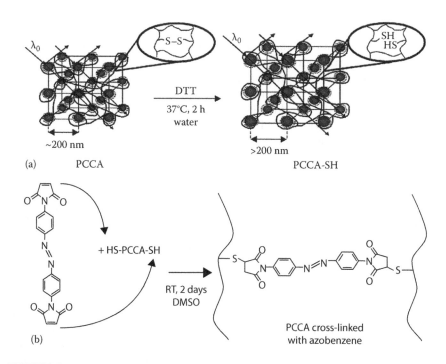

FIGURE 8.9
(a) Synthesis of a thiol-functionalized PCCA and (b) thiol-functionalized PCCA is cross-linked with an azobenzene derivative to create a photoresponsive PCCA. (From Kamenjicki, M. et al., *J. Phys. Chem. B*, 108, 12637, 2004.)

completely reversible; visible excitation converts the azobenzene back to the *trans*-form, and the diffraction redshifts back to the original diffraction wavelength.

A ~15 nm initial redshift at 200 ns was observed due to heating of the sample by the UV laser pulse (Kamenjicki et al. 2003). This thermally induced shift is threefold greater than that observed for the pendant azobenzene PCCA system. This larger redshift results from the larger temperature jump induced by the twofold increase in azobenzene concentration present in the azobenzene cross-linked PCCA. One UV pulse beam induces a short-lived ~40°C temperature jump in the sample which expands the PCCA volume, and increases the (111) spacing to redshift the diffraction.

The photochemical diffraction blueshift of the azobenzene cross-linked PCCA is ~11 nm with a characteristic time of 12 s, showing the opposite behavior to the pendant azobenzene PCCA derivatives. The origin of the blueshift is most likely the formation of less soluble hydrogel aggregates by the *cis*-azobenzene cross-linked species, as indicated by a study of the temperature dependence of the volume phase transition of PNIPAM hydrogels with azobenzene cross-links (Kang and Gupta 2002). The *cis*-azobenzene has a lower transition temperature compared to the *trans*-azobenzene derivative. The decreased transition temperature requires a more hydrophobic and less soluble hydrogel in the presence of the cross-linked *cis*-azobenzene, suggesting that the blueshift results from a less favorable free energy of mixing of the *cis*-derivative with the medium. This occurs despite an increased dipole moment, which in the pendant PCCA results in a redshift. The change in structure of the *cis*- cross-linked hydrogel species causes a more than compensating solubility decrease.

The azobenzene cross-link tethers two segments of the hydrogel chains at the cross-link sites, which possess separate sheaths of solvent at the length of the *trans*-derivative. In the *cis*-derivative, on the other hand, the distance between the hydrogel chains is too small for separate sheaths of solvent. As a result, the hydrogel locally collapses and the volume contracts. The smaller *cis*-azobenzene cross-link creates an excluded volume for solvent molecules in which the hydrogel chains form less soluble segments, which causes the hydrogel to shrink and blueshift (Kamenjicki et al. 2004).

UV excitation results in short- and long-time diffraction changes. In the fast nanosecond time regime, heating increases the PCCA volume and redshifts the diffraction, a change which relaxes in the microsecond time domain. At longer times, the volume of this PCCA hydrogel is controlled by the balance between the free energy of mixing of the polymer hydrogel with the solvent (dimethylsulfoxide) and the elastic restoring force of the hydrogel cross-links. The *cis*-azobenzene derivative exhibits a decreased free energy of mixing of the PCCA with the solvent. The diffraction of this photonic crystal can be photochemically controlled; a UV light pulse can write information and a visible blue pulse can erase it. The information is read out from the diffraction, which can occur at any desired wavelength.

Potential applications for this material include novel recordable and erasable memory and/or display devices. The active element of such a device is an azobenzene cross-linked hydrogel which contains an embedded CCA. UV excitation forms *cis*-azobenzene cross-links while visible excitation forms *trans*-azobenzene cross-links. The less favorable free energy of mixing of the *cis*-azobenzene cross-linked species causes the hydrogel to shrink and blueshift the photonic crystal diffraction.

8.3.2.3 Photoswitchable Spirobenzopyran-Based Photochemically Controlled Photonic Crystals

Another example of a photochemically actuated PCCA is one in which photoisomerization of a covalently attached spirobenzopyran modulates the PCCA's bandgap. Spiropobenzopyran can be photochemically cycled from its spiropyran (closed) form to its merocyanine (open, zwitterion) form. UV illumination increases the intensity of the 350 and 560 nm absorption bands, stabilizes the merocyanine form, swells the hydrogel by ~7%, and redshifts the diffraction peak by 13 nm (Figure 8.10a) (Maurer et al. 2005).

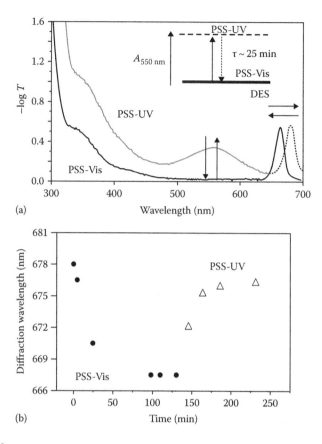

(a)

(b)

FIGURE 8.10
(a) Extinction spectra of a PCCA with spirobenzopyran linked to the hydrogel under UV and visible irradiation. Inset shows the relative absorptions of the UV photostationary state, the visible photostationary state, and the dark equilibrium state. (b) Changes in the diffraction peak of the PCCA over time after exposure to visible and then UV light. (From Maurer, M.K. et al., *Adv. Funct. Mater.*, 15, 1401, 2005. With permission.)

Visible irradiation results in the formation of the closed-form spiropyran, which shrinks the PCCA and blueshifts the diffraction. The characteristic time for both processes is 25 min at 20 mW/cm^2 UV and visible irradiation (Figure 8.10b). The characteristic thermal recovery time of the spiropyran-functionalized PCCA in dark is 35 h at 25°C.

Photoisomerization of covalently attached spirobenzopyran chromophores changes the hydrogel's free energy of mixing due to the large change in the charge distribution within the spirobenzopyran molecules. The resulting PCCA volume change alters the CCA lattice constant and shifts the bandgap and the diffracted wavelength. The diffraction shift results from changes in the free energy of mixing of the PCCA with dimethylsulfoxide solvent as the spiropyran is photoexcited to its different stable forms.

Spirobenzopyran-containing PCCAs are fabricated by covalently attaching a spirobenzopyran derivative to the hydrogel network. This is accomplished by reacting N-[p-maleimidophenyl]isocyanate (PMPI, a sulfohydryl-reactive and hydroxyl-reactive hetero-bifunctional cross-linker) with a spirobenzopyran derivative containing a hydroxyl group (Annunziato et al. 1993). The isocyanate end of PMPI reacts with the hydroxyl groups on the spirobenzopyran, forming urethane linkages. The resulting product contains maleimide ends which, when coupled to a thiol-containing PCCA, form stable thioether linkages (Maurer et al. 2005). The thiol-containing PCCA is made by dissolving N,N'-cystaminebisacrylamide, acrylamide, N,N'-methylenebisacrylamide and 2,2-diethoxyacetophenone (DEAP) in a 10 wt% polystyrene CCA dispersion. This dispersion was injected into a cell made of two quartz plates separated by a spacer and exposed to UV light. After 30 min irradiation, the cell was opened and the gel was washed with nanopure water. This PCCA was exposed to an aqueous solution of dithiolthreitol, which cleaved the disulfide bonds to leave reactive thiol groups within the PCCA. During this process, the hydrogel swelled due to the breaking of cross-links, and the diffraction wavelength redshifted 40 nm. This PCCA was then incubated with a 6 mM solution of maleimide-functionalized spirobenzopyran in dimethylsulfoxide for 2 h at room temperature, leading to a 5 mM concentration of spirobenzopyran within the PCCA.

Another approach to fabrication of spirobenzopyran-based photonic crystal hydrogels is through functionalization of the photonic crystal colloidal particles with a spiropyran derivative. This PCCA consists of spirobenzopyran-linked poly(styrene chloromethylstyrene) particles, which were synthesized as core-shell colloidal particles by emulsion polymerization. Styrene and chloromethylstyrene were added to a 10% weight suspension of previously synthesized 120 nm polystyrene colloidal particles used as seeds (Meyer et al. 2000, Tsoukatos et al. 2000, Park et al. 2002). The seed particles were allowed to swell for 1 h. The reaction temperature was then raised to 60°C, and potassium persulfate (initiator) and sodium bisulfite were added. After 4 h, the resulting crude dispersion was removed from the reaction vessel and dialyzed against nanopure water. The colloidal suspension was further cleaned by shaking the suspension with a mixed bed ion-exchange resin after which it became iridescent due to Bragg diffraction from the 135 nm diameter poly(styrene chloromethylstyrene) CCA. Elemental microanalysis of these particles showed 3% chlorine by weight. These particles were embedded into a polyacrylamide hydrogel network, gradually transferred to a pure dimethylsulfoxide solvent, and then exposed to a solution of 1-(2-hydroxyethyl)-3,3-dimethyl-6'-nitrospiro (indoline-2,2'-[2H-1]benzopyran) in dimethylsulfoxide. This spirobenzopyran derivative contains a hydroxyl group, which undergoes a nucleophilic substitution reaction by the chlorines on the polystyrene colloidal particle surfaces. After incubation, the concentration of spirobenzopyran in the PCCA was 2.5 mM.

Upon UV irradiation, the diffraction blueshifts 5 nm, and additional irradiation with visible light causes an additional 9 nm blueshift (Maurer et al. 2005). Starting at the dark equilibrium state and applying visible light results in the same photostationary state with a 14 nm diffraction blueshift. Irradiating this system with UV light redshifts the diffraction 9 nm (Figure 8.11). This suggests that the concentration of the closed form in the UV photostationary state is between the closed spiropyran-dominated visible photostationary state and the open merocyanine-dominated dark equilibrium state.

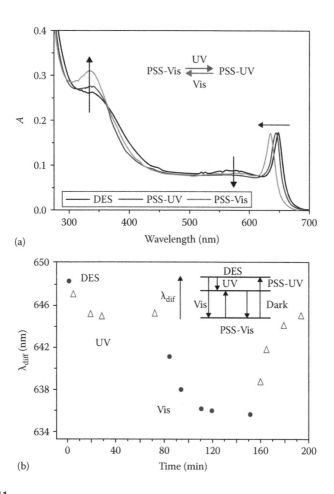

(a)

(b)

FIGURE 8.11
(a) Extinction spectra of a PCCA with spirobenzopyran linked to the colloidal particles. (b) Changes in the diffraction peak of the PCCA over time after exposure to visible and then UV light. (From Maurer, M.K. et al., *Adv. Funct. Mater.*, 15, 1401, 2005. With permission.)

This behavior, where UV irradiation can result in either a redshift or a blueshift, is very different than the one explained for the PCCA with the spirobenzopyran linked to the hydrogel, in which UV irradiation always results in a redshift. The blueshift can be explained by the "negative" photochromic behavior common to some spiropyran derivatives (Shimizu et al. 1969). It has been shown that in an acidic environment, the merocyanine form may steadily transform into the protonated form ($pK_a = 3.93$) (Zhou et al. 1995) which is negatively photochromic. The pH around the sulfonated colloidal particles is ~ 3, and therefore it is possible that the merocyanine form is completely converted to (or coexists with) the protonated form in this environment. UV or visible irradiation causes formation of the closed form, while dark relaxation has the opposite effect. The thermal dark relaxation characteristic time and the characteristic photoswitching times (on the order of seconds to minutes for macroscopic gels) are similar in the two systems explained here, and depend on the temperature and actinic power of the light. The kinetics of gel swelling and shrinking depend on the balance of the driving osmotic pressure associated with changes in the free energy of mixing and the collective diffusion constant of the polymer network relative to the solvent (Peters and Candau 1986, Matsuo and Tanaka 1988).

8.3.3 Nanogel Nanosecond Photonic Crystal Optical Switching

The timescale for the volume phase transition in all the above-mentioned macroscopic hydrogels is minutes and seconds (Peters and Candau 1988, Kawasaki et al. 1998, Shibayama and Nagai 1999, Tokita et al. 2000) due to the slow rate that water enters or leaves the hydrogel in response to osmotic pressure changes in the hydrogel. The kinetic response is diffusional, and the rate of volume change scales as l^{-2}, where l is the relevant length scale (Tanaka and Fillmore 1979, Peters and Candau 1986, Annaka and Tanaka 1992). Thus, small hydrogel particles should be able to undergo fast, microsecond and nanosecond phase transitions. As an example, individual ~ 200 nm PNIPAM colloids have shown small nanosecond to microsecond volume phase transitions upon laser-induced temperature increases (Wang et al. 2001). A robust submicrosecond photonic crystal switching material with nanogel PNIPAM colloidal particles that self-assemble into CCAs can be polymerized into a hydrogel, enabling the individual hydrogel-embedded PNIPAM particles to undergo thermally induced volume phase transitions (Reese et al. 2004).

PNIPAM particles are highly swollen (diameter is 350 nm) at 10°C; however, a laser temperature jump from 30°C to 35°C actuates shrinkage of the nanogel particle. As the temperature increases, the particles expel water and shrink (diameter is 125 nm at 40°C). The swollen particles scatter little light because their dielectric constant differs little from that of water. However, as the particles shrink, their scattering cross section, Q, increases with $\sim KD^{-2}$, where D is the particle diameter, resulting in an increased dielectric constant and efficient diffraction (Weissman et al. 1996). The resulting increased diffraction decreases the light transmission within 900 ns. Individual NIPAM sphere switching occurs in the ~100 ns time regime.

The CCA of PNIPAM nanogel particles are manufactured and embedded into an acrylamide/bisacrylamide hydrogel. The polymerization occurs under conditions at which the individual PNIPAM nanogel particles maintain their temperature-induced individual volume phase transitions. The hydrogel constrains the center of mass of the nanogel particles within the fcc lattice and prevents disorder upon the temperature jump.

At 10°C, the PNIPAM PCCA diffraction is weak, since the particles are mostly water, resulting in only a small dielectric constant modulation (Wang et al. 2001). As the temperature increases above the lower critical solution temperature (~32°C for PNIPAM), the particles collapse and expel water, increasing the dielectric constant modulation, which increases the diffraction efficiency. For example, at 45°C less than 10% of the light at 720 nm and less than 0.1% of the light at 400 nm is transmitted (Reese et al. 2004).

The rate of the optical switching was measured using a transient temperature-jump absorption spectrometer, which excites the sample with a 1.9 μm 3 ns pump pulse. This light was absorbed by residual water, causing the sample to heat. A delayed Xe flashlamp white light ~120 ns pulse probed the transmission, which was monitored with a fiber-optic diode spectrometer.

Figure 8.12 displays the extinction increase over time of a sample after a 5°C temperature jump pulse, with the largest extinction increase at

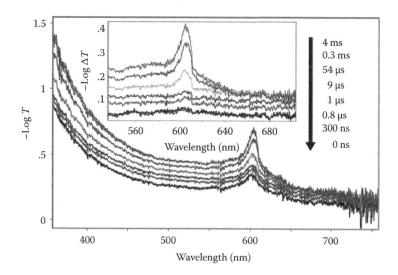

FIGURE 8.12
Temporal evolution of the extinction of a PNIPAM PCCA after a temperature jump from 30°C to 35°C. Inset shows difference spectra to highlight the spectral changes. (From Reese, C. et al., *J. Am. Chem. Soc.*, 126, 1493, 2004. With permission.)

the 111 plane diffraction peak (605 nm). Extinction changes show at least three different kinetic components. The shortest time kinetics (~900 ns) account for ~25% of the transmission change, while the longer time kinetics (~20 μs) account for another ~25%. The longest kinetics (~140 μs) account for the remaining 50% extinction change. The temperature-jump-induced particle size decrease results in an increased sphere scattering efficiency and a corresponding transmission decrease with a single-exponential time constant of ~100 ns. The scattering efficiency increase is readily observed at 600 nm, and occurs much faster than Wang et al. (Wang et al. 2001) observed for their PNIPAM particle shrinkage. The rate of the volume phase transition depends strongly on the osmotic pressure induced by the temperature jump and the cross-link density, which in turn depends on the polymer composition and architecture (Shibayama and Nagai 1999).

These nanogel nanosecond phenomena may be useful in the design of fast photonic crystal switches and optical limiting materials. These nanogels exhibit a coherent, synchronous optical response induced by a temperature jump. Smaller nanogels are expected to show even faster volume phase transitions.

8.4 Conclusion

CCA photonic crystal materials are promising systems for optical switching and optical limiting. They efficiently Bragg diffract UV, visible, or near-IR light depending on the colloidal particle array spacings. Robust semisolid polymerized materials are fabricated by embedding the CCA into a hydrogel network. They can be made from a stimuli-responsive polymer network, where appropriate stimuli alter the PCCA volume and the resulting CCA plane spacing and diffraction wavelength.

Temperature-actuated photonic crystals are fabricated by manufacturing temperature-sensitive, highly charged, monodisperse PNIPAM colloidal particles or by embedding self-assembled polystyrene colloidal particles into a PNIPAM hydrogel matrix. Alternatively, colloidal particles can be functionalized with Oil Blue N dye, which selectively absorbs light, heats the particles, and causes a change in the refractive index.

Photochemically actuated photonic crystals are fabricated by adding a photoresponsive monomer or cross-linker within the hydrogel matrix around the photonic crystal. Examples of these materials include an azobenzene-functionalized PCCA, where photoisomerization of covalently attached azobenzene derivative changes the hydrogel free energy of mixing. The resulting PCCA volume change alters the lattice constant and shifts the diffracted wavelength. Photoresponsive materials can also be developed using spiropyran-functionalized PCCAs, as well as spiropyran-based photonic crystal polystyrene particles. In addition, a photoresponsive interpenetrating network can be incorporated around the photonic crystal materials.

These processes create a diffracting structure that prevents transmission of light meeting the Bragg condition. Alternatively, polymer microphase separation self-assembly can be used to create layered periodic structures (Edrington et al. 2001). One-dimensional photonic crystals can also be created layer by layer through chemical vapor deposition techniques used to form dielectric mirrors and filters (Yariv and Pochi 1984). If the modulation amplitude of the dielectric constant could be altered quickly, these materials could be utilized for optical switching. The challenge is to develop optically nonlinear photonic crystal materials, which can efficiently control diffraction on fast timescales. This is a severe challenge since most materials have small optical nonlinearities.

References

Alfrey, T., Goldberg, A. I., and Price, J. A. 1950. Dilute-solution viscosity of polymethyl methacrylate and a methyl methacrylate-styrene copolymer. *J. Colloid Sci.* 5: 251.

Annaka, M. and Tanaka, T. 1992. Multiple phases of polymer gels. *Nature* 355: 430–432.

Annunziato, M. E., Patel, U.S., Ranade, M., Palumbo, P. S. 1993. p-Maleimidophenyl isocyanate: A novel heterobifunctional linker for hydroxyl to thiol coupling. *Bioconjugate. Chem.* 4: 212–218.

Asher, S. A. 1986. Crystalline narrow band radiation filter. U.S. Patent Nos. 4,627,689 and 4,632,517.

Asher, S. A. and Jagannathan, S. 1994. Method of making solid crystalline narrow band radiation filter. U.S. Patent No. 5,281,370.

Asher. S. A. and Pan, G. 1996. Crystalline colloidal array optical switching devices. *Nanoparticles in Solids and Solutions, NATO ASI Series* 18: 65–69.

Asher, S. A., Flaugh, P. L., and Washinger, G. 1986. Crystalline colloidal Bragg diffraction devices: The basis for a new generation of Raman instrumentation. *Spectroscopy* 1: 26–31.

Braun, P. V., Rinne, S. A., and Garcia-Santamaria, F. 2006. Introducing defects in 3D photonic crystals: State of the art. *Adv. Mater.* 18: 2665–2678.

Carlson, R. J. and Asher, S. A. 1984. Characterization of optical diffraction and crystal structure in monodisperse polystyrene colloids. *Appl. Spectrosc.* 38: 297–304.

Chang, S-Y., Liu, L., and Asher, S. A. 1994. Preparation and properties of tailored morphology, monodisperse colloidal silica–cadmium sulfide nanocomposites. *J. Am. Chem. Soc.* 116: 6745–6747.

Crocker, J. C. and Grier, D. G. 1994. Microscopic measurement of the pair interaction potential of charge-stabilized colloid. *Phys. Rev. Lett.* 73: 352–355.

Edrington, A. C., Urbas, A. M., DeRege, P. et al. 2001. Polymer-based photonic crystals. *Adv. Mater.* 13: 421–425.

Esquena, J., Pons, R., Azemar, N. et al. 1997. Preparation of monodisperse silica particles in emulsion media. *Colloids Surf. A.* 123–124: 575–586.

Filin, S. V., Puzynin, A. I., and Samoilov, V. N. 2002. Some aspects of precious opal synthesis. *Aust. Gemmol.* 21: 278–282.

Flaugh, P. L., O'Donnell, S. E., and Asher, S. A. 1984. Development of a new optical wavelength rejection filter: Demonstration of its utility in Raman spectroscopy. *Appl. Spectrosc.* 38: 847–850.

Fleming, J. G. and Lin, S. Y. 1999. Three-dimensional photonic crystal with a stop band from 1.35 to 1.95 mm. *Opt. Lett.* 24: 49.

Gates B., Qin, D., and Xia, Y. 1999. Assembly of nanoparticles into opaline structures over large areas. *Adv. Mater.* 11: 466.

Haacke, G., Penzer, H. P., Magliocco, L. G., and Asher, S. A. 1993. Narrow band radiation filter films. U.S. Patent No. 5,266,238.

Hachisu, S. and Kobayashi, Y., 1974. Kikrwood–Alder transition in monodisperse latexes. II. Aqueous latexes of high electrolyte concentration. *J. Colloids Interface Sci.* 46: 470–476.

Hiltner, P. A., Papir, Y. S., and Krieger, I. M. 1971. Diffraction of light by nonaqueous ordered suspensions. *J. Phys. Chem.* 75: 1881–1886.

Holland, B. T., Blanford, C. F., and Stein, A. 1998. Synthesis of macroporous minerals with highly ordered three-dimensional arrays of spheroidal voids. *Science* 281: 538–540.

Holland, B. T., Blanford, C. F., Do, T. et al. 1999. Synthesis of highly ordered, three-dimensional, macroporous structures of amorphous or crystalline inorganic oxides, phosphates, and hybrid composites. *Chem. Mater.* 11: 795–805.

Holtz, J. H. and Asher, S. A. 1997. Polymerized colloidal crystal hydrogel films as intelligent chemical sensing materials. *Nature* 389: 829–832.

Jiang, P., Bertone, J. F., Hwang, K. S. et al. 1999. Single-crystal colloidal multilayers of controlled thickness. *Chem. Mater.* 11: 2132.

Jiang P., Ostojic G. N., Narat R. et al. 2001. The fabrication and bandgap engineering of photonic multilayers. *Adv. Mater.* 13: 389–393.

Joannopoulos, J. D., Meade, R. D., and Winn, J. N. 1995. *Photonic Crystals*. Princeton, NJ: Princeton University Press.

Kamenetzky, E. A., Magliocco, L. G., and Panzer, H. P. 1994. Structure of solidified colloidal array laser filters studied by cryogenic transmission electron microscopy. *Science* 263: 207–210.

Kamenjicki, M. and Asher, S. A. 2004. Photochemically controlled crosslinking in polymerized crystalline colloidal array photonic crystals. *Macromolecules* 37: 8293–8296.

Kamenjicki, M., Lednev, I. K., Mikhonin, A. et al. 2003. Photochemically controlled photonic crystals. *Adv. Funct. Mater.* 13: 774.

Kamenjicki, M., Lednev, I. K., and Asher S. A. 2004. Photoresponsive azobenzene photonic crystals. *J. Phys. Chem. B* 108: 12637–12639.

Kang, M.-S. and Gupta, V. K. 2002. Photochromic cross-links in thermo-responsive hydrogels of poly(N-isopropylacrylamide): Enthalpic and entropic consequences on swelling behavior. *J. Phys. Chem. B* 106: 4127–4132.

Kawasaki, H., Sasaki, S., and Maeda, H. 1998. Effects of the gel size on the volume phase transition of poly(N-isopropylacrylamide) gels: A calorimetric study. *Langmuir* 14: 773–776.

Kesavamoorthy, R., Jagannathan, P., Rundquist, P. A. et al. 1991. Colloidal crystal photothermal dynamics. *J. Chem. Phys.* 94: 5172–5179.

Kesavamoorthy, R., Super, M. S., and Asher, S. A. 1992a. Nanosecond photothermal dynamics in colloidal suspension. *J. Appl. Phys.* 71: 1116–1123.

Kesavamoorthy, R., Tandon, S., Xu, S. et al. 1992b. Self-assembly and ordering of electrostatically stabilized silica suspensions. *J. Colloid Interface Sci.* 153: 188–198.

Klein, A. M., Nanohan, V. N., Pine, D. J. et al. 2003. Preparation of monodisperse PMMA microspheres in nonpolar solvents by dispersion polymerization with a macromonomeric stabilizer. *Colloid Polym. Sci.* 282: 7–13.

Krieger, I. M. and O'Neill, F. M. 1968. Diffraction of light by arrays of colloidal spheres. *J. Am. Chem. Soc.* 90: 3114–3120.

Li, W-H. and Stöver, H. D. H. 2000. Porous monodisperse poly(divinylbenzene) microspheres by precipitation polymerization. *J. Polym. Sci Polym. Chem.* 36: 1543–1551.

Lin, S. Y., Fleming, J. G., Hetherington, D. L. et al. 1998. A three-dimensional photonic crystal operating at infrared wavelengths. *Nature* 394: 251.

Maack, J., Ahuja, R. C., and Tachibana, H. 1995. Resonant and nonresonant investigations of amphiphilic azobenzene derivatives in solution and in monolayers at the air–water-interface *J. Phys. Chem.* 99: 9210–9220.

Matsuo, E. S. and Tanaka, T. 1988. Kinetics of discontinuous volume–phase transition of gels. *J. Chem. Phys.* 89: 1695–1703.

Maurer, M. K., Lednev, I. K., and Asher, S. A. 2005. Photoswitchable spirobenzopyran-based photochemically controlled photonic crystals. *Adv. Funct. Mater.* 15: 1401–1406.

McPhee, W., Tam, K. C., and Pelton, R. 1993. Poly(N-isopropylacrylamide) lattices prepared with sodium dodecyl sulfate. *J. Colloid Interface Sci.* 156: 24–30.

Meyer, U., Svec, F., and Frechet, J. H. 2000. Use of stable free radicals for the sequential preparation and surface grafting of functionalized macroporous monoliths. *Macromolecules* 33: 7769–7775.

Mittelman, D. M., Bertone, J. F., Jiang, P. et al. 1999. Optical properties of planar colloidal crystals: Dynamical diffraction and the scalar wave approximation. *J. Chem. Phys.* 111: 345–354.

Noda, S., Tomoda, K., Yamanoto, N. et al. 2000. Full three-dimensional photonic bandgap crystals at near-infrared wavelengths. *Science* 289: 604–606.

Norris, D. J. and Vlasov, Y. A. 2001. Chemical approaches to three-dimensional semiconductor photonic crystals. *Adv. Mater.* 13: 371–376.

Okubo, T. 1988. Extraordinary behavior in the structural properties of colloidal macroions in deionized suspension and the importance of the Debye screening length. *Acc. Chem. Res.* 21: 281.

Olson, W. L. and Liss, W. E. 1988. Preparation of monodisperse titania by titanium alkoxide hydrolysis. U.S. Patent No. 4,732,750.

Ozin, G. A. and Yang, S. M. 2001. The race for the photonic chip: Colloidal crystal assembly in silicon wafers. *Adv. Mater.* 11: 95–104.

Pan, G. S., Kesavamoorthy, R., and Asher, S. A. 1997. Optically nonlinear Bragg diffracting nanosecond optical switches. *Phys. Rev. Lett.* 78: 3860–3863.

Pan, G. S., Kesavamoorthy, R., and Asher, S. A. 1998a. Nanosecond switchable polymerized crystalline colloidal array Bragg diffracting materials. *J. Am. Chem. Soc.* 120: 6525–6530.

Pan, G. S., Tse, A. S., Kesavamoorthy, R. et al. 1998b. Synthesis of highly fluorinated monodisperse colloids for low refractive index crystalline colloidal arrays. *J. Am. Chem. Soc.* 120: 6518–6524.

Park, S. H. and Xia, Y. 1998. Fabrication of three-dimensional macroporous membranes with assemblies of microspheres as templates. *Chem. Mat.* 10: 1745.

Park, J., Kim, J., and Suh, K. 2002. Chloromethyl-functionalized polymer particles through seeded polymerization. *Colloids Surf. A* 191: 193–199.

Pelton, R. H. and Chibante, P. 1986. Preparation of aqueous lattices with N-isopropylacrylamide. *Colloids Surf.* 20: 247–256.

Peters, A. and Candau, S. J. 1986. Kinetics of swelling of polyacrylamide gels. *Macromolecules* 19: 1952–1955.

Peters, A. and Candau, S. J. 1988. Kinetics of swelling of spherical and cylindrical gels. *Macromolecules* 21: 2278–2282.

Pusey, P. N. and van Megen, W. 1986. Phase behaviour of concentrated suspensions of nearly hard colloidal spheres. *Nature* 320: 340–342.

Qiang, Y., Zhicheng, Z., Xuewu, G. et al. 2002. Formation of monodisperse poly(methyl methacrylate) particles by radiation-induced dispersion polymerization. II. Particle size and size distribution. *Colloid Polym. Sci.* 280: 1091–1096.

Reese, C. and Asher, S. A. 2002. Emulsifier-free emulsion polymerization produces highly charged, monodisperse particles for near infrared photonic crystals. *J. Colloid Sci.* 248: 41–46.

Reese, C., Guerrero, C., Weissman, J. et al. 2000. Synthesis of highly charged, monodisperse polystyrene colloidal particles for the fabrication of photonic crystals. *J. Colloid Sci.* 232: 76–80.

Reese, C., Mikhonin, A., Kamenjicki, M. et al. 2004. Nanogel nanosecond photonic crystal optical switching. *J. Am. Chem. Soc.* 126: 1493.

Roberts, J. M., O'Dea, J. J., and Osteryoung, J. G. 1998. Methods for determining the intrinsic and effective charges on spherical macroions. *Anal. Chem.* 70: 3667–3673.

Rotello, V. M. 2004. *Nanoparticles: Building Blocks for Nanotechnology.* New York: Springer.

Rundquist, P. A., Photinos, P., Jagannathan, S. et al. 1989. Dynamical Bragg diffraction from crystalline colloidal arrays. *J. Chem. Phys.* 91: 4932–4941.

Rundquist, P. A., Jagannathan, P., Kesavamorthy, R. et al. 1991a. Photothermal compression of colloidal crystals. *J. Chem. Phys.* 94: 711–717.

Rundquist, P. A., Kesavamoorthy, R., Jagannathan, S. et al. 1991b. Thermal diffuse scattering from colloidal crystals. *J. Chem. Phys.* 95: 1249–1257.

Sader, J. E. and Chan, D. Y. C. 1999. Long-range electrostatic attractions between identically charged particles in confined geometries: An unresolved problem. *J. Colloid Sci.* 213: 268–269.

Sakoda, K. 2004. *Optical Properties of Photonic Crystals.* New York: Springer.

Schmid, A., Fujii, S., and Armes, S. P. 2007. Polystyrene-silica colloidal nanocomposite particles prepared by alcoholic dispersion polymerization. *Chem. Mater.* 19: 2435–2445.

Schroden, R. C., Al-Daous, M., and Stein, A. 2001. Self-modification of spontaneous emission by inverse opal silica photonic crystals. *Chem. Mater.* 13: 2945–2950.

Seung-Man, Y., Shin-Hyun, K., Jong-Min, L. et al. 2008. Synthesis and assembly of structured colloidal particles. *J. Mater. Chem.* 18: 2177–2190.

Shibayama, M. and Nagai, K. 1999. Shrinking kinetics of poly(N-isopropylacrylamide) gels T-jumped across their volume phase transition temperatures. *Macromolecules* 32: 7461–7468.

Shimizu, I., Kokado, H., and Inoue, E. 1969. Photoreversible photographic systems. VI. Reverse photochromism of 1,3,3-trimethylspiro[indoline-2,2'-benzopyran]-8'-carboxylic acid. *Bull. Chem. Soc. Jpn.* 42: 1730–1734.

Soukoulis, C. M. 2002. The history and a review of the modelling and fabrication of photonic crystals. *Nanotechnology* 13: 420–423.

Stein, A. and Schroden, R. C. 2001. Colloidal crystal templating of three-dimensionally ordered macroporous solids: Materials for photonics and beyond. *Curr. Opin. Solid State Mater. Sci.* 5: 553–564.

Subramania, G., Constant, K., Biswas, R. et al. 1999. Optical photonic crystals fabricated from colloidal systems. *Appl. Phys. Lett.* 74: 3933.

Tanaka, T. and Fillmore, D. J. 1979. Kinetics of swelling of gels. *J. Chem. Phys.* 70: 1214–1218.

Thirumalai, D. 1989. Liquid and crystalline states of monodisperse charged colloidal particles. *J. Phys. Chem.* 93: 5637–5644.

Tokita, M., Miyamoto, K., and Komai, T. 2000. Polymer network dynamics in shrinking patterns of gels. *J. Chem. Phys.* 113: 1647–1650.

Tse, A. S., Wu, Z., and Asher, S. A. 1995. Synthesis of dyed monodisperse poly(methyl methacrylate) colloids for the preparation of submicron periodic light absorbing arrays. *Macromolecules* 28: 6533–6538.

Tsoukatos, T., Pispas, S., and Hadjichristicks, N. 2000. Complex macromolecular architectures by combining TEMPO living free radical and anionic polymerization. *Macromolecules* 33: 9504–9511.

van Blaaderen, A., Ruel, R., and Wiltzius, P. 1997. Template-directed colloidal crystallization. *Nature* 385: 321.

van der Beek, D., Radstake, P. B., and Petukhov, A. V. 2007. Fast formation of opal-like columnar colloidal crystals. *Langmuir* 23: 11343–11346.

Vlasov, Y. A., Bo, X.-Z., Sturm, J. C. et al. 2001. On-chip natural assembly of silicon photonic bandgap crystals. *Nature* 414: 289–293.

Vondermassen, K., Bongers, J., Mueller, A. et al. 1994. Brownian motion: A tool to determine the pair potential between colloid particles. *Langmuir* 10: 1351–1353.

Wang, J., Gan, D., Lyon, L. A. et al. 2001. Temperature-jump investigations of the kinetics of hydrogel nanoparticle volume phase transitions. *J. Am. Chem. Soc.* 123: 11284–11289.

Weissman, J. M., Sunkara, H. B., Tse, A. S. et al. 1996. Thermally switchable periodicities and diffraction from mesoscopically ordered materials. *Science* 274: 958–960.

Widoniak, J., Eiden-Assmann, S., and Maret, G. 2005. Synthesis and characterisation of monodisperse zirconia particles. *Eur. J. Inorg. Chem.* 2005/15: 3149–3155.

Wijnhoven, J. E. G. L. and Vos, W. L. 1998. Preparation of photonic crystals made of air spheres in titania. *Science* 281: 802–804.

Wong, S., Kitaev, V., and Ozin, G. A. 2003. Colloidal crystal films: Advances in universality and perfection. *J. Am. Chem. Soc.* 125: 15589–15598.

Wong, S., Deubel, M., Pérez-Willard, F. et al. 2006. Direct laser writing of three-dimensional photonic crystals with a complete photonic bandgap in chalcogenide glasses. *Adv. Mater.* 18: 265–269.

Wong, S., Kiowski, O., Kappes, M. et al. 2008. Spatially localized photoluminescence at 1.5 micrometers wavelength in direct laser written optical nanostructures. *Adv. Mater.* 20: 4097.

Yariv, A. and Pochi, Y. 1984. *Optical Waves in Crystals*. New York: John Wiley & Sons.

Ye, Q., Zhang, Z., Ge, X. et al. 2002. Formation of monodisperse poly(methyl-metha-crylate) particles by radiation-induced dispersion polymerization. II. Particle size and size distribution. *Colloid Polym. Sci.* 280: 1091–1096.

Yoshino, K., Lee, S. B., Tatsuhara, S. et al. 1998. Observation of inhibited spontaneous emission and stimulated emission of rhodamine 6G in polymer replica of synthetic opal. *Appl. Phys. Lett.* 73: 3506.

Zhou, J., Li, Y., Tang, Y. et al. 1995. Detailed investigation on a negative photochromic spiropyran. *J. Photochem. Photobiol. A* 90: 117–123.

9

Processing of MEMS and NEMS

Seyed Allameh

CONTENTS

9.1 Introduction

The recent exponential growth of microelectromechanical systems (MEMS) and nanoelectromechanical systems (NEMS) market has prompted a closer look at the processing and the reliability of micro- and nanoscale devices.[1] Silicon as the primary material for the fabrication of microscale devices is known to suffer fatigue failure in ambient atmosphere where moisture is present. A brief discussion of silicon properties is presented, followed by the fundamentals of MEMS and NEMS, including materials properties that are affected by scaling.

9.2 Scaling

Three main scales have been identified in today's technology: macro-, micro-, and nanoscale, corresponding to meter, micrometer, and nanometer, respectively. While macroscale properties are mostly known and documented, microscale and nanoscale properties are not. In fact, the properties may be very different from the macroscale. Consider carbon at macro- and nanoscale. While graphite is so soft that it can be used for writing, carbon nanotubes can be so strong that a strand of it with a size similar to human hair ($\sim$100 μm) can lift a semitrailer (strength is nearly 60 times the strongest carbon steel). At small scales, diffusion paths become small and, consequently, chemical reactions and other processes that deal with transport phenomena become much faster. Mechanical properties generally improve with the reduction in size. Smaller grains greatly increase the strength of Ni (e.g., $\sim$60 MPa for bulk compared to 410 MPa for 270 μm thick and 670 MPa for 70 μm thick LIGA thin films[2]).

Compared to a larger one, a small chunk of a material will have more of its atoms at the surface and less at the bulk. Based on this, surface properties become more important, and therefore new phenomena and new properties are obtained. These include transport properties (momentum, energy, and mass), optical, electrical, magnetic, physical, chemical, and mechanical among many others. New properties lead to new capabilities, new tools, and new technologies. As an example, Au changes its color based on its size, such that its colloid exhibits nearly all the colors of the rainbow: red ($\sim$20 μm), yellow ($\sim$30 μm), green ($\sim$80 μm), blue ($\sim$50 μm), and violet (at $\sim$40 μm). Mechanical properties change with the thickness of films. Microtesting and nanoindentation tests on thin films show a strengthening effect with the reduction of the film thickness.[3–9] LIGA Ni films with a thickness of 70 nm have shown to have a longer fatigue life at similar stresses when compared to similar films with a thickness of 270 nm.[2]

With smaller scale, mass production, the automation of the processes, a drop in the size of the devices, a decrease in the amount of raw materials used, a reduction in the processing time, and an increase in the accuracy and repeatability of the functions of the systems are possible, which translates into better, smarter, and faster devices at lower prices. Probably, the best of all is the multifunctionality of materials at small scale. For example, carbon nanotubes can be used as a structural component due to its high strength; however, it already has electronics application in field electron emitters in large displays. Additionally, many biomedical applications have been envisioned for the arrays of carbon nanotubes used for DNA recognition, sequencing, and fractionation.

9.3 Materials

Si and Ni are two commonly used materials for MEMS devices. The well-established micromachining techniques developed by silicon-based semiconductor industry makes silicon the material of choice. Ni-based MEMS components are produced by LIGA, which is a lithography-based electrode-position process. In addition to Si, and its compounds such as silicon oxides, nitrides, and carbides, other materials like metals (e.g., Al, W, and Cu) and polymers (e.g., polyimide) have use in MEMS.

The physical properties of Si make it attractive for MEMS in both single crystalline and polycrystalline or even amorphous forms. When grown as a topical layer, the thickness will be usually less than 5 μm. Single-crystal Si is usually doped with an n- or p-type dopants. When etched, the orientation of the crystal dictates the boundary planes. For example, (100)-oriented wafers etched with KOH will have etch pits bounded by {111} planes that make an angle of 54.7° with the surface of the substrate. The orientation and type of doping of silicon wafers marked by primary (longer) and secondary (shorter) flats are shown in Figure 9.1.

The mechanical properties of single-crystalline silicon vary with crystallographic direction. For example, elastic modulus is found to increase from 130 GPa in <100> to 169 GPa in <110> direction. Tensile strength coincides with fracture strength at 7 GPa and Poisson's ration is 0.22.[10] Unlike the single-crystalline form, polycrystalline silicon (polysilicon) is not stress free and must be annealed at nearly 900°C to reduce growth-induced stresses. The nature of stress depends on the deposition temperature and can be tensile or compressive in nature. The mechanical properties of silicon does

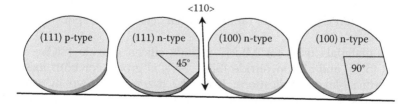

FIGURE 9.1
Orientation and type of doping of silicon wafers marked by primary (longer) and secondary (shorter) flats.

not change significantly with high temperature up to nearly 700°C, at which point the softening of silicon takes place.

The electrical properties of silicon make it desirable for integration with MEMS for electronic circuits, P–N junctions, and other electronic components necessary to control and drive MEMS. Important electrical properties of silicon include piezoresistivity, piezoelectricity, and thermoelectricity. Piezoresistivity effect is a change in the resistivity of silicon due applied mechanical stress. The effect is stronger in single-crystalline silicon; however, it varies with the direction, elastic modulus, temperature, and type and concentration of dopants. Silicon pressure sensors made of heavily doped ($>10^{19}/cm^3$) polycrystalline silicon are less sensitive to temperature variations. Piezoelectricity effect can be provided to Si-MEMS by applying a thin layer of piezoelectric material such as ZnO (sputtered), PZT (sputtered), PZT (sol–gel-processed), or polyvinylidene fluoride (PVDF, spin-coated). Seebeck, Peltier, and Thomson effects in silicon are related to its thermoelectric properties. The Seebeck coefficient of p-type polysilicon with a resistivity of 400 Ω cm is 270 $\mu V/K$, while n-type polysilicon with a resistivity of 2600 Ω cm exhibits a Seebeck coefficient of -400 $\mu V/K$.[10]

The thermal properties of silicon become important when the interface of Si and SiO_2 is exposed to variation in temperature. The coefficient of thermal expansion (CTE) of Si is close to that of SiC, but twice the CTE of SiO_2, and four times the CTE of SiN. Interfacial thermal stresses are therefore largest for the Si–SiO_2 interface during heating and cooling.

The optical properties of silicon make it suitable for light detector; however, due to its indirect bandgap, light emission is difficult. At thicknesses less than 1.1 μm, silicon is transparent, but reflects 60% of the light with wavelengths shorter than 400 nm.[11]

Germanium has been used in microelectronics since the development of the early transistors.[11] The use of Ge-based materials in MEMS includes low-pressure chemical vapor deposition (LPCVD)-deposited polycrystalline Ge, which is impervious to etchants such as KOH, tetramethylammonium hydroxide (TMAH), and buffered oxide etch (BOE). This makes poly-Ge a good etch-stop material for MEMS fabrication. The mechanical properties of Ge are comparable to those of Si.[11] While poly-Ge does not nucleate on SiO_2 (used as sacrificial layer), poly-Ge-Si can be deposited on SiO_2 by deposition processes such as LPCVD, atmospheric pressure chemical vapor deposition (APCVD), and rapid thermal chemical vapor deposition (RTCVD).[12]

Other materials used for MEMS include silicon dioxide (SiO_2), silicon carbide (SiC), and silicon nitride (SiN). SiO_2 is grown for both insulator- and sacrificial-layer applications. It can be thermally grown in oxidizing atmospheres at a temperature above 800°C. Alternatively, SiO_2 can be grown by CVD, sputtering, and spin-coating. SiN layers can be used as Na and K ion barrier in biological environments. Due to its high elastic modulus, the coatings of silicon nitride are used as suitable masks for alkaline etch processes. Metals and alloys used in MEMS include Ag, Al, Au, Cr, Cu, Ir, Ni, Pd, Pt, Ti, W, NiCr, TiNi, and TiW. Metals such as Au, Ag, and Ni do not directly adhere to the surface of Si and require a thin adhesion layer of Cr or Ti. For electrochemistry applications, Au and Ag are used due to their noble nature. Aluminum is used for the optical reflection of visible and infrared light. Both Al and Cu are used for low resistivity interconnects. Permalloy™ and Ni are used for magnetic transducers, while NiCr and SiCr are used for laser-trimmed resistors. The use of titanium alloys includes TiNi for shape memory alloy actuation. High temperature applications utilize W interconnects. The other applications of materials in MEMS include the use of SnO_2 for chemoresistance-based gas sensors, indium tin-oxide (ITO) for liquid crystal displays (LCD), and TaN for negative trickle charge resistor (TCR) laser-trimmed resistors.

9.4 Design and Modeling of MEMS

The subject of the design of MEMS and NEMS is closely tied to the fundamentals that dictate the properties and behavior of the devices and structures to be built. These fundamentals are related to the wide-ranging areas of mechanics, dynamics with and without control, electronics, and noise among others.[13] The most notable approach to the design of MEMS, as presented by Senturia,[13] emphasizes lumped-element models using either a network- or a block-diagram representation. Creating the right model elements though is critical to the success of such approach. The modeling of MEMS can be achieved using analytical as well as numerical methods. MATLAB®, Simulink®, and Maple have been successfully used for this purpose.[13]

9.5 Applications of MEMS

MEMS devices have applications in the various fields of electronics, optics, acoustics, mechanics, physics, chemistry, aerospace, automotive, military, and life sciences. Examples are microsensors including pressure sensors, accelerometers, radiation sensors, magnetic sensors, and smart sensors; microactuators including RF switches, piezoelectric MEMS; and drug delivery systems and lab-on-a-chip.

Sensing methods for mechanical sensors are based on motion-affecting physical processes, such as piezoresistivity, piezoelectricity, variable capacitance, and optical resonance. An example of the first method is the sensitivity of a conductive element in a strain gauge to the changes in its dimension. The resistivity change of the strain gauge element with its length is based on this sensitivity. The second mechanical sensing method is piezoelectricity at temperatures below Curie temperature. Piezoelectric materials can be used both as sensors (applied motion causes electric current) or actuators (applied current causes motion). The third sensing method is based on the capacitance change due to the motion of two electrodes can be used for mechanical sensors; however, it is affected by temperature changes in a nonlinear way. Corrections must be made by the integration of a temperature sensor with this type of mechanical sensors.

The fourth sensing method is optical, in which motion affects electromagnetic wave properties such as the intensity, phase, wavelength, spatial position, frequency, and polarization of light. Photodiodes, photoresistors, and LEDs are used with various techniques such as interferometry, frequency modulation, and laser Doppler frequency shift to measure the extent of motion. The fifth mechanical sensing method is based on the shifts in the resonant frequency of structures due to changes in their masses. Single-crystalline structures such as quartz (e.g., quartz tuning fork used as time base in watches) can detect smallest changes in their mass, their stiffness, or the shape of the resonator with an accuracy that is 5–10 times higher than that of capacitance-based mechanical sensors. The examples of resonant-structure-based mechanical sensors are force, pressure sensors, and accelerometers.

Nanotechnology has a wide range of applications from electronics to biomedical field. The biomedical application of nanomaterials includes drugs, drug delivery, photodynamic therapy, molecular motors, neuro-electronic interfaces, protein engineering, and nanoluminescent tags. The use of nanostructured materials in sensors has lead to the applications of technology in neural nanoscale sensors such as electromagnetic sensors, biosensors, and electronic noses. The other applications of NEMS include light energy, its capture and photovoltaics, light production, light transmission, light control and manipulation, electronics, carbon nanotubes, soft molecule electronics, memories, gates and switches.[14]

9.6 Processing of MEMS and NEMS

The processing of MEMS and NEMS in the terms of bulk and surface micromachining, additive techniques, wafer-bonding, self-assembly, and writing methods are briefly discussed here. Comparisons between competing techniques are made, and the implications of various techniques on the industry are elucidated.

9.6.1 Silicon Bulk Micromachining

Isotropic etching techniques have been in use in semiconductor industry since their introduction in early 1950s.[15] Wet bulk machining allows the carving of desired features in materials such as silicon, quartz, SiC, GaAs, InP, and Ge. Compared to conventional dry etching (0.1 μm/min etch rate), wet etching is faster (e.g., 1 μm/min for anisotropic etching[1]), and provides a wider variety of techniques for MEMS micromachining. In terms of the type of fluids used, liquids are used for wet bulk etching, while plasmas are used for dry etching. Higher etch rates are possible with enhanced dry-etching techniques such as inductively coupled plasma (ICP) etch (up to 6 μm/min). In isotropic etching, the use of acids such as HNA ($HF/HNO_3/CH_3COOH$) causes material removal in all directions at the same rate, leading to round features in single-crystalline silicon. While HNA etching utilizes near room temperatures ($<50°C$) since the early days of semiconductor processing,[15] basic etchants, discovered later, require higher temperatures ($>50°C$). Bases (e.g., KOH), with pH > 12, allow the anisotropic etching of single-crystalline Si, also known as orientation-dependent etching. The wet bulk machining of a membrane from a silicon single-crystalline wafer is shown in Figure 9.2.

Both isotropic and anisotropic etchings involve the oxidation of the silicon formation of silicate hydrates, and their transfer from the reaction front. In isotropic etching, the limiting factor is the diffusion of acid to, and the removal of hydrates from, the reaction front. In contrast, diffusion is fast for the anisotropic etching, with reaction rate being the limiting factor.

9.6.2 Isotropic Etching

Charge transfer, performed by the dopant atoms, facilitates the isotropic etching of silicon. Consequently, the dopant (n- or p-) type, and its concentration ($>10^{18}/cm^3$) become critical for the process. A milder electrostatic isotropic etching process permits the use of a simple photoresist mask. An electrical power supply allows the creation and transfer of electron holes to the chemical reaction front in a chemical reaction cell filled with etchant solution. A positive bias is established between the silicon (anode) and a counter (platinum) electrode as cathode. To create a membrane from lightly doped silicon (LDS), a highly doped silicon (HDS) wafer is etched to the depth where the reaction front meets LDS. The interface between the LDS and HDS acts as an etch stop.[16] Since electron holes are already supplied by the power supply, there is no need for HNO_3 to drive the oxidation of silicon. Immersion etching, spray etching, electrolytic etching, gas-phase etching, and molten salt etching are the variations of isotropic etching techniques.[17] Compared to anisotropic, the isotropic etching poses challenges that include the requirements of high precision masks, well-controlled agitation of the solution, and well-maintained temperatures. It also lacks lateral dimension control.

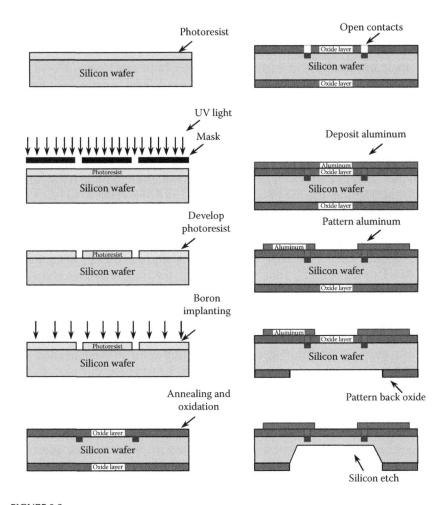

FIGURE 9.2
Wet bulk machining of a membrane from a silicon single-crystalline wafer. (Reproduced after Jia, G. and Madou, M., *MEMS Fabrication*, 2nd edn., CRC Press, Boca Raton, FL, 2006.)

9.6.3 Isotropic (Orientation-Dependent) Etching

Both single- and polycrystalline types of silicon are used for micromachining. Most applications of bulk micromachining use single-crystalline silicon wafers with diameters of 40 and 150 mm and thicknesses of 550 and 650 μm. The orientation of the crystalline structure makes a difference on the inclination of the walls created by anisotropic etching. The (111) wafers have either one flat (if p-type) or two flats oriented at 45° (if n-type). They are difficult to work with except for use with laser-assisted etching. Grooves and truncated pyramids that form during anisotropic etching on the (100) Si wafers will have inclined sidewalls of (111)-type with an inclination of 54.7° with respect to the wafer top surface. The edges of such recessed features will have <110> directions. If vertical sidewalls are desired, masks have to be aligned such that edges of the features coincide with the <100> directions of the wafer. However, this allows undercuts to be created. The extent of such undercut equals the depth of the recessed feature. The main reason for the orientation-dependent anisotropic etching of single-crystalline silicon is the much faster etching rate of low index planes such as {100} and {110}, which leave the slower etching {111} planes behind.[18]

Anisotropic etching provides the lateral dimension control of geometries on single-crystal planar substrates with an accuracy of 500 nm or better. The lateral dimension can be related to the vertical profile, hence providing a means of measuring the recessed feature depth. Different etch stops are used to reach desired depths for recessed features. The drawbacks of anisotropic etching include slow etching rates (1 μm/min) and relatively high solution temperatures (usually 80°C–110°C). A process[19] for the fabrication of a membrane by wet anisotropic etching is presented below:

1. Oxidize a single crystal wafer at 900°C–1200°C.
2. Spin the photoresist for 20–30 s at 5000 rpm.
3. Prebake the resist at 90°C for 10 min.
4. Expose to ultraviolet light at RT for 20 s.
5. Develop the resist at RT for 1 min.
6. Postbake the resist at 120°C for 20 min.
7. Strip the oxide in BHF at the rate of 1:7 at RT in 10 min.
8. Strip the resist in acetone at RT in 10–30 s.
9. Standard clean RCA1 [NH_4OH (27%):H_2O:H_2O_2 = 1:5:1] at boiling point for 10 min.
10. Standard clean RCA2 (HCl:H_2O:H_2O_2 = 1:6:1) at boiling point for 10 s.
11. HF dip at RT for 10 s.
12. Anisotropic etch at 70°C–100°C for an appropriate length of time.

Etchants used for the wet anisotropic etching of silicon include the alkaline aqueous solutions of KOH, NaOH, NH_4OY, RbOH, LiOH, and CsOH. The use of additives such as ethylenediamine, alcohol, choline, hydrazine, pyrocatechol, and pyrazine is common. The etching rate is dopant-insensitive due to the lack of a need for dopant-based charge transfer. The role of additives is to speed up the etch rate (e.g., pyrocatechol[20]) or decrease it (e.g., propanol, isopropanol, and butanol alcohols[21]). The important factors in etchant selection are the etch rate, topography of the bottom of the recesses feature, IC compatibility, etch stop, etch selectivity over other materials, mask material, and thickness of the mask. The most popular etchant is KOH used at near saturation at 80°C.

9.6.4 Electrochemical Etching

Doped n-type silicon can be etched by anodic dissolution if electron holes are supplied to its surface. Electrons in the conduction band of silicon react with the protons in the solution and reduce them to hydrogen. This requires a high reverse bias or the illumination of the surface of Si. For p-type silicon, a small forward bias will produce high anodic current even in the absence of illumination. In the illumination-assisted anodic etching, the dissolution of n-type silicon, associated with photocorrosion which is reported to be a barrier to the viability of photoelectrochemical cells.[22] Photocorrosion phenomenon is used in photoelectrochemical etching as an electrochemical polishing process.

9.6.5 Dry Etching

Plasma-assisted dry etching provides a better control of anisotropy and selectivity, with less contamination to the environment.[23] Silicon-based materials can be etched by plasmas including F_2, $CF_4 + O_2$, $CF_4 + N_2O$, $SiF_4 + O_2$, SF_6, $SF_6 + O_2$, NF_3, and ClF_3.[24] Silicon dry etching has become vital to the fabrication of MEMS structures.[23,25–28] An example is the application of dry-etching techniques to polycrystalline silicon sacrificial layer for the fabrication of RF MEMS switches, high-Q suspended inductors, and suspended-gate MOSFETs.[29,30] Some applications such as the fabrication of micromachined micromirrors[31] utilize both anisotropic wet-etching and isotropic dry-etching processes.

9.6.6 LIGA

Au,[32] Ni,[2,33,34] Cu, and alloys such as NiP[32] have been electrodeposited in polymethylmethacrylate (PMMA) molds in a lithography-based process called LIGA (the acronym stands for "Lithographie, Galvanoformung, Abformang," which means "lithography, electroplating, and molding"). While most silicon-based additive processes produce topical MEMS with thicknesses restricted to a few microns (e.g., 2 μm), high aspect ratio (e.g., 200 μm thick) microsamples can be made with practically vertical sidewalls. The feature size of LIGA greatly depends on variables such as the beam size (e.g., x-rays, UV) and film thickness; sidewall offset and slope[35]; and thermal expansion and swelling of PMMA.[36] The morphology of the sidewalls will affect the motion of mechanical LIGA components due to its implications for friction, adhesion, and wear of such components.[37]

Original LIGA process[38] uses x-rays from a synchrotron; however, its modifications include laser-LIGA,[39] and silicon-LIGA.[40] Laser (e.g., pulsed excimer, 248 nm) beam has been used to fabricate Ni serpentine microstructures with applications in microheaters.[39] In silicon-LIGA, deep reactive ion-etched silicon is used as preform. Silicon-LIGA process has been applied to underwater weapon safety and arming systems.[40]

9.7 Fabrication of NEMS Structures

9.7.1 Nanocharacterization Tools

Tools for the fabrication and characterization of nanostructures and NEMS vary from one field to another; however, there are a few tools, listed here, that are widely used in the rapidly growing dynamic area of nanofabrication. Scanning probe instruments with the scanning probe microscopy (SPM) tips performing surface modification as well as characterization have been used to manipulate or move small species (atoms or molecules) on surfaces. SPM tools can be used to make custom prototypes, but not commercial products that require low-cost fast processing. The variations of SPM include atomic force microscopy (AFM), scanning tunneling microscopy (STM), and magnetic force microscopy (MFM).

Other nanocharacterization tools include spectroscopy, which refers to the interaction of light with a sample leading to absorption, scattering, or other observable phenomena. X-ray machines allow the formation of an image from an object that is exposed to high-energy radiation. Magnetic resonance imaging (MRI) is well known for its medical applications. Due to the large wavelengths of visible light (400–800 nm), spectroscopy is only used for the characterization of nanostructures *en masse*.[14] Electrochemistry and electron microscopy are other common tools to characterize nanostructures in terms of topography, microstructures, chemical composition, and crystal structure. In electrochemistry, an applied current initiates or alters chemical reactions (e.g., a battery being charged by the passage of current) and vice versa (e.g., a battery producing electric current). Transmission electron microscopy and scanning electron microscopy (TEM and SEM) allow the interaction of electron beams (e-beams) with thin (TEM) and thick (SEM) samples generating various signals that lead to the characterization of the samples. Attachments to electron microscopes include energy dispersive spectroscopy (EDS), orientation imaging (OM), and backscatter diffractometry (BSD).

9.7.2 Nanofabrication Tools

Nanocharacterization tools mentioned above have use in the nanofabrication too. For example, nanostructures can be written by the tip of an SPM using the structural chemistry, electrical interactions, and magnetic behavior of the substrate and topical materials. Despite its ability to manipulate structures down to atoms, SPM is both expensive and slow. Nanoscale lithography includes dip-pen, e-beam, and nanoscale lift-off lithographies. These techniques will be described in more detail later. Nanoimprinting tools allow the transfer of patterns from a mold or die to soft and hard surfaces in a manner similar to printing.[41] The use of a solid-state electrochemical nanoimprinting process in the etching of metallic films has been demonstrated. This is achieved by passing an electric current between a patterned solid-electrolyte AgS$_2$ die and an Ag metallic film.[42] This process promises a large-scale single-step direct patterning of nanostructures. Molecular synthesis deals with crafting special-purpose molecules. An example is the fabrication of molecules with particular geometries. Probably the most promising tool for nanofabrication is self-assembly. This technique solves the slower-speed and higher-price problems associated with the other nanofabrication processes such as SPM techniques.

9.7.3 Nanolithography

While semiconductor industry uses optical and x-ray lithography, nano-scale lithography cannot use optical rays due to their long wavelengths. Micro-imprint[43] lithography and self-assembly[44-51] are well utilized for the fabricated nanostructures.[52-55] A review of the published work on the new approaches to nanofabrication including molding and printing is presented by Gates et al.[43] Some of the unconventional fabrication techniques cited in this review include those that use organic materials to replicate nanoscale patterns, scanning probe lithography, self-assembly, and edge lithography.[43] Attempts have been made to apply self-assembly to electrical connections in electronic circuits.[54]

An adaptation of lithography to nanoscale, micro-imprint lithography is based on inscribing a pattern onto a rubber surface, which is then coated with a molecular ink. This rubber stamp can then be used to print on metals, polymers, or ceramic surfaces. A different nanoscale lithography, dip-pen lithography, utilizes a reservoir of atoms or molecules (the ink), stored on the top of SPM tip.[56-58] When the tip is moved across a surface, such molecules are left behind in desired lines or patterns (Figure 9.3). Dip-pen

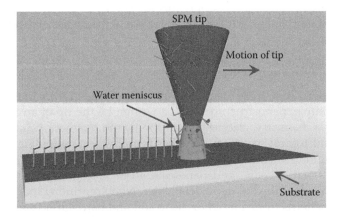

FIGURE 9.3
Schematic of transport of molecules from SPM into self-aligned array in dip-pen lithography.

nanolithography can be used for semiconductor patterning and biomedical and pharmaceuticals development. A commercial product of dip-pen nanolithography is available from Nanoink™ (Skokie, IL). The print arrays of this commercial dip-pen system (DPN 5000™) can be active or passive, and can contain up to 55,000 pens. Dip-pen inks include those with proteins, DNA, nanoparticles, and polymers. The main advantages of this technology are the ability to use any material as ink, and to write on any surface as substrate. On-wire lithography,[58] the patterning of oligonucleotides on metal and insulators,[59] and DPN etch resist[60] are variations of dip-pen nanolithography.

E-beam lithography is appropriate for nanoscale due to the shorter wavelengths of electrons at high voltages. Currently, e-beam lithography is used for the fabrication of microelectronics with nanoscale feature size. In simplest form, the process uses an electron microscope to focus and scan the beam in pathways dictated by a desired pattern from a computer. The beam will write the pattern into a layer of PMMA positive-resist. The exposed areas of the resist are then dissolved and the PMMA layer is used as mask for metal deposition. When the resist is dissolved, the patterned metal structure is revealed.

Nanosphere liftoff lithography technique uses nanoparticles to form a two-dimensional (2D) array with periodic holes in between them. If sprayed with a desired paint, the interstices between the nanoparticles will be filled with such paint material. When the nanoparticles are removed, a periodic 2D array of triangular dots will appear. The method allows formation of three-dimensional (3D) quantum dots by spraying repeatedly with the same or different types of paint, leading to monolithic or composite nanotextured surfaces.

9.7.4 Self-Assembly

Nanofabrication techniques, discussed above, require a design or a pattern to write, imprint, micromachine, or deposit structures onto substrates. Unlike these, self-assembly technique utilizes the atomic and molecular interactions for the nanofabrication of 2D and 3D structures. The driving force for self-assembly is the minimization of the energy associated with the atomic or molecular conglomerates. Intermolecular forces mainly arise from weaker columbic interactions between molecules with zero net charge. Called multipolar interactions, they bond positive and negative poles of structures that have nonuniform charge distribution, similar to that of water molecule. The process entails the introduction of the desired molecule onto a substrate surface. Once on the surface, the molecule will diffuse in various directions and align itself such that bonds form between this molecule and other molecules on the surface leading to the reduction of total energy. The examples of multipolar interactions include the hydrogen bonds that effect the folding of proteins into specific 3D structures and the hydrophobic forces that cause the separation of oil from water.

9.7.5 Ion-Beam-Induced Self-Assembly of Nanostructures

Writing techniques used for the fabrication of microstructures involve the use of ion beams to impinge desired adsorbed atoms onto the substrates surfaces from their precursors in predetermined patterns.[61–72] The use of a focused ion beam (FIB) in writing 3D Pt structures is an example of ion-beam-induced deposition method. Here, the differential pumping of a platinum organometallic compound allows its molecules to spread over the substrate surface. When large energetic Ga ions from the ion beam collide with the Pt-organometallic molecule on the surface, Pt gets affixed to the surface, while the accompanying carbon atoms leave. The continuation of this process leads to the formation of a column of Pt, for which the diameter will scale with the ion-beam diameter. If the beam is off-center and the stage is rotated while column deposition is in process, a spiral can result.[70] The repeated scanning action of the ion beam in a line (one-dimensional [1D]) or a plane (2D) leads to the formation of a wall or a solid block of platinum. A schematic of ion-induced deposition is presented in Figure 9.4. It illustrates the organometallic decomposition leading to the deposition of metal and evaporation of the carbon.

The self-assembly capability of FIB has been demonstrated for the fabrication of nanostructured silicon surfaces coated with Ir.[73] Figure 9.5 shows the image of the self-assembled nanofins grown to a height of 1 μm, a thickness of 70 nm, with a spacing of 140 nm. The random bifurcation of the fins is observed in Figure 9.5a. The cross section of the deposited Pt presented in Figure 9.5b shows the thickness of the deposited layer, along with the

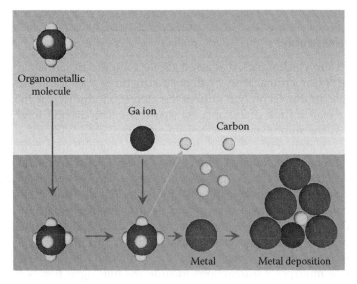

FIGURE 9.4
Schematic of ion-induced organometallic compound decomposition and metal deposition.

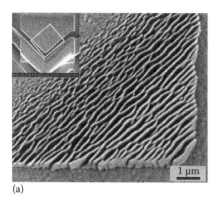
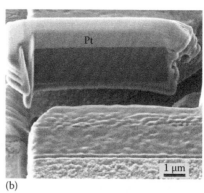

(a)   (b)

FIGURE 9.5
Nanofins self-assembled by ion-beam-induced CVD (a) nanofins with 1 μm height, 70 nm thickness, and 140 nm of spacing, the inset shows the location of the fins with respect to a MEMS device. (b) A cross-sectional cut using high FIB to examine the deposited Pt layer. Horizontal nanofins grown parallel to the surface are shown.

horizontal fins grown. The composition of the FIB-induced CVD-deposited structures usually shows large amounts of species other than the desired material. For the FIB-deposited nanofins, the x-ray diffraction (XRD) of the Pt film has shown 36% Pt, 48% C, and 16% Ga.

The mechanism of the self-assembly of the walls was postulated to be in the presence of two competing processes: a growth process that allows deposition in height, width, and thickness, and an etching process that suppresses the growth in the thickness direction more than in the other two directions. The latter was inferred from the simple geometry of the fins, and the orientation of the ion beam with respect to the fins.

The mechanism of the roughening of stressed solid surfaces have been discussed by Bradley and Harper[74] and further investigated by Erlebacher et al.[75–78] and others.[79–86] It is postulated that stressed solids exposed to etching or dissolution will not remain plane; rather, they develop the undulations of which the wavelength λ and growth rate, are given by

$$\lambda \sim \sqrt{FT} e^{\frac{-\Delta E}{k_B T}} \tag{9.1}$$

$$\dot{h} = S\kappa + B \frac{\partial^2}{\partial s^2} \kappa \tag{9.2}$$

where
    F is the ion flux
    κ is the wavelength of the undulations
    ΔE is the activation energy for the self-diffusion of the incoming energetic
        particles
    h is the height of the undulations
    T is the temperature

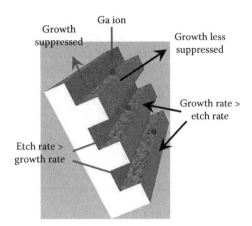

FIGURE 9.6
Growth mechanism of self-assembled nanofins for a substrate tilted greater than 45°.

There are two processes that compete, namely, roughening, incorporated by its coefficient, S, and a smoothing process, represented by its coefficient B. It is clear from these two equations that larger flux and higher temperatures lead to larger wavelengths. Figure 9.6 is an illustration of the effect of orientation on the growth mode of nanofins. At tilt angles greater than 45° (e.g., 52 for the case of Pt nanofins grown by FIB), the etch rate of the sidewalls (shown by grained texture) is greater than that of the topside of the walls. Consequently, growth rate is expected to be suppressed to a greater extent, leading to the growth favored in the direction perpendicular to the substrate. Obviously at smaller tilt angles, sideway growth rate is expected to be greater than the normal direction to the substrate, and solid films can grow.

9.8 Reliability of MEMS and NEMS Structures

Reducing the size to micro- and nanoscales imparts new properties to nanostructures. While some of these properties are superior to the bulk materials, there are others that can cause concerns about the reliability of the MEMS and NEMS devices.[87–91] Bulk silicon is not known to undergo fatigue under ambient conditions. However, microstructures made of silicon have been observed to undergo stress-assisted corrosion fatigue in ambient environments.[92] The culprit is postulated to be the presence of moisture in the air that facilitates the crack initiation in these structures.[83,93,94] Similarly, Ni is not known to suffer corrosion cracking under ambient conditions. It has been demonstrated that LIGA-deposited Ni microstructures can suffer from fatigue and failure by what was figured out to be stress-assisted oxidation.[2,95] Perhaps the most important difference between small structures and large bulk counterparts lies in the critical crack length that causes the failure of the component.

Since the time to grow an existing crack to its critical size is small in micro- and nanostructures, most of the life of the components falls within the crack initiation regime.[2] This fact prompts the surface protection of the load-bearing micro- and nanocomponents from aggressive media such as water vapor. Hermetic sealing addresses such a need for components that do not have to come in contact with ambient atmosphere (e.g., accelerometers). This is not the case, however, for sensors and detectors that sniff gases, probe fluids, and perform biomedical functions involving media transport. Materials selection becomes crucial for such applications; nevertheless, our knowledge of the behavior of materials at small scales is still in its infancy. The micro- and nanocharacterization of materials[80,96–103] for mechanical properties among other properties will help establish the suitability of materials for applications, where any combination of load, aggressive environment, electric current flow, and high temperatures coexist.

9.9 Conclusion

A brief overview of the fundamentals, processing, and applications of MEMS and NEMS has been presented in this chapter. The main ideas include the following:

1. A small scale enhances or affects mechanical, chemical, optical, and physical properties.
2. MEMS devices can be made using the established semiconductor industry practices.
3. There are promising large-scale nanostructure fabrication methods that can be commercially viable: the self-assembly, ink-jet printing of nanostructures, and nanolithography.

References

1. G. Jia and M. Madou, *MEMS Fabrication*, 2nd edn., CRC Press, Boca Raton, FL, 2006.
2. S. M. Allameh, J. Lou, F. Kavishe, T. Buchheit, and W. O. Soboyejo, *Materials Science and Engineering A* 371, 256–266 (2004).
3. I. Lou, P. Shrotriya, S. Allameh, N. Yao, T. Buchheit, and W. O. Soboyejo, *Plasticity Length Scale in LIGA Nickel MEMS Structures*, Materials Research Society Warrendale, PA, 2002, pp. 41–46.

4. J. Lou, S. Allameh, T. Buccheit, and W. O. Soboyejo, *Journal of Materials Science* 38, 4129–4135 (2003).
5. J. Lou, P. Shrotriya, S. M. Allameh, T. E. Buchheit, and W. O. Soboyejo, *Journal of Materials Science and Engineering A* 441, 299–307 (2006).
6. J. N. Ding, Y. G. Meng, and S. Z. Wen, Size Effect on the Mechanical Properties and Reliability Analysis of Microfabricated Polysilicon Thin Films, *Journal of Materials Research*, 16, 2223–2228 (2001).
7. D. Jianning, M. Yonggang, and W. Shizhu, *Chinese Science Bulletin* 46, 1392–1397 (2001).
8. W. N. J. Sharpe, *American Society of Mechanical Engineers*, Materials Division (Publication) MD 84, 153–155 (1998).
9. T. Tsuchiya, O. Tabata, J. Skakata, and Y. Taga, *Journal of Microelectromechanical Systems* 7, 106–113 (1998).
10. N. Maluf, *An Introduction to Microelectromechanical Systems Engineering*, Artech House, Boston, MA, 2004.
11. C. Zorman and M. Mehregany, *MEMS: Design and Fabrication*, 2nd edn., CRC Taylor & Francis, Boca Raton, FL, 2006.
12. S. Sedky, P. Fiorini, M. Cymax, S. Loreti, K. Baert, L. Hermans, and R. Mertens, *Journal of Microelectromechanical Systems* 7, 365–372 (1998).
13. S. D. Senturia, *Microsystem Design*, Springer Science, New York, 2000.
14. M. A. Ratner and D. Ranter, *Nanotechnology: A Gentle Introduction to the Next Big Idea*, Pearson Education, Inc., Upper Saddle River, NJ, 2002.
15. H. Robbins and B. Schwartz, *Journal of Electrochemical Society* 106, 505–508 (1959).
16. M. J. Theunissen, J. A. Apples, and W. H. C. G. Verkulen, *Journal of Electrochemical Society* 119, 351–365 (1970).
17. W. Kern and C. A. Deckert, in *Thin Film Processes*, Eds., J. L. Vossen and W. Kern, Academic Press, Orlando, FL, 411–413 (1978).
18. R. W. Ade and E. R. Fossum, *Journal of Electrochemical Society* 134, 3192–3194 (1987).
19. M. Elwenspoek, U. Lindberg, H. Kok, and L. Smith, *Wet Chemical Etching Mechanism of Silicon*, Oiso, Japan, 1994, pp. 23–28.
20. R. M. Finne and D. L. Klein, *Journal of Electrochemical Society* 114, 965–970 (1967).
21. H. Linde and L. Austin, *Journal of Electrochemical Society* 139, 1170–1174 (1992).
22. M. Madou, K. W. Frese, and S. R. Morrison, *Surface Science* 108, 135–152 (1980).
23. J. Bhardwaj, H. Ashraf, and A. McQuarrie, *Dry Silicon Etching for MEMS*, Electrochemical Society, Montreal, Quebec, Canada, 1997, pp. 118–130.
24. Y. Tzeng and T. H. Lin, *Journal of Electrochemical Society* 134, 2309 (1987).
25. A. R. Giehl, M. Kessler, A. Grosse, N. Herhammer, and H. Fouckhardt, *Journal of Micromechanics and Microengineering* 13, 238–245 (2003).
26. W. Liang, W. Jia-You, L. Dao-Guang, and Y. Yin-Tang, *Micronanoelectronic Technology* 41, 30–34 (2004).
27. I. W. Rangelow, *Vacuum Sixth International Conference on Electron Beam Technologies* (EBT '2000), June 4–7, 2000, Varna, Bulgaria, 62, 279–291 (2001).
28. H. Xie and G. K. Fedder, *IEEE Sensors Journal* 3, 622–631 (2003).
29. S. Frederico, C. Hibert, R. Fritschi, P. Fluckiger, P. Renaud, and A. M. Ionescu, *Silicon Sacrificial Layer Dry Etching (SSLDE) for Free-Standing RF MEMS Architectures*, IEEE, Kyoto, Japan, 2003, pp. 570–573.
30. R. Fritschi, C. Hibert, P. Fluckiger, and A. M. Ionescu, *Dry Etching Techniques of Amorphous Silicon for Suspended Metal Membrane RF MEMS Capacitors*, Materials Research Society, San Francisco, CA, 2002, pp. 69–74.

31. Y.-C. Cheng, C.-L. Dai, C.-Y. Lee, P.-H. Chen, and P.-Z. Chang, *Sensors and Actuators, A: Physical* 120, 573–581 (2005).
32. C.-W. Baek, Y.-K. Kim, Y. Ahn, and Y.-H. Kim, *Sensors and Actuators, A: Physical* 117, 17–27 (2005).
33. T. Bieger, *Microsystem Technologies* 3, 155–163 (1997).
34. T. R. Christenson, T. E. Buchheit, D. T. Schmale, and R. J. Bourcier, *Materials Research Society Symposium—Proceedings* 518, 185–190 (1998).
35. S. K. Griffiths, *Journal of Micromechanics and Microengineering* 14, 999–1011 (2004).
36. S. K. Griffiths, J. A. W. Crowell, B. L. Kistler, and A. S. Dryden, *Journal of Micromechanics and Microengineering* 14, 1548–1557 (2004).
37. A. C. Hall, M. T. Dugger, S. V. Prasad, and T. Christensen, *Journal of Microelectromechanical Systems* 14, 326–334 (2005).
38. E. W. Becker, W. Ehrfeld, D. Munchmeyer, H. Betz, A. Heuberger, S. Pongratz, W. Glashauser, H. J. Michale, and R. V. Siemens, *Naturwissenschaften* 69, 520–523 (1982).
39. H. Jin, E. C. Harvey, J. P. Hayes, M. K. Ghantasala, A. Dowling, M. Solomon, and S. T. Davies, *Proceedings of the SPIE—The International Society for Optical Engineering* 4592, 166–171 (2001).
40. L. Fan, H. Last, R. Wood, B. Dudley, C. K. Malek, and Z. Ling, *Microsystem Technologies* 4, 168–171 (1998).
41. S. Y. Chou, *MRS Bulletin* 30, 512–518 (2001).
42. K. H. Hsu, P. L. Schultz, P. M. Ferreira, and N. X. Fang, *Nano Letters* 7, 446–451 (2007).
43. B. D. Gates, Q. Xu, M. Stewart, D. Ryan, C. G. Wilson, and G. M. Whitesides, *Chemical Reviews* 105, 1171–1196 (2005).
44. A. Winkleman, L. S. McCarty, T. Zhu, D. B. Weibel, Z. Suo, and G. M. Whitesides, *Journal of Microelectromechanical Systems* 17, 900–910 (2008).
45. M. Hashimoto, P. Garstecki, and G. M. Whitesides, *Small* 3, 1792–1802 (2007).
46. L. S. McCarty, A. Winkleman, and G. M. Whitesides, *Angewandte Chemie—International Edition* 46, 206–209 (2007).
47. M. Hashimoto, B. Mayers, P. Garstecki, and G. M. Whitesides, *Small* 2, 1292–1298 (2006).
48. G. M. Whitesides, J. K. Kriebel, and J. C. Love, *Science Progress* 88, 17–48 (2005).
49. M. Boncheva and G. M. Whitesides, *MRS Bulletin* 30, 736–742 (2005).
50. A. Winkleman, B. D. Gates, L. S. McCarty, and G. M. Whitesides, *Advanced Materials* 17, 1507–1511 (2005).
51. G. M. Whitesides, *Chimia* 59, 65 (2005).
52. M. Boncheva, D. A. Bruzewicz, and G. M. Whitesides, *Pure and Applied Chemistry* 75, 621–630 (2003).
53. M. Boncheva, D. A. Bruzewicz, and G. M. Whitesides, *Langmuir* 19, 6066–6071 (2003).
54. M. Boncheva, R. Ferrigno, D. A. Bruzewicz, and G. M. Whitesides, *Angewandte Chemie—International Edition* 42, 3368–3371 (2003).
55. D. A. Bruzewicz, M. Boncheva, A. Winkleman, J. M. St. Clair, G. S. Engel, and G.M. Whitesides, *Journal of the American Chemical Society* 128, 9314–9315 (2006).
56. M. Jaschke and H.-J. Butt, *Langmuir* 11, 1061–1064 (1995).
57. F. D. Piner, J. Zhu, F. Xu, H. S., and C. A. Mirkin, *Science* 283, 661–663 (1999).
58. L. Qin, S. Park, L. R. Huang, and C. A. Mirkin, *Science* 309, 113–115 (2005).

59. L. M. Demers, D. S. Ginger, S.-J. Park, S. Li, and S.-W. Chung, Mirkin, C. A., *Science* 296, 1836–1838 (2002).

60. H. Zhang, S.-W. Chung, and C. A. Mirkin, *Nano Letters* 3, 43–45 (2003).

61. J. Cheng and A. J. Steckl, *IEEE Journal of Selected Topics in Quantum Electronics* 8, 1323–1330 (2002).

62. K. H. Church, C. Fore, and T. Feeley, *Commercial Applications and Review for Direct Write Technologies*, Materials Research Society, Warrendale, PA, 2000, pp. 3–8.

63. S. T. Davies, D. A. Hayton, and K. Tsuchiya, *Proceedings of the SPIE—The International Society for Optical Engineering* 2880, 248–255 (1996).

64. Y. Fu and N. K. Bryan, *IEEE Transactions on Semiconductor Manufacturing* 15, 229–231 (2002).

65. J. Glanville, *Solid State Technology* 32, 270–272 (1989).

66. C. Khan Malek, F. T. Hartley, and J. Neogi, *Proceedings of the SPIE—The International Society for Optical Engineering* 4075, 167–172 (2000).

67. H. Muessig, T. Hackbarth, H. Brugger, A. Orth, J. P. Reithmaier, and A. Forchel, *Materials Science & Engineering B (Solid-State Materials for Advanced Technology)* 35, 208–213 (1995).

68. Y. Ochiai, S. Matsui, and K. Mori, *Solid State Technology* 30, 75–79 (1987).

69. R. Puers and S. Reyntjens, *Advancing Microelectronics* 27, 33–36 (2000).

70. S. Reyntjens and R. Puers, *Journal of Micromechanical Microengineering* 10, 181–188 (2000).

71. H. Sawaragi, H. Manabe, H. Kasahara, R. Aihara, K. Nakamura, F. Nihei, Y. Ochiai, S. Matsui, and T. Nozaki, *Performance of a Focused-Ion-Beam Implanter with Tilt-Writing Function*, Japanese J. Appl. Phys., Tokyo, Japan, 1989, pp. 330–333.

72. A. Stanishevsky, V. Nagarajan, J. Melngailis, R. Ramesh, L. Kriachtchev, and E. McDaniel, *Journal of Applied Physics*, 92, 3275–3278 (2002).

73. S. M. Allameh, N. Yao, and W. O. Soboyejo, *Scripta Materialia* 50, 915–919 (2004).

74. R. M. Bradley and J. M. E. Harper, *Journal of Vaccine Science and Technology A* 74, 2390–2395 (1988).

75. E. Chason, J. Erlebacher, M. J. Aziz, J. A. Floro, and M. B. Sinclair, *Nuclear Instruments and Methods in Physics Research B* 178, 55–61 (2001).

76. J. Erlebacher, M. J. Aziz, E. Chason, M. B. Sinclair, and J. A. Floro, *Physical Review Letters* 82, 2330–2333 (1999).

77. J. Erlebacher, M. J. Aziz, E. Chason, M. B. Sinclair, and J. A. Floro, *Journal of Vacuum Science and Technology A* 18, 115–120 (1999).

78. J. Erlebacher, M. J. Aziz, E. Chason, M. B. Sinclair, and J. A. Floro, *Physical Review Letters* 84, 55–61 (1999).

79. S. Rusponi, G. Costantini, C. Boragno, and U. Valusa, *Physical Review Letters* 81, 2735–2738 (1998).

80. S. M. Allameh, P. Shrotriya, A. Butterwick, S. B. Brown, and W. O. Soboyejo, *Journal of Microelectromechanical Systems* 12, 313–324 (2003).

81. S. M. Allameh, P. Shrotriya, A. Butterwick, S. Brown, N. Yao, and W. Soboyejo, *Journal of Materials Science* 38, 4145–4155 (2003).

82. S. M. Allameh and W. O. Soboyejo, *Journal of Materials and Manufacturing Processes* 19, 883–897 (2004).

83. P. Shrotriya, S. Allameh, S. Brown, Z. Suo, and W. O. Soboyejo, *Proceedings of the Society for Experimental Mechanics* 50, 289–302 (2003).

84. K.-S. Kim, J. A. Hurtado, and H. Tan, *Physical Review Letters* 83, 3872–3875 (1999).

85. W. H. Yang and D. J. Srolovitz, *Journal of Mechanics and Physics of Solids* 42, 1551–1574 (1994).
86. D. J. Srolovitz, *Acta Metallurgica* 37, 621–625 (1989).
87. S. M. Allameh, *Journal of Materials Science* 38, 4115–4123 (2003).
88. C. L. Muhlstein, E. A. Stach, and R. O. Ritchie, *Applied Physics Letters* 80, 1532–1534 (2001).
89. C. L. Muhlstein, E. A. Stach, and R. O. Ritchie, *Applied Physics Letters* 80, 1532–1534 (2002).
90. S. B. Brown, W. van Arsdell, and C. L. Muhlstein, *Materials Reliability in MEMS Devices*, vol. 1, IEEE, New York, 1997, pp. 591–893.
91. S. B. Brown, C. Muhlstein, C. Chui, and C. Abnet, *Testing of MEMS Material Properties and Stability*, IEEE, Piscataway, NJ, 1998, pp. 161–162.
92. S. M. Allameh, B. Gally, S. Brown, and W. O. Soboyejo, *On the Evolution of Surface Morphology of Polysilicon MEMS Structures during Fatigue*, Materials Research Society, Warrendale, PA, 2001, pp. EE2.3.1–2.3.6.
93. C. L. Muhlstein, E. A. Stach, and R. O. Ritchie, *Acta Materialia* 50, 3579–3595 (2002).
94. P. Shrotriya, S. Allameh, A. Butterwick, S. Brown, and W. O. Soboyejo, *Materials Research Society Symposium—Proceedings* 687, 29–34 (2002).
95. P. Shrotriya, S. M. Allameh, and W. O. Soboyejo, *Mechanics and Materials* 36, 35–44 (2004).
96. P. Shrotriya, S. M. Allameh, J. Lou, T. Buchheit, and W. O. Soboyejo, *Mechanics of Materials* 35, 233–243 (2003).
97. J. Lou, P. Shrotriya, S. M. Allameh, N. Yao, T. E. Buchheit, and W. O. Soboyejo, *2001 Fall Meeting Symposium B*, Materials Research Symposium Proceedings, MRS, 687, B2.5, 41–46 (2001).
98. S. M. Allameh, Z. Suo, and W. O. Soboyejo, *Journal of Materials and Manufacturing Processes* 22, 170–174 (2007).
99. S. M. Allameh, B. Gally, S. Brown, and W. O. Soboyejo, Eds., *Surface Topology and Fatigue in Si MEMS Structures*, American Society for Testing and Materials, West Conshohocken, PA, 2001, Vol. STP 1413.
100. W. N. Sharpe Jr. and Bagdahn, *Mechanics and Materials* 36, 3–11 (2004).
101. W. N. Sharpe Jr., K. M. Jackson, K. J. Hemker, and Z. Xie, *Journal of Microelectromechanical Systems* 10, 317–326 (2001).
102. W. N. Sharpe, Jr. and K. J. Hemker, *Mechanical Testing of Free-Standing Thin Films*, Materials Research Society, Warrendale, PA, 2002, pp. 215–226.
103. K. J. Hemker and H. Last, *Materials Science and Engineering A* 319–321, 882–886 (2001).

10

System-Level Modeling of Nano-Electromechanical Systems

Kashif Virk

CONTENTS

10.1 Introduction

The growing complexity of micro/nano-electromechanical systems (MEMS/NEMS) devices and their increased use in embedded systems (e.g., automotive electronics, mobile handheld devices, wireless integrated sensor networks, etc.) demands a disciplined approach for MEMS/NEMS design as well as the development of techniques for the system-level modeling of these devices so that a seamless integration with the existing embedded system design methodologies is possible.

System-level modeling has a central role in systems engineering. The use of models can justifiably replace experimentation on actual systems with incomparable advantages such as the following:

- Enhanced modifiability of the system and its parameters
- Ease of system composition by integrating the models of heterogeneous components
- Generality through the use of genericity, abstraction, and behavioral nondeterminism

- Improved observability and controllability of the system
- Possibility of analysis and predictability by the application of formal methods

Devising system-level models that faithfully represent complex systems is not a trivial problem and is a prerequisite to the application of formal analysis techniques. Usually, system-level modeling techniques are applied at the early phases of system development and at a higher level of abstraction. Nevertheless, the need for a unified view of the various lifecycle activities and of their interdependencies have motivated the so-called model-based approaches that rely heavily on the use of modeling methods and tools to provide support and guidance for system development and validation [7].

Integrated micro/nano-systems and their subset—MEMS/NEMS—are inherently complex in nature. The level of their complexity can be realized from the fact that these systems involve coupled energy domains (e.g., optical, electrical, magnetic, mechanical, fluidic) and their signal conditioning units, typically, involve continuous-time (analog) and discrete-time (digital) electronic domains or a mixture of both (mixed-signal). The prototyping of these systems using the available manufacturing techniques is usually very expensive. Therefore, the commonly existing "build-and-test" approach for these systems has to be replaced by a systematic design methodology that introduces design hierarchy and information sharing across the domain dichotomies. As a part of design methodology, the modeling and simulation of MEMS/NEMS-based systems play an important role in reducing the number of design iterations and their time-to-market [2].

The design methods of MEMS/NEMS have traditionally been viewed from either a bottom-up or a top-down perspective.

In the bottom-up approach, which currently is the most common design approach among the MEMS/NEMS design communities, the idea of an MEMS/NEMS device is conceived and the necessary **physical-level** modeling on the device design is conducted to establish its physical characteristics. However, the computational resource requirements associated with physical-level modeling render it an impractical approach for modeling the entire system. Therefore, the physical-level modeling techniques are only employed to analyze the physical characteristics of MEMS/NEMS device structures and to generate the data necessary to create a reduced-order model of the device. The reduced-order modeling of the MEMS device, along with the necessary signal conditioning and control electronics, is then conducted to determine its proper functioning at the **device level**. Then, **system-level** modeling is carried out to determine the potential impact the device will have on the whole system.

On the other hand, in the top-down approach, the critical system parameters are first determined from the **system-level** (reduced-order) analytical equations governing the system behavior regardless of the implementation and the fabrication options to be used. After determining the critical system parameters, the implementation details (device structure) are considered through the use of **device-level** (reduced-order) models. The specific fabrication details (process technologies) are considered later on.

Modeling at the device level involves a MEMS/NEMS structure with or without the signal conditioning and control electronics. At the device-level, reduced-order modeling allows the designers to determine what boundary and load conditions will be placed on the individual components. After device-level modeling, more detailed **physical-level** modeling (3D modeling) allows the designer to examine a structure's response to a particular physical environment in finer detail [3].

In this chapter, we focus mainly on the system-level modeling of MEMS/NEMS-based systems.

10.2 Modeling MEMS/NEMS

A large variety of MEMS/NEMS-based systems include complex interacting components that must be carefully designed for proper operation. Designing MEMS/NEMS structures necessitates an a priori understanding of their behavior. The way to gain such understanding is to formulate, analyze, and interpret proper mathematical models. One of the primary goals of system-level modeling is to formulate a model within which a broad class of designs can be developed and explored.

The modeling of MEMS/NEMS devices is about the formulation, analysis, and interpretation of proper mathematical models of MEMS/NEMS. A study of these models is intended to illuminate micro/nano-scale phenomena and to aid in the design and optimization of MEMS/NEMS devices as well as the systems in which these devices are embedded, such as wireless microsensor networks. The main challenge in the successful modeling of such systems at the device level is not determining what to include but determining what to exclude.

As mentioned above, to specify, design, and implement a complex MEMS/NEMS-based system, it is modeled on four levels of abstraction: the process level, physical level, device level, and system level. The process-level modeling is out of the scope of this chapter and will not be discussed.

10.2.1 Physical Level

As discussed above, the vast majority of problems of interest to MEMS/NEMS designers comprise multiple energy domains such as optical, thermal, magnetic, electrical, mechanical, etc.

The design of MEMS/NEMS devices, therefore, often depends on large-scale transient simulation of coupled physical domains. This requires the solution of very large systems of ordinary differential equations (ODEs), resulting from the spatial discretization of a computational domain. However, instead of a common "brute force" approach to integrate a large system of ODEs, one can use modern mathematical methods to drastically reduce the problem dimension and, thereby, achieve dramatic speedup of the calculation time. Hence, nowadays, it is possible to simulate models that only several years ago were too large (due to lack of time, computer memory, or processor speed). Indeed, it has been shown that for many MEMS/NEMS devices, such as accelerometers, gyroscopes, and many different electrothermal devices, the number of ODEs obtained from finite-element modeling can be reduced by several orders of magnitude almost without sacrificing precision.

The physical-level of modeling allows the designer to examine a particular micro/nano-structure's response in great detail and it involves numerically solving the equations of physics governing the device behavior using numerical field solvers such as the finite-element method (FEM), finite volume method (FVM), boundary-element method (BEM), etc.

The first step in physical-level modeling is to create a 3D model that represents the physical shape of the MEMS/NEMS device structure to be analyzed. This can be done by a variety of methods. Once the 3D structural model of the device is created, the material properties of the structure can be provided and the type of analysis (i.e., nodal, electromechanical, etc.) to be performed on the structure is defined. The next step is meshing, which is the process of breaking the complex 3D structure into an arrangement of simple shapes that represent the structure in a manner required by the field solver being used for the particular analysis. The BEM requires meshing only on the outer surfaces of the model while the other two methods require the whole volume of the model to be meshed. After the mesh has been created, the appropriate boundary and load conditions are applied to the model and the analysis is performed after which the results are examined to determine how well the structure performed for the particular load and boundary conditions. In order to refine the mesh and rerun the analysis, adaptive re-meshing techniques are employed where the mesh density is automatically adjusted so that critical areas have a finer mesh than noncritical areas. The computed results can either be used for iterating the design of the structure until it meets the specifications or to refine the fabrication process to minimize residual stresses during fabrication. The system designer can use the results to create a reduced-order model of the structure required to design the control circuitry for the device or determine its functional behavior in a subsystem or system.

10.2.2 Device Level

Device-level modeling involves reduced-order modeling through the generation of macro-models from the physical-level models using the macro-modeling solvers. The interface from the device level to the system level is not simply the change of data format but the change of degree of freedom for the device models and is usually called macro-modeling.

A macro-model is a low-order behavioral representation of a device. Usually, it has attractive attributes such as correct and explicit energy conservation and dissipation behavior, covering both quasistatic and dynamic behavior, expressible in a simple-to-use form such as an equation or a network analogy or a small set of coupled ODEs that are easy to connect to the system-level simulator, etc. Compared with an FEM/BEM analysis, the macro-model-based device-level simulation could consume much less time with approximate accuracy [9].

In electrical engineering, the common approach to device-level modeling is to find a "compact model" of the device in an analytical form. Whereas there is almost no problem in writing down a relationship in the form of an equation for simple passive circuit elements, such as resistors, capacitors, and inductors, the modeling of active (semiconductor) devices has been a challenge right from the start. In principle, to accurately describe the transistor operation, the transport partial differential equations (PDEs) for electrical carriers coupled with a Poisson–Boltzmann equation have to be solved. This is possible in analytic form for some special cases. However, as technology develops, the previously developed compact model cannot be applied anymore to a newly developed device, and newer models must be employed [8].

For MEMS/NEMS, analytic solutions for describing PDEs of each component are only available for simple geometries. For complex geometries, due to the large number of possible devices, working principles, and design freedoms for the designer, the conventional approach, i.e., simplifying a system to an equivalent circuit by manual or semiautomatic "compact models," is not a viable solution for the long term. Therefore, either approximations or numerical methods have to be used. However, the numerical treatment of the PDEs of thousands of interconnected individual devices with each exhibiting a complex behavior is almost impossible without a reduction of the order of unknowns to a lower-dimensional system.

Automatic model order reduction (MOR) aims at providing reduced-order models only with minimal intervention by the designer. The goal is to provide a software tool that—based on a spatial discretization of the PDE, e.g., by the FEM—is capable of returning ODEs with a far lower number of state variables than the previous discretized system without sacrificing too much accuracy. These ODEs can then be used in SPICE-like simulators, allowing for system simulations in acceptable time. The designer does not need to worry about the details of the order reduction process, and the software tool should be robust enough for use in industrial applications. Thus, MOR provides a form of "compact modeling on demand."

10.3 Compact Modeling

Traditionally, size reduction or MOR is performed via compact modeling, which was developed in electrical engineering long before MEMS/NEMS design. The goal of compact modeling is to create a small-size equivalent network of resistors, capacitors, inductors, etc., which accurately describes the dynamics of the device and can be directly inserted into SPICE-like simulators. Naturally, MEMS/NEMS designers try to use the same methodology. Mathematically speaking, compact modeling starts by choosing the topology of a small-dimensional equivalent circuit. During the second step, parameters within this network (resistivities and capacitances) are found by fitting the model parameters to measured or simulated curves. This approach requires the designer to choose the correct network topology intuitively, i.e., without strict guidelines, and then to perform a model parameterization. It should be noted that although the second step requires time-consuming data fitting, the first step, usually, takes even more time in practice as it is based on intuition [1].

10.4 Model Order Reduction

An alternative to compact modeling is mathematical MOR (also called the approximation of large-scale dynamic systems). The simulation of a single device starts with the governing PDEs. The next step is the discretization in space of the original PDE using, for example, the FEM, which integrates the PDE over a number of small nonoverlapping subsets of the complete domain. This results in a system of ODEs whose dimension is proportional to the number of introduced nodes. The finer the spatial discretization required, the more nodes are produced. Due to the complex nature of MEMS/NEMS devices, their discrete models are usually large (100,000 equations are the engineering standard nowadays). The second step is mathematical MOR, which is based on the transformation of a high-dimensional system of ODEs to a low-dimensional one, done by projection. This conversion is formal, robust, and can be fully automated. If necessary, the reduced order system model can also be represented as an equivalent electrical circuit and inserted into a system-level simulator. Hence, mathematical MOR can be considered as "compact modeling on demand." Compact modeling requires the designer to intuitively choose the topology of a small-sized equivalent network, which is not a trivial task. It further requires either simulation of the original large-scale system or the use of experimental data. These are then used for the parameterization (via data fitting) of the compact model. No mathematical properties of the original system matrices or their connection to the matrices of the reduced ODE system are taken into account. MOR, on the other hand, does not require the simulation of the original system. Instead, it reduces the original large-scale system matrices using the concept of mathematical projection. Hence, the reduced system is obtained completely formally and without relying on intuition.

MOR for linear systems and systems with nonlinear input functions is much more effective than compact modeling.

Currently, the development of MOR is strongly driven by diverse MEMS/NEMS engineering applications that would benefit significantly if the dimension reduction could be done in a completely automatic fashion. In general, one can say that the development of efficient MOR methods for automatically creating accurate low-order dynamic models is about to become a major subject of MEMS/NEMS simulation and modeling research.

10.5 System Level

Embedded computer systems are, simultaneously, growing more complex in design and shrinking in physical size, resulting in an increasing number and diversity of components being fabricated on a single chip. The system-on-chip (SoC) design paradigm exploits increases in transistor count to deliver simultaneous benefits in performance, power dissipation, reliability, footprint, and cost. Modern SoCs can incorporate not only digital but also nondigital (analog and MEMS/NEMS) components on the same silicon substrate. Extensive research has been done on analog and MEMS/NEMS component fabrication techniques, with the result that many such components can now be fabricated using processes compatible with standard digital CMOS process technologies.

In typical MEMS/NEMS application areas, such as inertial sensors, microfluidics, bio-NEMS, and communication systems—devices ranging from accelerometers, gyroscopes, nano-mirrors, DNA sensors, resonators, filters, and RF switches—the designer is often faced with the challenge of predicting the performance of the MEMS/NEMS device when it is integrated with the electronics and software. To exploit the wide range of components and to perform the hardware/software codesign and validation for accurate optimization and reduction in design cycle, system-level simulation of the micro/nano-devices along with the transistor-level integrated electronic circuits should be performed. The system-level models used must accurately represent all the components comprising the SoC. Therefore, in addition to the initial device modeling, the system-level behavior of the device with other signal processing electronics and software needs to be performed for accurately predicting its overall performance.

In practice, the requirement to model all the components on a SoC faithfully can possibly be relaxed under certain circumstances—for example, if the communication between a nondigital and a digital component is predominantly unidirectional or deterministic. During system-level modeling, such analog components as clock generators or pad drivers can be abstracted away without significant loss of accuracy because they do not, usually, impact system-level behavior in complex ways. However, this approach—abstracting away nondigital behavior entirely—becomes invalid when there is feedback in the system such as in the case of microprocessors running control programs that interact with analog or MEMS/NEMS-based sensors and actuators. The components with complex time-dependent behavior cannot be abstracted away because the behavior of the digital system can depend on both time and the state of the nondigital component.

Unfortunately, current system-level SoC design tools only allow digital components to be modeled. There is, thus, a gap between the system-level event-driven simulation methodology used by the SoC designer and the FEM, SPICE, or MATLAB®-based differential-equation-solving approach used for the design and analysis of nondigital components.

The accurate modeling of systems with feedback containing heterogeneous components requires bridging this gap. The alternative—waiting for a hardware prototype before performing software development and verification—is undesirable for reasons of cost, complexity, and time-to-market. Therefore, the SoC design flow should have such capabilities that the complete system (hardware and software) be modeled, tested, debugged, and verified well before the expensive fabrication stage, where design modification costs become prohibitive.

There have been several attempts at generalized modeling of mixed-signal elements for SoC design, which have included VHDL-AMS and Verilog-AMS—aimed at extending the VHDL and Verilog language definitions to include analog and mixed-signal regimes [6]. These have been moderately successful for mixed domain component modeling; however, they are designed for implementation and end-of-design verification late in the design flow, not for system-level design and verification. Effective system-level design involves representing the entire system at high levels of abstraction and modeling it at high simulation speeds. These requirements are not adequately met by HDL frameworks that primarily target component-level design, creating the need for higher-level techniques and tools that are more efficient at a system-level design.

A recent key advance in system design has been the development of higher-level languages and tools for expressing hardware constructs. In particular, SystemC—a freely available C++-based library that provides a variety of hardware-oriented constructs and an event-based simulation kernel—has gained rapid acceptance. It is now supported by a variety of EDA tools and IP vendors and has been widely adopted as a standard system-level modeling platform that enables the development and exchange of fast system-level models and supports system-level software development, top-down design, IP core integration, hardware/software codesign, and system verification. Designs can be expressed at a variety of levels of abstraction in SystemC. In particular, accurate and high-speed simulation at high levels of abstraction is a key feature, enabling designers to model the behavior of large workloads on complex systems.

The SystemC 2.0 standard addresses purely digital simulation. However, increasing on-chip heterogeneity has led to the demand for modeling both digital and nondigital components within an integrated framework. Ongoing research efforts, such as SystemC-AMS, propose extensions to the SystemC language definition and additions to the SystemC kernel to incorporate analog and mixed-signal devices into the simulation framework. In contrast, the techniques and models that use a standard, unmodified SystemC kernel and library to model nondigital components and represent some of the earliest applications of SystemC to a MEMS SoC design have also been demonstrated [4].

As described above, the designers of SoCs with nondigital components, such as analog or MEMS/NEMS devices, can currently use system-level design languages, such as SystemC, to model only the digital parts of a system. This is a significant limitation, making it difficult to perform key system design tasks—design-space exploration, hardware/software code-sign, and system verification—at an early stage. However, lumped analytical models of complex nondigital MEMS/NEMS devices can now be integrated into a SystemC simulation of a heterogeneous SoC. This approach makes the MEMS component behavior visible to a full-system simulation at higher levels, enabling realistic system design and testing.

The problem of modeling coupled energy domains is addressed by setting up a lumped-parameter model that correctly models the coupling between the various energy domains. Another modeling problem is posed by the fact that the behavior of analog and MEMS/NEMS components is best represented by differential equations, not by the discrete-time event-based state machines used for digital simulation in SystemC. This is solved by expressing the MEMS/NEMS device behavior in discrete time, so that numerical methods can be applied, and then integrating this efficiently into SystemC's *network-of-communicating-processes* model of computation. In addition, the values of the various simulation parameters must be known to enable accurate system-level modeling.

A SystemC simulation consists of a hierarchical network of parallel processes that exchange messages and concurrently update signal and variable values under the control of a simulation kernel. Signal assignment statements do not affect the target signals immediately, and the new values become effective only in the next simulation cycle. The kernel resumes when all the processes become suspended, either by executing a wait statement or upon reaching the last process statement. On resuming, the kernel updates the signals and variables and suspends itself again while scheduled processes resume execution. If the time of the next scheduled event is the current simulation time, the processes execute a delta cycle in which signal and variable values are updated without incrementing the current time.

The ability to model the complete system in detail enables designers to easily and quickly find answers to questions about overall system behavior. Such questions can range from how a MEMS/NEMS device responds to a given input to finding out whether a given piece of code running on a microcontroller can meet desired performance parameters while controlling one or more MEMS/NEMS devices.

The integrated simulation of both digital and MEMS components can be an extremely useful tool in the hardware/software codesign of SoCs as:

- Full-system simulation results can be used to make critical design decisions.

- Integrated simulations can be used to assess system robustness while facing process variations in device parameters.

- Running the software stack under realistic conditions enables more thorough testing, leading to better defect detection before the system is fabricated.

- Interrupt routines, timer settings, operating frequency, I/O, and control algorithm parameters can be better optimized when realistic simulation results are available. In the absence of these, designers need to allow larger margins of error to account for uncertainty in the final performance of the system.

Complex system-level interactions, such as those illustrated above, need to be taken into account by system, software, and component designers and the integrated modeling of both microcontroller and MEMS/NEMS device behavior in SystemC has enabled precisely that.

10.6 Conclusion

System-level modeling of MEMS/NEMS-based systems is an ongoing area of research with the aim to model, as accurately as possible, the micro/nano-system device behavior at the system level as well as at the lower levels of abstraction. The capabilities of the existing design tools for designing micro/nano-systems are still limited to some extent, either in the diversity of components in the parametric components library, accuracy of simulation, or in terms of simulation speed and ease of use. A lack of standards and interoperability are additional limitations. Information about macro-modeling or lumped-parameter modeling of micro/nano-systems is ample but patchy. Moreover, the automated synthesis of micro/nano-systems from system-level models seems to be a far-off dream [5].

References

1. J. A. Pelesko and D. H. Bernstein, *Modeling MEMS & NEMS*, Chapman & Hall/CRC, Boca Raton, FL, 2003, 357pp, ISBN 1-58488-306-5.
2. K. Virk, J. Madsen, M. Shafique, and A. Menon, System-level modeling & simulation of MEMS-based sensors, in *IEEE INMIC*, New York, 2005.
3. J. van Kuijk, G. Schropfer, and M. da Silva, Design automation for MEMS/MST, *Design, Automation, and Test in Europe (DATE)*, Munich, Germany, 2005.
4. A. Varma, M. Yaqub Afridi, A. Akturk, P. Kein, A. R. Hefner, and B. Jacob Modeling heterogeneous SoCs with SystemC: A digital/MEMS case study, *Proceedings of the International Conference on Compilers, Architecture and Synthesis for Embedded Systems* 2006, Seoul, Korea, pp. 54–64.
5. Z. Fan, K. Seo, J. Hu, R. C. Rosenberg, and E. D. Goodman, System-level synthesis of MEMS via genetic programming and bond graphs. Genetic & Evolutionary Computation - GECCO 2003, Lecture Notes in Computer Science (LNCS) 2724, pp. 2058–2071, Springer, Germany.
6. L. Khine and M. Palaniapan, Behavioural modelling & system-level simulation of micromechanical beam resonators, *International MEMS Conference, Journal of Physics: Conference Series*, 34, 1053–1058, 2006.
7. K. Virk and J. Madsen, Computing platforms—Modeling & simulation, in *Embedded Systems Design—The ARTIST Roadmap for Research & Development*, B. Bouyssounouse and J. Sifakis (Eds.), LNCS 3436, Springer, New York, 2005.
8. J. Lienemann, D. Billger, E. B. Rudnyi, A. Greiner, and J. G. Korvink, MEMS compact modeling meets model order reduction—Examples of the application of Arnoldi methods to microsystem devices, in *Technical Proceedings of the 2004 NSTI Nanotechnology Conference and Trade Show, NSTI-Nanotech 2004*, Vol. 2, Danville, CA, 2004.
9. H. Chang, J. Xu, J. Xie, C. Zhang, Z. Yan, and W. Yuan, One MEMS design tool with maximal six design flows, in *Proceedings of DTIP 2007*, Stresa, Italy, 2007.
10. www.systemc.org

11

MEMS-Based Wireless Communications

Salahuddin Qazi

CONTENTS

11.1 Introduction

Wireless communications continues to grow with new applications in the consumer, medical, and military arena by employing radio frequency (RF) and microwave circuits and systems. The future wireless communication system necessitates the use of highly integrated RF front ends containing low power, small in size and weight, high-performance, and low-cost components. The current RF integrated devices use GaAs FET and positive intrinsic negative (PIN) diode switches in the front end of wireless communication systems such as cell phones for switching antenna bands and transmitter/receivers. In multiband cell phones, the GaAs FET switches, however, suffer from a lack of isolation to minimize cross interference and signal jamming from other channels. PIN diodes consume considerable power to operate, which decreases battery life. Similarly, the off-chip passive components such as high-Q inductors, ceramic and surface acoustic wave (SAW) filters are bulky and limited in frequency operation. According to Clark T.-C. Nguyen, microelectromechanical systems (MEMS) can be used at subsystem levels in cellular handsets, PDAs, low-power networked sensors, and ultrasensitive radar and jam-resistant communicators designed for hostile environments. The properties of MEMS, such as low power consumption, small size and volume, added functionality, and reconfigurability, are making wireless communications a ubiquitous connectivity. Industry experts predict that several RF MEMS devices in each of the approximately 700 million cellular phones sold each year and a new wave of RF MEMS will offer the lowest cost solution for 3G phones [1,2].

MEMS technology, a term evolved in the United States in the 1990s, has led to the development of complex microsystems for applications including biomedicine, environmental sensing, infrastructure monitoring, wireless communications, and others. MEMS can be defined as mechanical structures fabricated with integrated circuit processing mainly on silicon wafers [3]. Currently the micro- and nanotechnology-based systems have become the standard with progress in the integration and introduction of nanotechnology into systems. Microtechnology employs a top-down approach to investigate the limits of fine processing and nanotechnology employs a bottom-up approach to develop new functions by designing materials from the atomic and molecular level. These two different approaches have come together on a MEMS substrate. As a result, MEMS provides the medium for implementing nanotechnology into practical applications.

A mobile phone consists of $\sim$100 components of which 75% are passive elements such as inductors, variable capacitors, filters, and others. MEMS version of these components promises to make phones smaller in size, power efficient, and more reliable with added functionality [4]. A schematic of a wireless transmitter and receiver is shown in Figure 11.1, which shows the shadowed components that can be realized by MEMS equivalent. It is

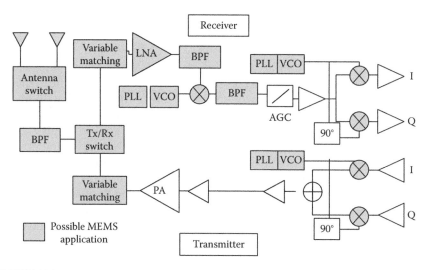

FIGURE 11.1
Simplified conventional transceiver architecture. (Reprinted from De Los Santos, H.J. and Richards, R.J., *Microw. J.*, 44(7), 2001. With permission.)

estimated that MEMS will become a standard component for the new generation of cell phones with more than 40 RF MEMS devices integrated into the system. According to industry analysts at Cahners In-Stat Group, the market for RF MEMS is expected to grow to nearly $350 million in 2006. The prediction by the analysts at Mega Tech Resources for the quantitative model forecasts of worldwide shipment of MEMS is to reach $100 billion in 2010 and the long-term potential is estimated at $2 trillion by 2025. It is further reported that MEMS devices have been shipping mobile handsets since 2002, which is a large single opportunity for component suppliers after personal computers [5].

Despite the many advantages of MEMS, the reliability and packaging issue of RF MEMS is complex and presents many challenges because of the combination of circuits and micromachinery, which includes both the electronics and mechanical parts and their interactions. Many solutions have been presented to overcome these challenges and improve on the existing technology before it can be commercialized and make a major impact in the field of wireless communications.

This chapter is divided into six sections. Section 11.1 deals with an introduction and importance of MEMS in wireless communications. Section 11.2 deals with RF MEMS and discusses various components including MEMS-based inductors, variable capacitors, tuning of capacitors, and RF switches. Section 11.3 deals with the problems of switches and reliability issue of RF MEMS. Section 11.4 deals with the packaging technology of RF MEMS and discusses wafer-level packaging. Section 11.5 discusses the methods of fabrication of RF MEMS. Section 11.6 concludes the chapter and provides the references.

11.2 RF MEMS

In wireless communications, research efforts are currently aimed at developing a single-chip RF circuit to increase functionality and meeting the demand of low power consumption, lower cost, and lower weight. This will allow MEMS devices to integrate directly on RF chips and replace numerous discrete components on the RF chips. The ultimate goal in applying RF MEMS is to propagate the device-level benefits all the way up to the system level to achieve high system performance, as shown in Figure 11.2 [3].

This will lead to the replacement of all passive RF chips with on-chip devices and will offer considerable benefits such as smaller form factors for cell phones with added functionalities including Internet connectivity.

RF MEMS technology enables superior passive devices, which can be used as monolithic alternatives in the following areas [6].

- Reactive components, such as inductors and variable capacitors for integrated voltage-controlled oscillators
- Switches to build impedance networks in front of power amplifiers
- Resonators and filters

RF components based on MEMS have superior performance and tunability at much broader range of operating frequencies. RF MEMS switches, for example, provide improved insertion loss, isolation, and linearity at the same time.

11.2.1 MEMS-Based Inductors

Inductors are important components in the front-line circuitry of wireless communications for filters, voltage-controlled oscillators, impedance matching networks, and amplifiers. These applications require inductance, typically in the range of 1–5 nH with a high Q factor, and a self-resonance frequency (SRF) of more than 10 GHz. For high performance on board inductors, the requirements are high Q, small size, and low power consumption. In standard, low-cost CMOS or bipolar technologies, the Q factor of spiral inductors is limited to about 10 and SRFs of 5–20 GHz. Conventional planar on-chip inductors are made with a single layer metallization in which a conducting layer is etched on

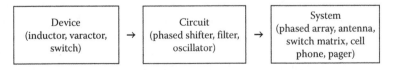

FIGURE 11.2
A typical RF MEMS chain. (Reprinted from De Los Santos, H.J. and Richards, R.J., *Microw. J.*, 44(7), 2001. With permission.)

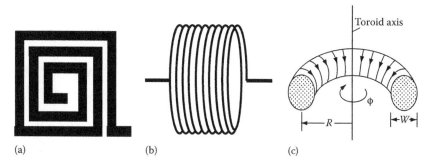

FIGURE 11.3
Types of inductors (a) Spiral, (b) solenoid, and (c) toroid. (Reprinted from Huang, X. and Cao, Y., 3D-solenoid MEMS RF inductor design in standard CMOS technology, Department of EECS, University of California, Berkeley, CA. With permission.)

a dielectric substrate. The finite conductivity of the metal and the loss in dielectric substrate introduces parasitic capacitance and loss due to capacitance and inductive coupling into the substrate, which results in low Q factor and low self-resonant frequency. MEMS technology is one of the ways to improving inductor characteristics by exploiting the power of surface and bulk micro-machining fabrication techniques. This is achieved by suspending the inductor structure either within the bulk, or elevating it above the wafer surface, which helps to overcome the major sources of performance degradation [7–9].

The geometrical designs of planar inductors are spiral, solenoid, and tor-oidal-meander-type inductors. The conventional geometries for integrated inductors have been mostly meander types or spiral types and solenoid types for macroscale inductors. This is mainly because the fabrication of a coil wrapped around a core, which has been more difficult using conventional IC processes than the fabrication of meander or spiral types. Figure 11.3 is a simplified schematic of three types of inductors (a) spiral-type inductor (b) solenoid and (c) meander-type inductor [10]. A fundamental problem of inductor design is to obtain the greatest possible inductance with a piece of round wire. Although meander-type inductor is simple to fabricate, it suffers from low overall inductance because of negative turn-to-turn mutual induct-ance. A spiral-type inductor has a relatively high inductance, but has several disadvantages including larger size, presence of stray capacitance between the conductor and lead wire, and the perpendicular direction of flux to substrate, which can interfere with the underlying circuit in multiple modules [11,12].

11.2.1.1 Planar Spiral Inductor

Planar spiral inductor, shown in Figure 11.4, has been widely used in on-chip and PCB applications because of their fabrication compatibility with stand-ard CMOS technology. The spiral inductor is usually composed of the top metal spiral and the center tap underpass using a lower metal layer.

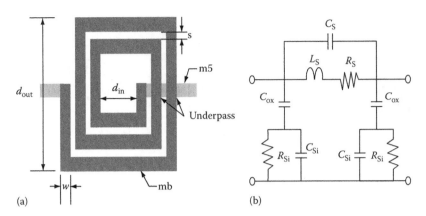

FIGURE 11.4
(a) Schematic diagram of planar spiral inductor and (b) equivalent circuit.

The layout parameters are the number of turns n, the linewidth (w), the line spacing (s), the outer dimension (d_{out}) and the inner dimension (d_{in}). These parameters plus process parameters determine the electrical behavior of the inductor. In low frequencies, when the substrate effects and the metal eddy current can be neglected, the inductance (L) of a square spiral inductor can be regarded as the function of the layout.

The Q factor of this inductor can be expressed with the spiral conductor resistance of R as

$$Q = \frac{WL}{R}$$

When the thickness of the conductor lines is increased, the conductor resistance will be reduced. Increase in Q factor resulting from increasing the conductor thickness will be caused mainly by the decrease in the conductor resistance. Thus, the Q factor increases almost linearly as the thickness of the conductor is increased.

In MEMS-based inductors, research efforts have been applied to separate the current-carrying coil from the lossy substrate to reduce parasitic. Some of the structures used to achieve this include fabrication of inductors on thick airbridged inductors and various techniques based on removal of the substrate underneath the coil [13,14].

A spiral-type inductor, shown in Figure 11.5, was introduced by Ahn and Allen [15] to improve the usefulness of spiral geometry in micropower applications. This is achieved by making use of micromachined fabrication of electroplated thick nickel-iron permalloy to complete the closed magnetic circuits to reduce magnetic reluctance and to minimize magnetic field interference. It also helps to minimize the resistance to reduce power consumption in the conductors. A schematic diagram is shown in Figure 11.5 for

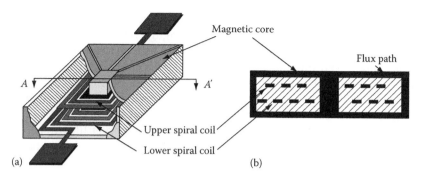

FIGURE 11.5
Spiral-type inductive structure. (a) Schematic diagram that has a closed magnetic circuit and (b) simulation model. (Reprinted from Ahn, C.H. and Allen, M.G., *IEEE Trans. Ind. Electron.*, 45(6), 866, 1998. With permission.)

modeling and determining its inductance value. The total inductance L is the summation of the inductance for path 1, path 2, and the internal self-inductance of the coil.

Ramachandran and Pham [16] in 2002, developed micromachined spiral inductor on lossy substrates by using thick SU-8 dielectric layer to virtually eliminate substrate parasites, which also act as supporting mechanism for realizing an inductor on a solid ground plane. It is a single turn, one-port inductor that uses organic micromachining process to elevate the inductor above a substrate, as shown in Figure 11.6. The inductor is supported by a 200 μm high SU-8 substrate, which has been partially removed to allow for the creation of air gaps between the traces of the inductor to enable air to be the main substrate. The experimental results were obtained from a micromachined integrator on a Si substrate to achieve Q factor of 19.3 at 2.1 GHz.

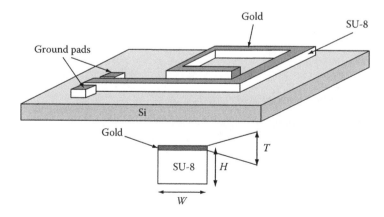

FIGURE 11.6
Schematic diagram of the micromachined inductor. (Reprinted from Ramachandran, R. and Pham, A.-V.H., *IEEE Trans. Adv. Packag.*, 25(2), 244, 2002. With permission.)

For high Q factor and self-resonance frequency, a number of suspended spiral inductors are used, relying on improving the performance by reducing the influence of substrate losses. The fabrication technique is based on micro-molding and electroplating of thick/high aspect ratio microstructures and removal of sacrificial layers similar to the one used in MEMS micromachining. It is a low-cost, low-temperature manufacturing technique, which is compatible with the fabrication of a complete system on package module. Pinel et al. [17] implemented different designs using a typical air gap of 100 μm between the conductor line and the substrate for application in C-band, X-band, and KU-band. The result of physical modeling and characterization showed very high performance of Q over 100 and self-resonance frequency of 50 GHz.

Yoon et al. [18] also proposed a new process allowing for the fabrication of arbitrary-shaped highly suspended metal microstructure, fully compatible to CMOS manufacturing for process stability and structural robustness. The design involved fabrication of inductor structures, which are suspended by two signal posts of 20 μm in diameter without the support of any additional mechanical supporting posts, as shown in Figure 11.7. This is done in order to minimize substrate coupling without sacrificing structural robustness. The suspension of inductor over the substrate has been 50 μm, which was passivated by 1 μm thick thermal oxide for electrical isolation. In order to reduce metal ohmic loss and ensure structural robustness, inductors are made of copper of 10 μm thickness. A special three-dimensional (3D)

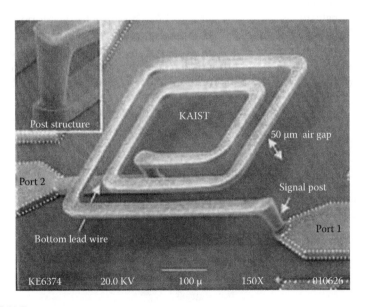

FIGURE 11.7
SEM microphotograph of the highly suspended MEMS spiral inductor fabricated on standard silicon. The white dotted line illustrates the open-pad pattern utilized in de-embedding procedure. (Reprinted from Yoon, J.-B. et al., *Electron. Dev. Lett.*, 23(10), 591, 2002. With permission.)

photoresist made of single positive photoresist layer (AZ9260) is used to fabricate the processing of suspended inductor.

11.2.1.2 Solenoid-Type Inductor

A conventional inductor usually has a solenoid shape, and can be easily fabricated by wrapping coils manually around the magnetic cores. However, fabrication of a coil wrapped around a core has been more difficult using conventional IC processes or realizing such a 3D winding structure in planar shape on a wafer. Solenoid structures generally have more than twice the inductance compared to the toroidal structure from stronger mutual coupling. But when induced electromagnetic interference is a problem, toroidal inductors may be preferred to solenoids because of very small external magnetic field [19]. For on-chip applications, the design choice lies in the process compatibility, resonant frequency, and Q values. Solenoid structures have a significant advantage because they have lower ground capacitance, and are much more area efficient. Recent advances in micromachining technology have provided alternative processes for fabricating integrated solenoid-type inductors.

Ahn and Allen [15] proposed an integrated solenoid inductor shown in Figure 11.8, which was designed and fabricated using polymer/metal multi-layer processing and surface micromachining techniques for high-frequency applications. This inductor uses an air core in which an air gap between the substrate and the conductor is created by use of a surface micromachining technique and an organic sacrificial layer. This inductor can be fabricated using low-temperature fabrication processes, allowing the possibility of fabrication on either ceramic or organic (laminate) substrates.

Thus, the calculation of the inductance of a solenoid-type structure is very simple and the inductance L of the solenoid inductor structure is expressed as

$$L = \frac{\mu_0 \mu_r N_2 A_c}{l_c}$$

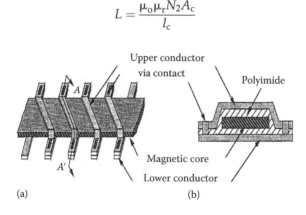

(a)   (b)

FIGURE 11.8
A schematic of the solenoid-type inductive components: (a) schematic view and (b) A-A cut view. (Reprinted from Ahn, C.H. and Allen, M.G., *IEEE Trans. Ind. Electron.*, 45(6), 866, 1998. With permission.)

where

A_c is the cross-sectional area of the magnetic core

l_c is the length of the closed magnetic core

μ_o and μ_r are the permeability of the vacuum and the relative permeability of the magnetic core, respectively

Kim and Allen [20] designed a new technique to overcome these limitations using planarization with polyimides and inserting a magnetic bar core in place of the air core for high-frequency applications. To fabricate this inductor on a planar substrate, quasi-3D micromachining techniques that are compatible with IC processes are required. This planar solenoid-type inductor shown in Figure 11.9 with a closed magnetic circuit can be one of the favorable structures as a magnetic flux generator in microscale. As the wires wrapped around closed magnetic cores result in low leakage flux, the generated flux can be flexibly guided to the required points. In this technique, such structures have been fabricated using multilevel metal schemes to "wrap" a wire.

Chomnawang et al. [21] designed a solenoid inductor (shown in Figure 11.10) that has flat bottom conductors and arch-shape top conductors with only two contact points per turn instead of four contact points per turn, as in conventional micromachined solenoid inductors. This inductor uses a low-resistivity material such as copper as a conductor, which helps decrease the series resistance. As low-resistivity substrate limits the Q factor of an inductor, a high-resistivity material as a substrate, removing the substrate bulk beneath the inductor, or suspending the inductor above the substrate with surface micromachining was used to reduce this problem. In the design of this inductor surface, micromachining technique is utilized to raise the inductor's body above the substrate with metallic spacers at both ends to reduce substrate losses. Materials inside inductor core can also degrade inductor performance.

Huang and Cao [10] proposed a 3D solenoid inductor structure, shown in Figure 11.10, which utilizes the existing multiple-level interconnect in the

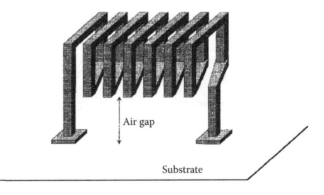

FIGURE 11.9

A schematic of an integrated solenoid-type inductor with an air gap. (Reprinted from Kim, Y.-J. and Allen, M.G., *IEEE Trans. Compon. Packag. Manuf. Technol. C*, 21(1), 26, 1998. With permission.)

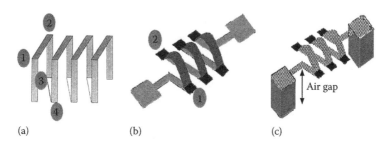

FIGURE 11.10
3D on-chip air-core solenoid inductors: (a) conventional 3D inductor design, (b) nonsuspended arch-shape solenoid inductor, and (c) suspended arch-shape solenoid inductor. (Reprinted from Chomnawang, N. et al., *J. Microlith. Microfab. Microsyst.*, 2, 275, 2003. With permission.)

standard CMOS technology. The design maintains the process compatibility of planar inductor. To realize area efficiency and minimize the substrate capacitive coupling, the solenoid structure is laid out on top of the normal Si resulting in a smaller footprint and less ground capacitance. There is an optional post-micromachining step in order to remove the substrate to eliminate the substrate loss under high frequency, which will further boost the Q factor and resonance frequency. The net result is that such a fabrication gives rise to higher Q and self-resonant frequency and less interference between inductors if placed orthogonal because of its 3D geometry (Figure 11.11).

11.2.1.3 Toroidal-Meander-Type Inductor

An inductor with toroidal geometry shown in Figure 11.3c has most of the magnetic field well confined within because of loop structure. This makes it

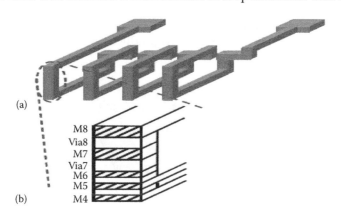

FIGURE 11.11
Solenoid inductor in standard CMOS process: (a) schematic of the inductor and (b) details of the cross-section layer (not to scale). (Reprinted from Huang, X. and Cao, Y., 3D-solenoid MEMS RF inductor design in standard CMOS technology, Department of EECS, University of California, Berkeley, CA. With permission.)

optimal in electromagnetic characteristics with higher Q compared to planar coils and lower interferences with surrounding circuits. However making 3D toroidal structures with conventional integrated circuit processes have been difficult to build on a chip. A toroidal-meander inductor consists of toroidal-shaped multilevel magnetic core wrapped around a meander planar conductor. A meander-type inductor is simple to fabricate but because of negative turn-to-turn mutual inductance, it suffers from low overall inductance.

The metal interconnection used to construct the wrapping coils usually includes metal via contact, which has relatively high contact resistance. This causes heat dissipation in the via contact, which is a practical problem. For high inductance, more turns of solenoid coils are required, which increases the total coil resistance because more vias are needed. In order to solve this problem, Ahn et al. [22], in 1994, used electroplating to fabricate the conductor lines and the vias to reduce the resistance of metal contacts. This method of fabricating toroidal inductor minimizes the coil resistance by increasing the thickness of conductor lines.

Nickel is the most commonly used material to produce 3D microstructures. Nickel-iron permalloy is one of the soft magnetic materials, which has been used in a variety of magnetic film applications because of its favorable magnetic and magnetic properties.

Ni-Cu alloys and nanocomposites are investigated as possible replacements for nickel in microsystems. Ni-Cu alloys are attractive for their corrosion resistance, magnetic, and thermophysical properties. Alumina nanoparticles included into metal matrices improve hardness and tribology of deposits.

In 1998, Kim and Allen [20] proposed an integrated toroidal-meander inductor in which a multilevel magnetic core is wrapped around a planar meander conductor, as shown in Figure 11.12. It uses a relatively short planar conductor and has the advantage of reducing the total conductor resistance. As the magnetic core in this structure is produced on two levels, it makes it

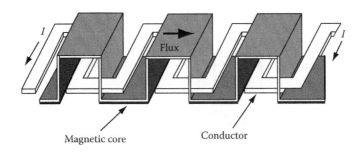

Magnetic core   Conductor

FIGURE 11.12

Schematic diagram of the toroidal-meander-type inductor. (Reprinted from Huang, X. and Cao, Y., 3D-solenoid MEMS RF inductor design in standard CMOS technology, Department of EECS, University of California, Berkeley, CA. With permission.)

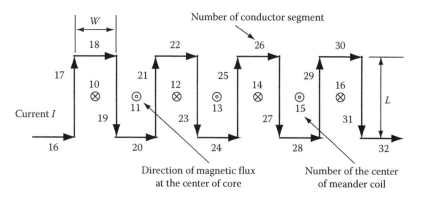

FIGURE 11.13
Meander conductor model for coordinate and meander elements for Biot–Savart law calculations. (Reprinted from Huang, X. and Cao, Y., 3D-solenoid MEMS RF inductor design in standard CMOS technology, Department of EECS, University of California, Berkeley, CA. With permission.)

readily available for surface micromachining of movable core actuators for realizing in the applications of microactuators. The magnetic flux density at the center of each meander coil can be calculated by evaluating magnetic fields at center points, which are generated from the current flowing through all meander conductor elements, as shown in Figure 11.13.

The inductance of meander-toroidal conductor can be calculated from the total flux linkage (both self and mutual flux linkage) as

$$L = \sum \frac{A}{I}$$

where
  A denotes the total flux linkage that occurs between the closed multilevel meander magnetic circuit and the flux density generated from the current flowing through all meander conductor elements
  I is the current flowing through the coil

Ermolov et al. [23] reported the fabrication of truly 3D high-quality conductor using polymer replication process. It uses electroforming to form a negative copy of master structure, which is produced in silicon or glass of a desired pattern. The master is then used to make mold inserts, which are used to produce polymer replicas using different replication methods such as casting, hot embossing, and injection molding. This method resulted in inductance of 6.0 nH and a peak Q of 50 and is not compatible with conventional integrated circuit process.

Kim et al. [24] fabricated toroidal inductors on low-resistivity (10–20 Ω·cm) and high-resistivity (10 kΩ·cm) silicon substrates using recently introduced stressed metal technology. This technology, as shown in Figure 11.14, is fully

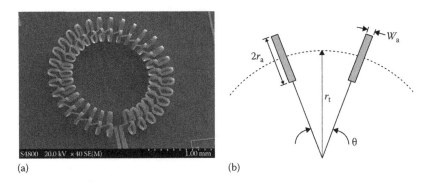

(a)   (b)

FIGURE 11.14
Self-assembled inductor: (a) SEM microphotograph (31 turns) and (b) design parameters. (Reprinted from Kim, D. J.-I. et al., Design of toroidal inductors using stressed metal technology, 2005 IEEE MTT-S International, *Microwave Symposium Digest*, Long Beach, CA, p.4, 2005. A report for DARPA Technology for Efficient and Agile Microsystems (TTEAM), Arlington, VA, DAAB07-02-1-L430. With permission.)

compatible with Si and compound-semiconductor fabrication technologies. It can be either implemented as a post-processing step or as a part of a vertical chip to interposer packaging scheme. The appropriate bending moment due to an induced stress gradient is obtained by the use of Au and Cr layers, which are both compatible with integrated circuit fabrication.

11.2.1.4 Tunable Inductors

Inductors are important part of RF front-end architecture and their tunability can give the advantage of optimizing the performance of RF front-end circuits by adjusting the center frequency of a band-pass filter, changing the impedance offered by a matching network, and tuning the oscillation frequency of a voltage control oscillators (VCO). However, there are only a few tunable inductor configurations, which have been developed in comparison to tunable capacitors. Most of the approaches that are developed employ MEMS switches to activate different static inductive sections. In one approach, bimorph switches are used to vary their overall inducting depending on the on/off state of the switches by dividing the inductor into four segments. This technique gives a wide range of inductance between the maximum and minimum, but suffers from lack of continuous tunability [25].

Balachandran et al. [26] designed a variable inductor using a series-type DC-contact MEMS switch to operate from 5 to 30 GHz. The DC-contact switches are fabricated on a 500 μm thick quartz, which uses cantilever beams suspended on 1.5 μm thick posts located on the center conductor, as shown in the schematic of tunable inductor of Figure 11.15. A small length of high-impedance line is used to achieve high inductance and a low inductance is realized by reconfiguring the same circuit by DC-contact switches to yield

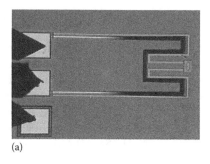 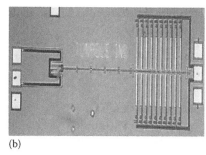

(a) (b)

FIGURE 11.15
(a) Inner and outer inductor and (b) two inductors with beam and the actuator. (Reprinted from Alexander, D. and Suyama, K., *IEEE Trans. Microw. Theor. Tech.*, 46(12), 2587, 1998. With permission.)

a low-impedance line. In the nonactuated state of the beams, the signal is carried only on the thin center conductor of the CPW line and a high value of characteristic impedance is obtained. As the length of this section is electrically small, the topology effectively emulates an inductor with high inductance value. In the actuated state of the beams or when the beams make contact, the effective width of the center conductor increases and the characteristic impedance with respect to the high-impedance state is less, which represents a low-inductance state. The inductance ratio is directly related to the change in the impedance state.

Zine-El-Abidine et al. [27] designed a new method of achieving tunability by fabricating and connecting two inductors in parallel and using thermal actuators that control spacing between the two inductors, the inner and the outer, as shown in Figure 11.15a. Because of the residual stress between the metal and the polysilicon layer, the inner inductor moves away from the substrate. The outer inductor is attached to a beam that is connected to an array of actuators for thermal actuation. On actuation of array, the beam bends and lifts up the outer inductor. This changes the angle separating the two conductors and allows tuning the mutual component of the total inductance. The maximum tuning range of 13% was obtained in the first prototype design using multi user MEMS (MUMS) process, which allows for only one thin metal layer deposited on polysilicon.

11.2.2 MEMS Variable Capacitor

Variable capacitors are important control elements and have the potential to replace conventional varactor diodes in many applications such as phase shifters, oscillators, and tunable filters. MEMS technology has enabled new possibilities for creating and controlling capacitance with wide tuning range and high Q due to micromechanical tuning and the use of low-loss manufacturing material. There are many types of MEMS-tunable capacitors, which have been proposed essentially focusing on new architecture solutions to increase their tuning range. This includes two or three parallel plate capacitors with vertical electrostatic actuation (air gap variation) and others using the effective area or dielectric constant of capacitors to increase the tuning range [28,29].

A variable parallel plate capacitor consists of two electrodes with the lower electrode fabricated on the surface of the IC and a thin aluminum membrane suspended over the electrode [7]. The metal membrane is connected directly to ground on either side of the electrode while a thin dielectric layer covers the lower electrode, as shown in Figure 11.18a. MEMS capacitor off-capacitance is determined by the air gap between the two conductors. The residual tensile stress of the membrane keeps it suspended above the RF path in the absence of applied actuation potential. When actuation potential in the form of a DC electrostatic field is applied to the lower electrode, it causes the formation of positive and negative charges on the electrode and membrane conductor surfaces. The strong attractive force produced as a result of these charges causes the suspended metal membrane to snap down onto the lower electrode and dielectric surface. This forms a low-impedance RF path to ground, as shown in Figure 11.16b, and is the actuated state of the MEMS capacitor. The characteristics of the dielectric layer primarily determine the MEMS' on-capacitance in this state. This is an example of a variable MEMS capacitor where the capacitance of the MEMS device can be changed over a significant capacitance range by changing the position of the membrane with an applied DC voltage (electrostatic actuation).

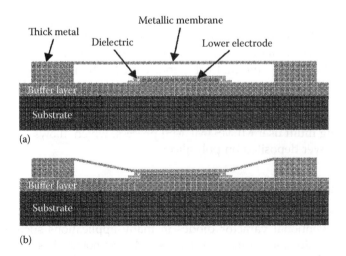

FIGURE 11.16
Cross section of an RF MEMS capacitor: (a) unactuated and (b) actuated positions. (Reprinted from Goldsmith, C.L. et al., *RF MEMS Variable Capacitors for Tunable Filters*, John Wiley & Sons, New York, 1999. With permission.)

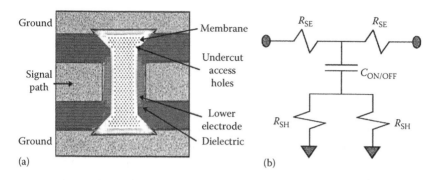

FIGURE 11.17
(a) Top view of a shunt MEMS capacitor and (b) RF model of a MEMS capacitor. (Reprinted from Goldsmith, C.L. et al., *RF MEMS Variable Capacitors for Tunable Filters*, John Wiley & Sons, New York, 1999. With permission.)

The top view of shunt RF MEMS capacitors is given in Figure 11.17a and the RF model is given in Figure 11.17b. The fabrication of this capacitor is achieved by utilizing surface micromachining. The capacitor is built on high-resistivity silicon substrate with a thick layer of silicon dioxide, which acts as a buffer layer. The capacitor circuitry is fabricated on top of the silicon dioxide using 4.0 μm thick aluminum interconnects. This fabrication is CMOS compatible and presents low conductor losses at high frequencies. The construction of bottom electrode uses a film of refractory material, which provides good conductivity for low loss and has smooth finish. A film of silicon nitride on top of the lower electrode blocks the DC control signal from shorting out during capacitor actuation but allows RF signals to capacitively couple from lower electrode to upper membrane. The membrane made of thin aluminum has high conductivity for low RF resistance and good mechanical properties. Two grounded posts are used for signal path and the suspended membrane spans between them. The upper membrane is completely patterned with a series of 2 μm holes to accelerate the removal of the sacrificial polymer from beneath the membrane. The freed membrane released as a result of removing mechanical material moves up and down onto the lower electrode in response to applied electrostatic forces. The size of the membrane is approximately 120 μm in width and spans 280 μm from ground to ground. The size of signal path line is 120 μm [28].

11.2.2.1 Tuning of MEMS Variable Capacitor

MEMS tuning avoids the high-resistive capacitive losses associated with semiconductor varactor diodes at high frequencies. MEMS tuning is obtained by the movement linearity of MEMS devices, which tune the capacitors linearly. MEMS tune the capacitance by adjusting the physical dimension or parameters electromechanically using electrostatic or thermal actuation.

By ignoring the fringing field, the capacitance of two electrodes of area (A) separated by a distance (d) is given by

$$C = \varepsilon \frac{A}{d}$$

where ε is dielectric constant of the medium.

The capacitor can be tuned by varying one of three parameters such as distance between the electrode, area, or dielectric of the material. It can be classified as gap tuning, area tuning, or dielectric tuning. Gap and area tuning is commonly used by employing electrostatic or thermal actuation.

The behavior of a MEMS varicap can be modeled by a mass–spring–damper. A voltage over the varicap results in an electrostatic force between the top and the bottom plates. This force pulls the top plate down to the position where it is in equilibrium with the elastic force in the top plate. When the voltage is higher than a critical voltage, called the pull-in voltage, there is no equilibrium anymore and the top plate collapses on the bottom plate. This is unwanted for a varicap, but it is the normal operation of a capacitive switch. Mechanical inertia makes the MEMS varicap behave as a low-pass filter on the tuning signal. This is an important difference when compared to a MOS varicap, which does not show this low-pass behavior at all.

11.2.2.2 Electrostatic Actuation

Electrostatic actuation uses direct drive to control the gap area of the dielectric layer by varying the distance between two parallel plates with one variable and the other fixed plate. Parallel plate configuration of the capacitor exhibits high Q and is easy to fabricate. The range for this variable capacitance, however, is limited by the maximum theoretical tuning range of C_{max}/C_{min} of 50% due to the collapse of the membrane as the voltage is increased beyond the pull-in voltage [29,30].

Bakri-Kassem and Mansour [31] introduced parallel plate nitride-loaded MEMS variable capacitor with a wide tuning range of more than five times improvement over the tuning range of fixed capacitor. It consists of two movable plates with an insulation dielectric on the top of the bottom plate, as shown in Figure 11.18a. The design of two flexible plates makes the two plates possible to attract each other and decreases the maximum distance before the pull-in voltage occurs. The micromachining of the capacitor used two structural layers, three sacrificial layers, and two insulating layers of nitride. The top plate, as shown in Figure 11.18a, is fabricated from nickel with a thickness of 26 μm covered by a gold layer of 2 μm of thickness. The bottom plate is made of polysilicon covered by a nitride layer of a 0.35 μm thickness. An SEM picture of the fabricated capacitor is shown in Figure 11.18b.

Electrothermal actuation uses indirect drive to control the gap variation. The capacitor plates used are designed from the substrate, which is obtained

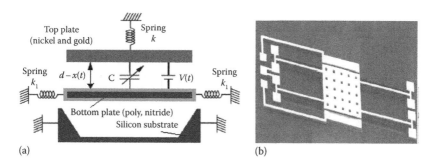

FIGURE 11.18
(a) Schematic diagram of the capacitor and (b) SEM picture of the fabricated capacitor. (Reprinted from Bakri-Kassem, M. and Mansour, R.R., Wide tuning range MEMS variable capacitor, CWMEMS 2003 Abstract. With permission.)

by embedding a polysilicon heater inside a beam that is one of the highest thermal expansion materials. Motion is induced because of the different temperature coefficient of the metal and oxide layers inside the beams. Electrothermal actuation has the advantage of linear capacitance tuning and lower driving voltage compared to electrostatic actuation. Electrothermal actuation, however, suffers from the disadvantage of consuming more power in continuous operation. Polysilicon produces high insertion loss, and stray capacitance between the poly-layer of the capacitor plate and the silicon [32,33].

11.2.2.3 Comb Drive Actuators

The parallel plate-tunable capacitors based on gap-closing electrostatic offers relatively short response times, low power consumption with very little heat generation and can be easily integrated with electronic components. Despite these advantages, it suffers from the fundamental limitation of pull-in phenomena, which limits its tuning range to 50% (1.5). This inherent nonlinearity makes them better suited for large force, small displacement applications.

Tuning ratio and linearity can be improved significantly by using lateral comb drivers in which capacitance tuning is based on lateral motion of the movable fingers and does not suffer from pull-in limitation. Lateral comb drive actuators exhibit better stability characteristics over longer range of displacement and can be easily designed for linear operation over large operating ranges. Such an actuator is suited for large force and large displacement applications. The comb drive consist of one top membrane, a number of comb drives and has an interdigitated shape, as shown in Figure 11.19a and b [34–36]. The capacitance generated by the comb drive is given by

$$C = \frac{2n\varepsilon(L - x)l}{g}$$

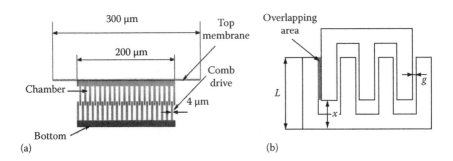

FIGURE 11.19
(a) Cross-sectional schematic of comb drive capacitor unit and (b) comb drive array. (Reprinted from Seo, D.-B. and Shandas, R., Design and simulation of a MEMS based comb-drive pressure sensor for pediatric post-operative monitoring applications, in *Summer Bioengineering Conference*, Sonesta Beach Resort, FL, June 25–29, 2003. With permission.)

where
    ε is the dielectric constant
    L is the finger height
    n is the number of fingers
    g is gap width (gap between top and bottom fingers)
    l is length
    x is the displacement

Actuation force on the top membrane creates a displacement, which changes the capacitance based on the change in the overlapping area.

Oz and Fedder [37] designed a large tuning range CMOS-compatible tunable capacitor based on interdigitated beams using electrothermal actuation. In this method of actuation, polysilicon resistors act as heaters inside the inner frame and the interdigitated beams curl down vertically and sideways upon heating the structure. This curling changes the area between interdigitated beams, which is used for tuning the variable capacitor. Curling takes place because of different temperature coefficient of expansion (TCE) of the metal and oxide layer inside the beams. Gap tuning is also obtained by this method and it is essentially a tunable capacitor, which uses both gap and area tuning. The capacitors are fabricated using Austria Microsystem (AMS) 0.6 μm CMOS, Agilent 0.5 μm CMOS, and TSMC 0.35 μm CMOS. A SEM picture of a released tunable capacitor in AMS 0.6 μm CMOS process is shown in Figure 11.20.

11.2.3 RF MEMS Switch

MEMS switches have been developed to provide more efficient switching features including low insertion loss and higher isolation, nearly zero power consumption, small size and weight, and very low intermodulation

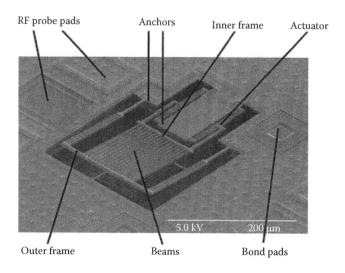

RF probe pads   Anchors   Inner frame   Actuator

Outer frame   Beams   Bond pads

FIGURE 11.20

SEM picture of a released tunable capacitor in AMS 0.6 μm CMOS process. (Reprinted from Oz, A. and Fedder, G.K., RF CMOS capacitor having large tuning range, MEMS Laboratory, Carnegie Mellon University, Pittsburgh, PA, 2003. With permission.)

distortion. The RF MEMS switches, however, have some disadvantages like low switching speed, higher actuation voltages, higher cost, and the challenges of packaging and reliability issue. Work has been undertaken to improve on the existing MEMS in new fields of applications such as reconfigurable antennas, high-Q passives and resonators, filters and tuners, low-pass planar THz waveguide components and low-pass shifters.

An RF MEMS switch operates by making use of physical motion to achieve a short circuit or an open circuit in the transmission or power line. The physical movement can be produced by actuation of electrostatic, piezoelectric, magnetostatic, or thermal forces. There are two types of switches: series switch and shunt switch. These switches can be further categorized by the way they are coupled or by contact methods: Capacitive (metal–insulator–metal) and resistive (metal-to-metal).

11.2.3.1 Series Switch

A series switch is in series with the transmission or power line and either closes or opens the line to turn it on or off. The contact surface in the switch is located at the end of a singly supported cantilever beam with a control electrode located under the beam. The control beam is pulled down when voltage to the control beam is applied and completes the connection between two conductors.

Series switches are classified into in-line and broad-line depending on the plane of actuation. In broad-side switches, the plane of actuation is perpendicular to the transmission line, as shown in Figure 11.21a and b. The actuation plane in in-line switches is colinear with the transmission line, as shown in Figure 11.21c [38].

In an in-line shunt switch, the beam is clamped at both ends and the control plane pulls down the beam when the potential is applied, which ensures that the signal finds a shorter path to the ground and does not pass on to the following network.

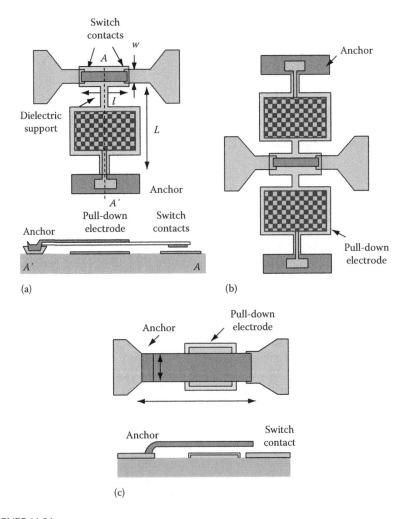

FIGURE 11.21
Broadside series MEMS switch and in-line series MEMS switch. (Reprinted from Rebeiz, G.M. and Muldavin, J.B., *IEEE Microw. Mag.*, 2(4), 59, 2001. With permission.)

11.2.3.2 Shunt Capacitive Switch

A shunt switch is connected between ground lines where the switch turns on to short the power on the signal line to the ground and as a result prevents the power from going past the switch. A shunt switch is usually based on fixed–fixed beam design in which the anchors are connected to the coplanar waveguide (CPW) ground plane and the membrane is grounded. The electrostatic actuation as well as the RF capacitance between the transmission line and the ground is provided by the center electrode. When the switch is in the up-state, it provides the low capacitance to the ground, and does not affect signal on the transmission line. In the down-state of the switch, the capacitance to the ground becomes higher, and this results in an excellent short circuit and high isolation at microwave frequencies [38].

11.2.3.3 Electrostatic Actuation of the MEMS Switch

The actuation mechanism is achieved using an electrostatic force between the top and bottom electrodes, and is given by [39,40]

$$F = \frac{QE}{2} = \frac{CVE}{2} = \frac{\varepsilon A V^2}{2(g + t_{\mathrm{d}}/\varepsilon_{\mathrm{r}})^2}$$

where
   V, g, and C are the voltage, distance, and capacitance between the lower and upper electrodes, respectively
   A is the area of electrode in Figure 11.21
   The bottom electrode is often covered by a dielectric layer to prevent a short circuit between the top and bottom plates

DC actuation voltage or the pull-in voltage of the fixed–fixed beams or air bridges is given by [38–40]

$$V_{\mathrm{P}} = \sqrt{\frac{8K_z g_0^3}{27\varepsilon_0 A}}$$

where
   K_z is the equivalent spring constant of the moving structure in the direction of desired motion (typically the z-direction)
   g_0 is the gap between the switch and the actuation electrode
   ε_0 is the free-space permittivity
   A is the switch area where the electrostatic force is applied

The above equation shows that there are several ways that may decrease the required actuation voltage. The first approach is to reduce the gap between the switch and the actuation electrode (g_0); this reduction, however, affects the high-frequency off-state switch performance by compromising the switch isolation or insertion loss. A second approach in lowering actuation voltage is by increasing the actuation area (A). This approach is not feasible because the area has to remain small to be able to create miniaturized circuits. The third approach, which offers the maximum design flexibility for a low-to-moderate actuation voltage, is to lower the switch spring constant [38].

11.3 Problems and Solutions

Despite many advantages of RF MEMS switches over PIN diode or FET switches, operating at RF to millimeter wave frequencies, it also has some drawbacks including high actuation voltage, low speed, low-power handling, packaging problem and low long-term lifetime. For a reliable operation of RF MEMS switches, a voltage of 20–80 V is required to produce a large electrostatic force, which means that a voltage up-converter must be used in the application of telecommunication systems [38]. This approach takes a lot of space and does not help in reducing the size of the wafer. Furthermore, it is difficult to integrate such MEMS switches with CMOS technology.

11.3.1 Low Actuation Design

For low actuation voltage design, the choice of the membrane material and support design is critical. Although increasing the area of membrane should reduce the actuation voltage, it also increases the size of device. As a result, the stiffness of the support and size of the air gap must be suitably chosen to reduce the actuation voltage of the switch and keep the device size small. Many designs and techniques have been proposed to reduce the actuation voltage.

Balaraman et al. [40] in 2002 proposed a design and fabrication of capacitive copper RF MEMS switch with various hinge geometries fabricated on high-resistivity silicon substrate. Lee et al. [41] presented a novel RF MEMS switch, which uses an actuation voltage below 5 V and has high long-term lifetime. The working of this switch is based on the freely moving structure, where mechanical movement of the movable lower contact pad is achieved at very low voltage due to the lack of elastic deformation involved in the actuation. Finally, research has shown that actuation voltages in the order of 1.5 V can be achieved with capacitive-based shunt switches. Specifically, the meander cantilever structure and meander bridge design structures show potential to reduce the actuation voltage enough for integration into SoCs [42].

11.3.2 Problem of Stiction and Solutions

Stiction or static friction is the adhesion of contacting surfaces due to surface forces failure associated with both capacitive (metal–insulator–metal) and resistive (metal-to-metal) RF MEMS switches. In electrostatically actuated capacitive switches, stiction is caused by dielectric charging effects such as control-voltage drift. In a capacitor RF switch, a dielectric is typically made quite thin (less than 0.3 µm) of silicon dioxide or nitride formed by plasma-enhanced chemical vapor deposition (PECVD). As the capacitive membrane switches generally require 30–50 V of actuation voltage, the electric field across the dielectric during switch operation can be of the order of 10^6 V/cm causing electrons or holes to be injected into the dielectric and become trapped. During the repeated operation, charge gradually builds up in the dielectric resulting in control voltage drift resulting into stiction [43]. Number of techniques have been proposed to overcome stiction including selecting contact materials with less adhesion, applying surface treatment or eliminating contamination [44,45].

Mercado et al. [46] in 2003 proposed a mechanical approach to provide enough restoring force to overcome the adhesion force generated at the interface. By using an iterative solution methodology and coupled electrostatic–structural analysis, the authors studied beam geometry changes in three ways. The first method is to double the beam area by doubling the beam width. The second method used was to shorten the cantilever beam arm to one-third of its original length and the third is to increase the dielectric layer thickness by 60%. The result from the simulations showed that the most favorable modification is to double the area by making the structure short and wide. This structure doubled the restoring force and lowered the actuation voltage to 25 V. In addition, sandwich design of the cantilever can reduce the actuation voltage compared to bilayer designs.

Fukushige et al. [47] proposed a new method to prevent the moving electrode from sticking to the substrate electrode. It is achieved by applying only a voltage of certain waveform and not by changing the structure or dimensions of the materials, as in some other solutions.

11.3.3 Reliability Issues of MEMS Switches

As MEMS devices are a combination of circuits and micromachinery, the reliability aspect of MEMS includes both the electronics and mechanical parts, complicated by the interactions. MEMS reliability is particularly a challenge [48].

- MEMS technologies are new; therefore, many new failure mechanisms are poorly understood.
- MEMS technology is evolving rapidly with the introduction of new processes, materials, and structural geometries.

- Many applications of MEMS are in safety critical systems where the cost of failure is catastrophic and the requirements for qualification/ certification are stringent and expensive to satisfy.

- In both safety critical and noncritical applications alike, design trade-offs must account for reliability in order to control product warranty costs.

The reliability issue of RF MEMS switches is a major concern in wireless communications especially when it is introduced in cellular nanosatellites (space communications) or in biomedical applications. In case of DC-contact switches, the failure mechanisms are resistive, and for capacitive switches, the mechanism is due to dielectric discharge leakage leading to stiction.

The reliability of resistive or ohmic switches is related to the metal contact used and is limited by damage, pitting, hardening, or welding of the contact area due to the impact of force between the beam and bottom contact area. This can lead to failure in the form of permanent open or short after number of switching cycles. The resistive failure occurs suddenly when the contact resistance increases from 1–2 to 4–10 Ω and above. The University of Michigan and Rockwell Science Center improved the reliability of DC contact switch by varying the drive voltage and reducing the impact energy on the contact area [49].

At lower RF power, stiction in these switches is not a problem, as for most design the pull-down electrode does not need to touch the micromachined beam. For higher RF power of 100–1000 mW, microwelding can be a serious problem because the contact dissipates 2% of the incident power for a contact resistance of 1 Ω, which amounts to 2 mW for an RF power of 100 mW. This results in microwelding of the contact area and a failure of the switch in the closed position [38].

11.4 Packaging of RF MEMS

The packaging of MEMS devices is one of the most difficult parts of product development process because, unlike ICs, it contains movable fragile parts that must be packaged in a clean and stable environment. The objectives of MEMS packaging are twofold. The first is to provide support and protection to the delicate core elements like the dyes of the sensors, the associate wire bonds and transduction units from mechanical or environmentally induced damages. The second is to provide protection to these elements requiring interface with working media, which may be environmentally hostile to these elements [50].

RF MEMS switches are very sensitive to environment and must be packaged in hermetic or near hermetic seals to protect from moisture, dirt, as well as from mechanical and radiation loads. The technique of packaging MEMS switches will determine the cost, as it is the most expensive step in the production line. The well-proven hermetic packages with ceramic/glass feedthroughs in clean room can be used for RF MEMS switches but are complicated and expensive. MEMS packaging using epoxy seals, glass-to-glass anodic bonding and gold-to-gold bonding techniques give problems, which affect the reliability of RF MEMS switches. The first problem arises from outgassing of organic materials inside the MEMS cavity during the bonding process due to wetting compounds in the glass, gold, or epoxy layers. The second problem arises due to high temperature of 300°C–400°C needed to make a good seal in the bonding process. This high temperature is likely to bend the cantilever or membrane making the switch unusable [38].

11.4.1 Wafer-Level Packaging

MEMS packaging costs account for 70%–90% of the device and 90% of the size. One way to reduce the cost of packaging and size for applications requiring large number of units is to use wafer-level packaging. Wafer-level packaging is increasing in importance because of low cost due to the batch character of the process and small size. In this packaging technique, structures are protected early in the process and testing can be done on the wafer. Wafer-level packaging also gives rise to smaller packages, better reliability, and better electrical properties due to the shorter paths used [51–53]. In wafer-level packaging, a cavity is fabricated during wafer processing prior to die singulation, as shown in Figure 11.22. This is achieved by creating an on-wafer enclosure around the MEMS device, which serves as a first protective interface. The widely used bonding techniques with silicon micromachining for packaging of RF MEMS are limited. An ideal bonding technique should yield a hermetic seal that has a dielectric constant equal to the substrate, which can be processed at low temperatures, and can tolerate a large degree of nonplanarity/roughness [54].

One major drawback of most wafer-level packaging techniques is the requirement for a seal ring, which, with the inclusion of appropriate bonding

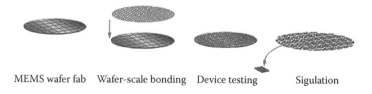

MEMS wafer fab   Wafer-scale bonding   Device testing   Sigulation

FIGURE 11.22

Sequence of steps in wafer-level packaging. Wafer-level packaging enables the whole MEMS wafer to be packaged at once, while in the controlled environment of the clean room. (Reprinted from RF MEMS packaging, MEMtronics Inc., Publications, Plano, TX, http://www.memtronics. compage.aspx?page_id = 36. With permission.)

pads, increases the area of RF MEMS. There are four areas that need to be considered for this: (1) RF MEMS circuit, (2) seal ring, (3) interconnect area, and (4) saw kerf. The regions required for the seal ring, interconnect area, and the saw kerf increase the final size of the RF circuit, thereby reducing the available number of circuits per wafer.

MEMtronics [54] used an innovative approach to packaging called wafer-level microencapsulation, which is designed to be compatible with RF MEMS switch fabrication. This wafer-level packaging does not use any seal ring, which makes it smaller in size and less costly. It is achieved by constructing individual micropackages on top of each RF switch using the same MEMS process as used to construct the switch. It does not use bonding of a separate glass wafer to the RF MEMS wafer and microencapsulation process provides a protective, low-loss package with RF friendly interconnects. This innovative approach of microencapsulation process decreases the cost of packaging to only 28% of total packaged switch cost compared to conventional packaging techniques of 75%–90% of total cost. These packaging processes require only moderate temperature of 200°C–250°C.

11.5 Fabrication of RF MEMS

A significant part of fabrication technology for MEMS is adopted from 2D fabrication techniques used in semiconductor industry. The third dimension in the realization of MEMS is created by layering where micromachining techniques are used to produce 3D mechanical structures. There are essentially three main approaches of micromachining, namely, the surface, bulk, and LIGA.

11.5.1 Surface Micromachining

The surface micromachining technology makes thin micromechanical devices on the surface of a silicon wafer on which large numbers of devices can be made inexpensively. The moving parts are made by alternating layers of thin films of a structural material (typically silicon) and a sacrificial material (typically silicon dioxide). The structural material will form the mechanical elements, and the sacrificial material creates the gaps and spaces between the mechanical elements. At the end of the process, the sacrificial material is removed, and the structural elements are left free to move and function. This technology integrates well with electronics. Different structures of surface micromachining include monolithic solenoid inductors, suspended spiral inductors, and vertical spiral inductors. The highest Q factors on silicon substrate were achieved by making use of spiral inductors [7,18,20].

11.5.2 Bulk Micromachining

Bulk micromachining builds micromechanical devices by etching deeply into the silicon, removing unwanted parts, and being left with useful mechanical devices. Etching is termed as anisotropic when different crystallographic directions are etched at different rates using etchants like KOH (potassium hydroxide). Isotropic etching etches all directions in the silicon wafer with nearly the same rate, and produces rounded depressions on the surface of the wafer that usually resemble hemispheres and cylinders. Anisotropic etching usually produces Vee grooves, pyramids, and channels into the surface of the silicon wafer. Deep reactive ion etching, RIE or DRIE, uses a plasma to etch straight-walled structures on the wafer This is a relatively simple and inexpensive fabrication technology, and is well suited for applications that do not require much complexity, and that are price sensitive. Bulk micromachined inductors are realized by removing the substrate under the inductor spiral via top-side etching [7].

11.5.3 LIGA

LIGA is a German acronym that stands for LI (Roentgen–lithography meaning x-ray lithography), G (Galvanik, meaning electrodeposition), and A (Abformung, meaning molding). In this technique, a thick photoresist is used in the form of a sacrificial layer, which is selectively etched through a mask by synchrotron x-ray radiation to produce a mold that is filled by electroplating. The remaining photoresist is then etched away leaving the electroplated parts attached to the substrate. This technique allows manufacturing of structures with sidewalls of high aspect ratio of over 100:1. Although not well suited to mass production, it is possible to electroplate a metal master that can then be used for injection molding mass production. Ultraviolet-based LIGA can be used to directly expose deep structures in resist such as SU8 employed in the fabrication of RF MEMS.

11.6 Conclusion

- Boom in wireless communication has intensified the quest to develop low-cost, ultra-low-loss RF MEMS including switches, switchable capacitors, varactors, and inductors.
- RF MEMS promise to replace the traditional FET switches for reduced loss and improved linearity in key components.
- RF MEMS in addition will help realizing frequency-agile RF/wireless systems capable of serving multiple frequency bands.

- Emphasis of research is shifting toward system integration, reliability, and packaging. This the most critical issue as it determines the cost, size, and reliability, as reviewed in this chapter.
- Many key issues remain for realizing full potential. Some of these issues include impedance, drift stability, and transistor integration. Work is being done to resolve these issues.

References

1. J. Bryzek, Principles of MEMS, in *Handbook of Measuring System Design*, John Wiley & Sons, New York, 2005.
2. C.T.-C. Nguyen, Vibrating RF MEMS for next generation wireless application, in *Proceedings, 2004 IEEE Custom Integrated Circuits Conference*, Orlando, FL, pp. 257–264, October 3–6, 2004.
3. H.J. De Los Santos and R.J. Richards, MEMS for RF/microwave wireless applications: The next wave—Part II, *Microwave Journal*, 44(7), July 2001. http://www-ece.rice.edu/~jdw/432/cache/microwave_journal_partII.pdf
4. M. Chapman, The impact of MEMS on cellular phone architecture, WiSpry Inc., Special Report, *Microwave Journal*, 49(5), 256, May 2006.
5. MEMS Markets & Technologies, MEMS-making their mark in mobile handsets, In-Stat 2006 Research report, http://www.instat.com/catalog/scatalogue.asp?id = 47
6. S. Raman, RF MEMS for wireless communications applications, Session D3, June 2006.
7. H.J. De Los Santos, *Introduction to Microelectromechanical (MEM) Microwave Systems*, Artech House, Norwood, MA, 1999.
8. Y. Sun et al., Suspended membrane inductors and capacitors for application in silicon MMICs, in *IEEE Microwave and Millimeter-Wave Monolithic Circuits Symposium Digest of Papers*, San Francisco, CA, pp. 99–102, 1996.
9. H. Jiang et al., Fabrication of high-performance on-chip suspended spiral inductors by micromachining and electroless copper plating, in *2000 IEEE IMS Digest of Papers*, Boston, MA, pp. 279–282, 2000.
10. X. Huang and Y. Cao, 3D-solenoid MEMS RF inductor design in standard CMOS technology, Department of EECS, University of California, Berkeley, CA.
11. A.C. Reyes et al., Coplanar waveguides and microwave inductors on silicon substrates, *IEEE Transactions on Microwave Theory and Techniques*, 43, 2016–2022, September 1995.
12. M. Yamaguchi et al., Analysis of the inductance and the stray capacitance of the dry-etched micro inductors, *IEEE Transactions on Magnetics*, 27, 5274–5276, November 1991.
13. L.E. Larson, Integrated circuit technology options for RFIC's—Present status and future directions, *IEEE Journal of Solid-State Circuits*, 33(3), 169–176, March 1998.
14. G. Ternent et al., Coplanar waveguide transmission lines and high Q inductors on CMOS grade silicon using photoresist and polyimide, *Electronics Letters*, 35(22), 1957–1958, October 1999.

15. C.H. Ahn and M.G. Allen, Micromachined planar inductors on silicon wafers for MEMS applications, *IEEE Transactions on Industrial Electronics*, 45(6), 866–876, December 1998.

16. R. Ramachandran and A.-V.H. Pham, Development of RF/microwave on-chip inductors using an organic micromachining process, *IEEE Transactions on Advanced Packaging*, 25(2), 244–247, May 2002.

17. S. Pinel et al., Very high-Q inductor using RF-MEMS technology for system-on-package wireless communication integrated module, *IEEE MTT-S Digest*, 3, 1497–1500, 2003.

18. J.-B. Yoon et al., CMOS-compatible surface-micromachined suspended spiral inductors for multi-GHz silicon RF ICs, *Electronic Device Letters*, 23(10), 591–593, October 2002.

19. P.N. Murgatryod, The optimal form for coreless inductors, *IEEE Transactions on Magnetics*, 25(3), 2670–2677, May 1989.

20. Y.-J. Kim and M.G. Allen, Surface micromachined solenoid inductors for high frequency application, *IEEE Transactions on Components, Packaging, and Manufacturing Technology, Part C*, 21(1), 26–33, January 1998.

21. N. Chomnawang, J.-B. Lee, and W.A. Davis, Surface micromachined arch-shape on-chip 3-D solenoid inductors for high-frequency application, *Journal of Microlithography, Microfabrication, and Microsystems*, 2(4), 275–281, October 2003.

22. C.H. Ahn, Y.J. Kim, and M.G. Allen, A fully integrated planar toroidal inductor with a micromachined nickel-iron magnetic bar, *IEEE Transactions on Components, and Manufacturing Technology—Part A*, 17(3), 463–469, September 1994.

23. V. Ermolov et al., Microreplicated RF toroidal inductor, *IEEE Transactions on Microwave Theory and Techniques*, 52(1), 29–37, January 2004.

24. D. J.-I. Kim et al., Design of toroidal inductors using stressed metal technology, A report for DARPA Technology for Efficient and Agile Microsystems (TTEAM), Arlington, VA, DAAB07-02-1-L430.

25. X.Q. Sun, S. Zhou, and W.N. Carr, A micro variable inductor chip using MEMS relay, in *IEEE International Conference on Solid-State Sensors and Actuators*, Chicago, IL, p. 1137, 1140, 2001.

26. S. Balachandran et al., MEMS tunable inductors using DC-contact switches, 34th European Microwave Conference, pp. 713–716, Amsterdam, 2004.

27. I. Zine-El-Abidine, M. Okoniewski, and J.G. McRory, A new class of tunable MEMS inductors, in *Proceedings of the International Conference on MEMS and Smart Systems* (ICMENS'03), Banff, Alberta, Canada, pp. 114–115, 2003.

28. C.L. Goldsmith et al., *RF MEMS Variable Capacitors for Tunable Filters*, John Wiley & Sons, New York, 1999.

29. D. Alexander and K. Suyama, Micromachined capacitors and their application to RF IC's, in *IEEE Transactions on Microwave Theory and Techniques*, 46(12), 2587–2596, 1998.

30. D. Alexander and K. Suyama, 2.4 GHz CMOS LC VCO using micromachined variable capacitors for frequency tuning, in *Microwave Symposium Digest, IEEE MTT-S International*, Vol. 1, Anaheim, CA, pp. 79–82, 1999.

31. M. Bakri-Kassem and R.R. Mansour, Wide tuning range MEMS variable capacitor, CWMEMS 2003 Abstract.

32. Z. Feng et al., Design and modeling of RF MEMS tunable capacitors using electro-thermal actuators, in *Microwave Symposium Digest, 1999 IEEE MTT-S International*, Vol. 4, Anaheim, CA, pp. 1507–1510, 1999.

33. K. Harsh et al., Flip-chip assembly for Si-based RF MEMS, in *MEMS'99 Twelfth IEEE International Conference on Micro Electro Mechanical Systems*, Orlando, FL, pp. 273–278, 1999.

34. D.-B. Seo and R. Shandas, Design and simulation of a MEMS based comb-drive pressure sensor for pediatric post-operative monitoring applications, in *Summer Bioengineering Conference*, Sonesta Beach Resort, FL, June 25–29, 2003.

35. R. Legtenberg, A.W. Groeneveld, and M. Elwenspoek, Comb-drive actuators for large displacement, MESA Research Institute, University of Twente, Enschede, Germany, 1996.

36. W. Ye, S. Mukherjee, and N.C. MacDonald, Optimal shape design of an electrostatic comb-drive in micro-electro-mechanical systems, *Journal of Micro-Electro-Mechanical Systems*, 7(1), 16–26, 1998.

37. A. Oz and G.K. Fedder, RF CMOS capacitor having large tuning range, MEMS Laboratory, Carnegie Mellon University, Pittsburgh, PA.

38. G.M. Rebeiz and J.B. Muldavin, RF MEMS switches and switch circuits, *IEEE Microwave Magazine*, 2(4), 59–71, December 2001.

39. D. Peroulis et al., Electromechanical considerations in developing low-voltage RF MEMS switches, *IEEE Transactions on MTT*, 51(1), 259–270, January 2003.

40. D. Balaraman et al., Low cost actuation voltage copper RF MEMS switches, in *2002 IEEE MTT-S International Microwave Symposium Digest*, IF-WE-20, Seattle, WA, pp. 1225–1228, 2002.

41. S.-D. Lee et al., An RF-MEMS switch with low-actuation voltage and high reliability, *Journal of Micromechanical Systems*, 15(6), 1605–1611, December 2006.

42. Multiband and multimode RF-MEMS, building blocks for wireless world WiSpry Publications, December 2005, http:www.perfectdisplay.com/feature_articles/rfmems/rf_mems.htm

43. C. Goldsmith et al., Lifetime characterization of capacitive RF MEMS switches, in *IEEE international Microwave Symposium*, Phoenix, AZ, pp. 227–230, 2001.

44. J. Schimkat, Contact materials for microrelays, in *Proceedings of the 11th Annual International Workshop on Micromechanical Systems*, Heidelberg, Germany, pp. 190–194, 1998.

45. R. Maboudian, Anti-stiction coatings for surface micromachines, *Proceedings of SPIE*, 3511, 108–113, 1998.

46. L. Mercado et al., A mechanical approach to overcome RF MEMS switch stiction problem, in *IEEE Electronic Components and Technology Conference*, New Orleans, LA, 2003.

47. T. Fukushige, S. Hata, and A. Shimokohbe, A new method for electrostatic MEMS actuators to prevent sticking, in *Proceedings of the 4th Euspen International Conference*, Glasgow, Scotland, May–June 2004.

48. P. Sandborn, Position paper: MEMS packaging and reliability, CALCE Electronic Products and System Center, University of Maryland, College Park, MD.

49. Private Communications, University of Michigan and Rockwell Science Center, June–July 2001.

50. T.-R. Hsu, Reliability in MEMS packaging, in *44th International Reliability Physics Symposium*, San Jose, CA, pp. 398–402, March 2006.

51. K. Najafi, Micropackaging technologies for integrated microsystems: Applications to MEMS and MOEMS, *Proceedings of the SPIE*, 4982, 9–27, 2003.

52. T. Harder et al., Wafer-level encapsulation of MEMS using solder sealing, in *Micro System Technologies*, Franzis Verlag GmbH, Poing, Germany, 2003.
53. M. Madou, *Fundamentals of Microfabrication*, CRC Press, New York, 1997.
54. RF MEMS packaging, MEMtronics Inc., Publications, Plano, TX, http://www. memtronics.compage.aspx?page_id = 36

12

Quantum-Dot Cellular Automata:
The Prospective Technology for Digital
Telecommunication Systems

Shahram Mohammad Nejad and Ehsan Rahimi

CONTENTS

12.1 Introduction

This chapter is organized into several sections. Section 12.2 accounts for the very general comparison of the conventional CMOS technology to that of the quantum-dot cellular automata (QCA). More about quantum-dots, quantum mechanical concepts, and quantum cells, which are the basic elements for the implementation of QCA circuits and systems will be discussed in

Section 12.3. It is also important to know how digital systems are designed in this technology. Useful information about this process will be given in Section 12.4.

Section 12.5 provides the reader with basic knowledge about modeling and simulation of QCA circuits. One of the interesting applications of QCA is in developing high-density memories. In today's world of digital communication, high-capacity and high-performance digital data storage provide an ultimate solution to preserve sheer volume of data. Nanotechnology emerges to bring new digital media with high capacity and high performance, where digital data are stored at nanoscale. In Section 12.6, a novel ROM as a digital data storage for QCA implementation has been presented. The QCA technology enables us to store 250 Gb/cm^2 of digital data. Also, architectures in digital data storage have been compared in terms of latency, size, and complexity in this section.

It is also thought that QCA is more secure than CMOS technology from the viewpoint of implementation of secure devices and systems, such as cryptographic processors in secure telecommunication systems and networks [1]. In Section 12.7, how the QCA technology is useful for implementation of secure systems is described.

Since the introduction of side-channel attacks, cryptographic devices have been highly susceptible to power and electromagnetic analysis attacks. This is because these attacks require relatively inexpensive equipments. Unless adequate countermeasures are implemented, side-channel attacks allow an unauthorized person to get access to the private key of a cryptographic module. As a new countermeasure, a novel logic approach to implement cryptographic processors, known as secured clocked quantum-dot cellular automata (SCQCA) logic, has been presented in Section 12.7. The proposed logic takes advantage of low power consumption QCA together with complicated clocking schemes as a paradigm of nanotechnology advances in cryptography engineering. Section 12.8 finally concludes this chapter.

12.2 CMOS versus QCA

The development in microelectronics has been mostly due to the industry's ability to continuously scale down the transistor. CMOS technology of conventional basic transistor follows Moore's law. According to this law, the number of transistors doubles every 18 months. Obviously the Moore's law cannot continue for microelectronic device manufacturing forever. The leakage currents through the gate oxide resulting from the quantum mechanical tunneling of electrons from the gate electrode through the oxide and into the transistor channel are one clear reason. As the size of transistors scale down, more and more of these quantum mechanical effects, such as, nondeterministic behaviors in low currents and the technological restrictions including, power consumption and design complexity, appear and affect the normal operation of transistors [2,3]. Therefore, a new paradigm beyond current switches is needed for future digital telecommunication systems. QCA takes advantage of quantum mechanical effects such as electron tunneling in nanoscale. It was introduced by Lent et al. [4] and has recently become one of the promising candidate paradigms for nanocomputing. A significant amount of theoretical and practical research has been done in this field [5–8]. Although QCA is still in research stages, some experimental nanoelectronic devices have been manufactured with expected functionality [9–11].

12.3 Quantum-Dot Cellular Automata

12.3.1 What Is a Quantum-Dot?

Quantum-dots are semiconductors or conductors in nanosize. These dots consist of a few to several hundreds of atoms, which are set beside each other with various arrangements. In fact, they are islands that are confined by another material. According to quantum confinement theory, the energy difference between two materials implies that free electrons of these islands have just certain energies. A set of these allowable energies specify absorption and emission spectrum of quantum-dots. The important point is that the wavelength of absorption and emission spectrum generally depends on the size of quantum-dots. The bigger the size of a quantum-dot, the wider the energy gap will be, and the wavelength of the spectrum will be smaller as a consequence. For this reason, quantum-dots are called artificial atoms, too. Just like atoms, where electrons can be present in certain orbits with the particular energies, in quantum-dots, free electrons have certain energies but they have greater chance of presence in particular areas in dots.

12.3.2 Quantum Mechanical Concepts

The basic idea of the quantum-dot comes from quantum mechanical concepts. A quantum-dot can be visualized as a three-dimensional (3D) well.

The behavior of an electron in this well can be well described by Schrödinger equation:

$$\frac{-\hbar^2}{2m} \nabla^2(\psi) + V(r)\psi = E\psi \tag{12.1}$$

where ψ and V represent wave function of the confined electron and potential of the barrier, respectively.

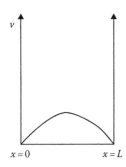

FIGURE 12.1
One-dimensional infinite potential well.

Solving this equation for one-dimensional (1D) infinite well as in Figure 12.1 would result in

$$\psi(x) \approx \sin\left(\frac{n\pi}{L}\right) \tag{12.2}$$

and the total energy of the electron would be

$$E_T = \frac{n^2\hbar^2}{8ml^2} + \frac{p_y^2}{2m} + \frac{p_z^2}{2m} \tag{12.3}$$

In a similar manner, for a −3D confinement, we have

$$\psi(x,y,z) \approx \sin\left(\frac{n\pi x}{L_x}\right) \sin\left(\frac{s\pi x}{L_y}\right) \sin\left(\frac{q\pi x}{L_z}\right) \tag{12.4}$$

$$E_T = \frac{n^2\hbar^2}{8ml^2} + \frac{s^2\hbar^2}{8ml^2} + \frac{q^2\hbar^2}{8ml^2} \tag{12.5}$$

In (12.2) through (12.5) *n*, *s*, and *q* are integer quantum numbers where *p* represents electron momentum and *m* denotes electron mass.

We notice that if the confinement is only in one dimension (i.e., in *x*) then the energy of the electron is continuum in other two directions, and when the electron is in a −3D well, the energy of the electron is quantized in three dimensions. This in turns forms a quantum-dot. Table 12.1 summarizes

TABLE 12.1

Structures and DoS

Structure	Degree of Confinement	DoS
Bulk material	0D	$\sqrt{E}$
Quantum well	1D	1
Quantum wire	2D	$1/\sqrt{E}$
Quantum-dot	3D	$\delta(E)$

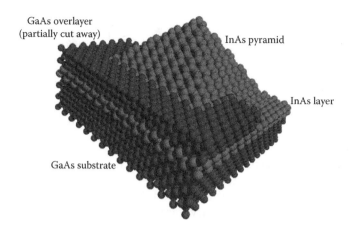

FIGURE 12.2
InAs/GaAs quantum-dot structure. (Photo from University of Newcastle, Newcastle, New South Wales, Australia.)

density of states (DoS) and degree of confinements for different structures. Implementation of such dots is practical using standard semiconductor materials such as InAs/GaAs. Figure 12.2 depicts a pyramid quantum-dot.

12.3.3 Quantum Cells

QCA offers a novel alternative to the transistor paradigm [12,13]. Quantum cells are the basic elements of QCA circuits. A simple quantum cell consists of four quantum-dots and two loaded electrons [7], as shown in Figure 12.3. It is clear that the dots are the places where the charge can be localized. The two electrons will tend to occupy antipodal sites as a result of their mutual electrostatic repulsion. However, they can change their positions within dots as a result of tunneling effect. This phenomenon takes place when potential barrier, which separates the dots, is low. However, tunneling process into or out of a cell will be blocked severely. Consequently, two configurations are possible, which can be used to encode binary information, as shown in Figure 12.4.

The numbering of dots starts clockwise from the top-right dot. Polarization (P), which represents the distribution form of electric charge in the four dots in each cell is defined as follows:

$$P = \frac{(\rho_1 + \rho_3) - (\rho_2 + \rho_4)}{\rho_1 + \rho_2 + \rho_3 + \rho_4} \tag{12.6}$$

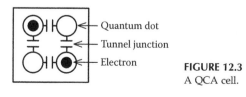

Quantum dot
Tunnel junction
Electron

FIGURE 12.3
A QCA cell.

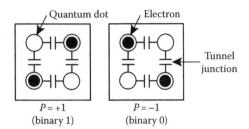

FIGURE 12.4
QCA cells with different polarities representing different binary information.

where ρ_i indicates electric charge density at dot i. Because of Coulomb repulsion, electrons rest on the two extremes of diagonals in each cell. Considering ρ_i values in (12.6), it is concluded that P can only take values $P = 1$ and $P = -1$, which represent binary values "0" and "1" respectively. These two states are used for encoding binary data. When a polarized cell is placed in a line close to another cell, Coulomb repulsion between them makes the second cell to be in the same state as the first one and this causes electrostatic energy to get minimized in the charge configuration of the cells. This is how the state is propagating in a line of cells. Based on these Coulomb interactions between the cells, basic QCA devices can be developed. In addition to semiconductor dots, magnetic dots, metal dots, and molecular dots have also been implemented. In [14,15], the development of a basic silicon QCA with four silicon dots has been reported. The performance of this QCA has been measured at 4.2 K. The implementation of metal dot QCA cells working at about 0.5 K has been presented in [11]. In magnetic QCA, the logical states are shown through magnetic polarization of single magnetic dots. Coupling of these dots to the neighboring dots is done through magnetic interaction. The development of a sample room temperature magnetic QCA is in [16] while theoretical and laboratory research on magnetic QCA is presented in [17,18]. But an interesting possibility is the implementation of QCA cells on the molecular level [19]. By using single molecules as the cells and regions within the molecule as sites for electrons, molecular QCA holds the promise of densities upward of 10^{13} devices per cm^2. Clock speeds for these cells could be in the 1–10 THz range at room temperature. Power consumption should be far less than end-of-line high-performance transistors in CMOS.

12.4 QCA Digital Logic Implementation

When placing quantum cells adjacent to each other, due to the electrostatic force and Coulomb's law they will interact in the way that the polarization of one cell will be directly affected by the polarization of its neighboring cells. One can use an array of QCA cells as a wire to transmit information from one end to another, as shown in Figure 12.5.

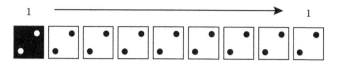

FIGURE 12.5
A QDCA wire.

A very basic QCA gate is called majority gate. We can easily prove that all other logical gates like AND, OR are implementable by majority gate [7,12], Figure 12.6. The truth table of this gate is shown in Table 12.2. Figure 12.7 shows how AND, OR gates are created by majority gate. Figure 12.8 also shows the structure of a NOT gate. One important thing about QCA information flow is the clocking scheme; that is to say for adjacent cells, in order to control the polarizations reactions and effects, one should hold the polarization of the first cell fixed and lower the potential barrier of its adjacent cell in order to let the electrons of the adjacent cell relocate. This phenomenon should repeat over and over again to pass the information through cells. It has been shown that for a QCA circuit to function correctly, only four clocking zones are necessary. Each clock signal lags 90° in phase with respect to the previous clocking [12,20]. The four clock zones are shown in Figure 12.9.

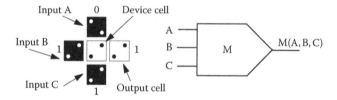

FIGURE 12.6
The basic QCA majority gate.

TABLE 12.2

Truth Table of QDCA Majority Gate

A	B	C	M(A, B, C)
0	0	0	0
0	0	1	0
0	1	0	0
0	1	1	1
1	0	0	0
1	0	1	1
1	1	0	1
1	1	1	1

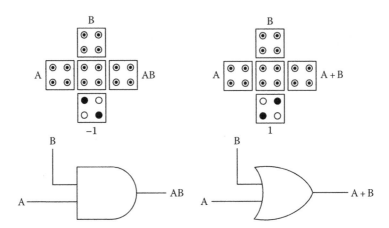

FIGURE 12.7
Creating AND, OR gates using majority gate.

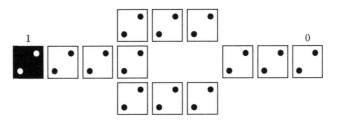

FIGURE 12.8
QCA NOT gate.

There is an interesting possibility in QCA arrays, that is, when dots in a cell are rotated by 45°, as shown in Figure 12.10 (vertical QCA cell array), an "inversion chain" forms. When an inversion chain crosses a wire of 45° cells, the two wires do not interact. This feature in QCA cells gives us the ability to cross signals directly over each other as in Figure 12.10. This in turn helps the designer to optimize the layout and to avoid extra layers, metal or via cross connections.

12.5 Modeling and Simulation of QCA Cells

The "orthodox theory" of Coulomb blockade can be used to model and simulate metal-dot QCA [21]. The number of electrons on each of the metal islands and charge configurations help to model the circuit. Although

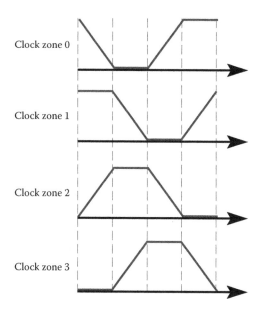

FIGURE 12.9
Clocking scheme of QCA circuits.

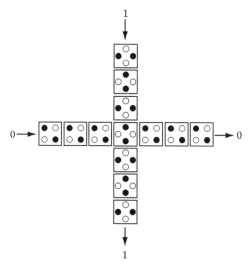

FIGURE 12.10
Inversion chain and QCA array crossover.

metal islands are surrounded by thin layers of insulators, they are coupled to other metal islands and metal electrodes via tunnel junctions and capacitors. A class of metal electrodes whose voltages can be fixed via external sources is called leads.

Figure 12.11 shows the schematic of islands and leads. Dot charges are best described as on—island charges and lead charges as on—electrodes charges. The free energy of charge configuration and lead charges is calculated by

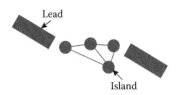

FIGURE 12.11
Schematic of islands and leads.

subtracting the electrostatic energy of the capacitors and junctions from the work done by leads [22].

$$F = \frac{1}{2}\begin{pmatrix} q \\ q' \end{pmatrix} C^{-1} \begin{pmatrix} q \\ q' \end{pmatrix} - v^{\mathrm{T}} q' \tag{12.7}$$

where
   C stands for the capacitor matrix that includes all the junctions and capacitors
   v is the column vector of lead voltages
   q and q' are the column vectors of dot and lead charges

The number of electrons on each island is an integer at zero temperature. The charge configuration with the minimum of free energy is the equilibrium one. Only when the amount of free energy decreases, a tunneling occurs. The number of charges is not always an integer, for example, at some finite temperatures it would be a thermal average of the charges over all possible configurations. It is always possible for a tunneling event to occur, even if the free energy increases. To obtain the transition rate of the tunneling between two charge configuration states that is deduced from first-order perturbation theory, the following equation can be used:

$$\Gamma_{ij} = \frac{1}{e^2 R_{\mathrm{T}}} \cdot \frac{\Delta F_{ij}}{1 - e^{-\Delta F_{ij}/(k_{\mathrm{B}} T)}} \tag{12.8}$$

where
   R_{T} indicates the tunneling resistance
   F stands for the energy difference between the initial state i and final state j
   k_{B} is the Boltzmann constant
   T is the temperature

As co-tunneling can be suppressed in QCA system, the second-order co-tunneling is ignored.

If the following two prerequisites are met, then this single electron-tunneling phenomenon will occur. First, the system requires metallic islands that are linked to other metallic regions through tunnel barriers; the tunneling resistance R_{T} must be much greater than the resistance quantum, $h/e^2 \approx 26$ kΩ, to make sure about the localization of electrons on islands.

Second, in order to suppress the thermal fluctuation, the thermal energy $k_B T$ must be much less than the charging energy $(e^2/2C)$ of the tunnel junction. Meeting these two conditions guarantees that the transportation of charge from island to island is dominated by Coulomb charging energy [23].

12.5.1 Master Equation

A tunneling event is defined as an instantaneous and stochastic process during which each successive tunneling event is uncorrelated and constitute a Poisson process.

A master equation is used to define the tunneling events of many electrons. For the temporal change of the probability distribution function of a physical quantity, the conservation law used is as follows:

$$\frac{dP}{dT} = \Gamma P \tag{12.9}$$

A solution to (12.9) can be in the form:

$$P(t) = e^{\Gamma t} p(0) \tag{12.10}$$

where

$$e^{\Gamma t} = \sum_{K=0}^{\infty} \frac{(\Gamma T)^K}{K!} \tag{12.11}$$

In the above equations, P stands for the state probabilities of charge configurations and is the ensemble average of the charge in the islands. In (12.9), Γ denotes a time-dependent transition matrix. In this matrix, the diagonal elements are $-\sum_{i \neq j} \Gamma_{ij}$ and other elements are the transpose of $\Gamma_{i,j}$ [23]. All the states of the system have to be tracked in order to solve (12.9) indirectly in master equation method. One of the drawbacks of this method is that when the number of the states is too large, it is difficult and totally impossible to exploit it. The interesting point is that the QCA generally operates near the ground state, thus the master equation is tractable considering just a few states.

Rui et al. have proposed a Spice model for simulation of QCA circuits in [24]. Their model is based on the verification of the behavior cells. Figure 12.12a depicts a possible realization of a basic QCA with four metal dots. The left pair of metal dots has been separated by a tunneling barrier but is coupled to the right pair of metal dots through capacitors [25]. The left half and the right half of a QCA cell are exactly symmetrical. This will result in a simple model for each pair of metal dots in a QCA cell. Figure 12.12b shows the schematic diagram of the simplified half-QCA cell [24,25]. In this figure, the two black dots represent quantum-dots and T_1 is a tunneling junction.

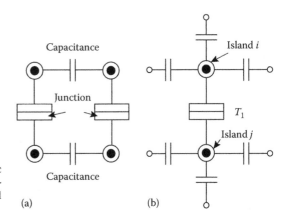

FIGURE 12.12
(a) A possible realization of a basic QCA with four metal dots. (b) Schematic diagram of the simplified half-QCA cell.

The required energy to move an electron from island i to island j can be computed as [24]:

$$\Delta E = -e(v_j - v_i) + \frac{\left(C_{ii}^{-1} - 2C_{ij}^{-1} + C_{jj}^{-1}\right)e^2}{2} \tag{12.12}$$

In (12.12), C_{ii}, C_{jj}, and C_{ij} are capacitances of nodes i and j with respect to ground, and the capacitance between them. v_i and v_j are the voltages on island i and j before the electron tunnels from node i to node j. The single tunneling event can be modeled based on the birth-and-death Markov chain, as shown in Figure 12.13. The probability that one island holds k electrons, p_k, is given by the following equation [24]:

$$p_k = p_{k-1}\left(\frac{\lambda_{k-1}}{\mu_k}\right) \tag{12.13}$$

where λ_k and μ_k are transition rate from state k to state $k-1$. In a simple case, there is only one extra electron trapped in island i, and the other state is island j, holding the extra electron. Thus, it is clear that [25]:

$$p_{i=1, j=0}, \quad \Gamma_{i \to j} = p_{i=0, j=1}\Gamma_{j \to i} \tag{12.14}$$

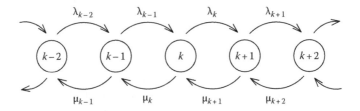

FIGURE 12.13
Birth-and-death Markov chain.

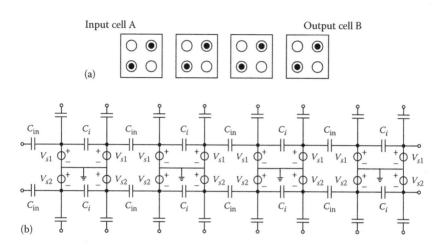

FIGURE 12.14
Spice model of a QCA wire.

in which, $\Gamma_{j \to i}$, is the tunnel rate for an electron tunnel from island j to island i. $p_{i=0,j=1}$ stands for the probability that there are b electrons available on node j and a electrons on node i (a, b can be either 0 or 1). The tunnel rate is calculated based on the orthodox theory as in (12.8). The Spice model proposed by Rui et al. [25] for a QCA wire has been shown in Figure 12.14. The Spice model presented can be used for logic simulations of digital QCA gates and circuits.

12.5.2 Monte Carlo Approach

Single electron tunneling events can be modeled through Monte Carlo approach, which is a stochastic method of simulation. The probability of tunneling event, for times $t > \tau$ out of state zero, can be described by Poisson distribution as

$$P_o(\tau) = e^{-\Gamma \tau} \tag{12.15}$$

Γ is the tunneling rate as in (12.8).

One can write the inverse of (12.15) to generate random numbers as follows:

$$\tau = -\frac{\ln(r)}{\Gamma} \tag{12.16}$$

In (12.16), r is a number that has even distribution in the interval [0,1]. This method requires an efficient random number generator. Some rare events may lengthen the simulation time, which is one of the drawbacks of this method. SIMON [26] is an interesting software, which is widely used for the simulation of single electron systems based on Monte Carlo method.

There is an interesting software, the QCADesigner, which has been initially developed at the ATIPS Laboratory with a number of tools for QCA layout designing and logic simulation of digital QCA circuits [27]. The software has three built-in simulation engines, namely, digital simulation engine, non-linear approximation engine, and two-state simulation engine. The digital simulation engine considers the polarization of the neighboring cells to compute the polarization of the cell, which is about to switch for each change in the clock. Therefore, this engine provides the designers with quick analysis of the functionality of their circuits. The kink energy $E_{i,j}^k$ represents the energy cost of cells i and j provided that they have opposite polarization. The electrostatic interaction between all the charges (mobile/immobile) can be used for its calculation. The electrostatic interaction between each dot in cell i and other dots in cell j are calculated as follows:

$$E_{\text{total}} = \sum_{i,j} \frac{1}{4\pi\varepsilon_0\varepsilon_r} \frac{q_i q_j}{|r_i - r_j|} \tag{12.17}$$

where
ε_0 represents the permittivity of free space
ε_r is the relative permittivity of the material system

This is summed over all i and j. The first step to calculate the kink energy, is calculating this electrostatic energy when the cells are in the opposite polarization state and repeating the same computation procedure when the cells have the same polarization, and then subtracting them. In the following sections, we briefly introduce the coherence vector simulation method.

12.5.3 Coherence Vector Simulation

The cell-to-cell response function can be approximated excluding quantum mechanical correlation between cells to compute the polarization of the cells. The coherence vector λ is a vector representation of the density matrix ρ of a cell, projected onto the basis spanned by the Identity and the Pauli spin matrices $\sigma_x, \sigma_y,$ and σ_z. The components of λ are found by taking the Trace of the density matrix multiplied by each of the Pauli spin matrices; that is [28],

$$\lambda_i = \text{Tr}\{\hat{\rho}\hat{\sigma}_i\}, \quad i = \{x, y, z\} \tag{12.18}$$

The polarization of cell i, P_i is just the z-component of the coherence vector:

$$p_i = \lambda_{z,t}$$

The Hamiltonian must also be projected onto the spin matrices as

$$\Gamma_i = \frac{\mathrm{Tr}\{\hat{H}\hat{\sigma}_i\}}{\hbar}, \quad i = \{x, y, z\} \tag{12.19}$$

The vector Γ represents the energy environment of the cell, including the effect of neighboring cells. We can evaluate the explicit expression for Γ by substituting it into the related Hamiltonian. This explicit expression is

$$\vec{\Gamma} = \frac{1}{\hbar}\left[-2\gamma, 0, \sum_{j \in S} E_{i,j}^k P_j\right] \tag{12.20}$$

where S is the effective neighborhood of cell i. The equation of motion for the coherence vector including dissipative effects is

$$\vec{\Gamma} = \frac{\partial}{\partial t}\vec{\lambda} = \vec{\lambda} \times \vec{\gamma} - \frac{1}{\tau}(\vec{\lambda} - \vec{\lambda}_{ss}) \tag{12.21}$$

where τ is the relaxation time and is a time constant representing the dissipation of energy into the environment. λ_{ss} is the steady state coherence vector defined as

$$\vec{\lambda}_{ss} = -\frac{\vec{\Gamma}}{|\vec{\Gamma}|}\tanh(\Delta) \tag{12.22}$$

Δ is the temperature ratio defined as

$$\Delta = \frac{\hbar|\vec{\Gamma}|}{2k_B T} \tag{12.23}$$

where
   T is the temperature in Kelvin
   k_B is Boltzmann's constant

The simulation engine evaluates the equation of motion (a partial differential equation) using an explicit time marching algorithm. For each time step, Γ and λ_{ss} for each cell are evaluated and then the coherence vector for each cell is stepped forward in time [29].

The density matrix approach is used to manipulate the coherence vector simulation. It is used to model the dissipative effects and also perform a time-dependent simulation of the design [30]. It assumes each cell as a simple-state system. The following Hamiltonian is constructed for this two-state system [28]:

$$H = \sum_j \begin{bmatrix} -\dfrac{1}{2}P_j E_{ij}^k & -\gamma_j \\ -\gamma_j & \dfrac{1}{2}P_j E_{ij}^k \end{bmatrix} \tag{12.24}$$

where E is the kink energy between cell i and j. This kink energy and the energy cost of two cells with opposite polarization are associated with each other. P represents the polarization of cell j and γ is the tunneling energy of electrons within the cell and is directly related to the clock, i.e., the clock value is the tunneling energy. The summation is over all cells within an effective radius of cell i, and can be set prior to the simulation.

With the aid of this Hamiltonian, Schrödinger time-invariant equation can be written as

$$H_i \psi_i = E_i \psi_i \tag{12.25}$$

A solution to this equation that helps us to find the polarization of a cell will be [27]:

$$P_j = \frac{\dfrac{E_{ij}^k}{2\gamma} \sum_j P_j}{\sqrt{1 + \dfrac{E_{ij}^k}{2\gamma} \sum_j P_j}} \tag{12.26}$$

12.6 Digital Data Storage

Conventional passive digital data storage medias such as compact disks (CDs) and digital versatile disks (DVDs) have been developed so far to increase the amount of digital data that can be stored in a storage medium. A standard single-sided CD is capable of storing approximately 800 MB of data while a single-sided DVD can store 4.4 GB of data. The area of a typical CD or DVD is about 86 cm. Consequently, the rate of data that could be stored on a CD is 74 Mb/cm². With a simple rule of thumb, this rate for a single-sided DVD is 0.4 Gb/cm². Active digital storage mediums, such as CMOS ROMs, have also been developed to be used as single-chip digital data storage mediums for many microprocessor applications. Typical NAND

TABLE 12.3

Different ROMs and Their Capacities

Media	Technology	Cell Area	Data Rate
CD ROM	Laser	$1.34\ \mu m^2$	$74\ Mb/cm^2$
DVD ROM	Laser	$0.24\ \mu m^2$	$0.4\ Gb/cm^2$
NOR ROM	CMOS—0.25 μm	$0.62\ \mu m^2$	$161\ Mb/cm^2$
NAND ROM	CMOS—0.25 μm	$0.25\ \mu m^2$	$0.4\ Gb/cm^2$
QDC ROM	Nanotechnology	$400\ nm^2$	$250\ Gb/cm^2$

ROMs use cell size of $4F^2$, where F stands for feature size of the used technology. That is to say, exploiting 0.25 μm process technology parameters, storing one digital bit requires area of about 0.25 μm^2, yielding data bit rate of $400\ Mb/cm^2$. Table 12.3 shows bit cell area for different storages [31].

Here, our focus is primarily on capacity. However, other parameters such as access time, performance, cost, and lifetime should be taken into account. Several circuits based on the capabilities of the QCA have been proposed so far. Many of them take advantage of coplanar wire crossing, which makes the design independent of metal layer interconnects. Moreover, with the aid of nanotechnology, nanoscale implementation of digital circuits with high performance and ultralow power has become possible. Table 12.3 compares different storage mediums in terms of data rate.

Any digital data storage consists of two parts, namely, the storage part and the readout part. The readout part may consist of decoders, tristate buffers, flip-flops, etc. In this section, we discuss some of the important elements of QCA digital data storage from the viewpoint of design, simulation, and the contributing factors such as latency, area, access time, etc.

12.6.1 Decoders

It is very easy to implement decoders in QCA technology. However, special considerations have to be taken into account when realizing such circuits in QCA. Figure 12.15 depicts the conventional three-layer 2 × 4 decoder implemented in QCA. This decoder consists of main cell layer, two via layers, and a cross contact layer. The radius of effect of each cell in this decoder is 65 nm, which results in at least two-cell spacing between different signals running on adjacent wires. Figure 12.16 depicts a single layer 2 × 4 QCA decoder layout. The simulation results of this decoder in QCA designer [29] for both coherence vector and bistable approximation methods have been shown in Figure 12.17.

Taking advantage of 45° rotated cells, it is also possible to implement the decoder in one layer as shown in Figure 12.16. However, the radius of effect of each cell in this decoder should be less than the previous one (in our

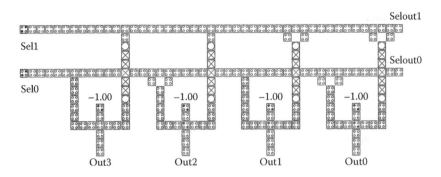

FIGURE 12.15
QCA 2 × 4 three-layer decoder.

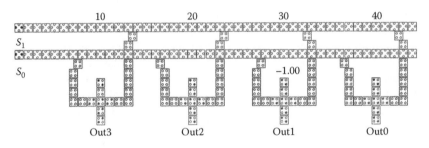

FIGURE 12.16
QCA 2 × 4 one-layer decoder.

example 45 nm). The specifications of these decoders have been described in Table 12.4. In this table, the delay has been computed according to four zones of a clock period of T.

12.6.2 QDCROM

Typical ROMs consist of data bit cells, decoders, and tristate buffers. Each bit cell has a select input line and an output. When the select input line is active, binary information appears at the output line. Figure 12.18 shows a simple bit cell and its QCA equivalence. The select input has been added to the cell by using an AND operation. An array of bit cells can be selected based on the given address by a decoder.

The block diagram of a simple ROM has been shown in Figure 12.19. Current semiconductor memories achieve random access by connecting the memory cells to the bit lines in parallel. In this kind of ROM, the cell outputs

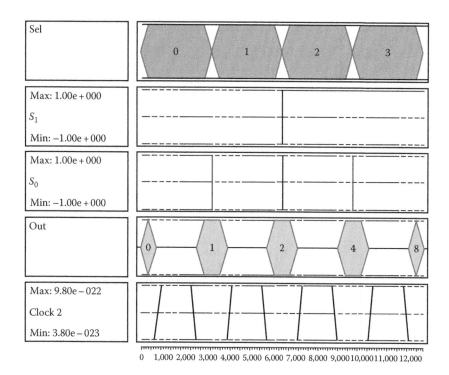

FIGURE 12.17

2×4 QCA decoder simulation results.

TABLE 12.4

Specifications of the Decoders

Decoder Type	2×4 Four Layer	2×4 One Layer
Number of cells	226	172
Delay (T)	3/4	3/4
Area (μm^2)	0.33	0.22

connected to the bit lines act similar to wired OR gates. As a result, when the bit cell has not been selected, the output of that cell is "0" or "1" depending on the binary bit stored in the cell. Consequently, the bit value stored in the selected cell will come on the output bit line of the ROM. This is a kind of serial OR that is very easy to implement using QCA cells. Figure 12.20 shows how the serial OR can be carried out on multiple inputs with QCA technology. Since there is no tristate buffer available in QCA technology, the serial OR gate can be used as a substitute.

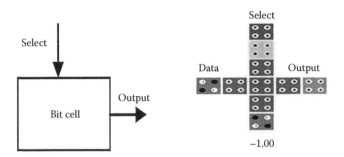

FIGURE 12.18
QCA ROM bit cell.

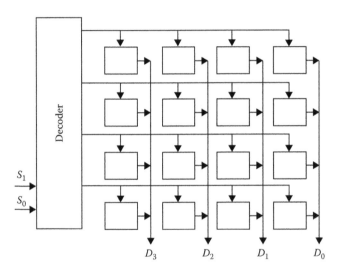

FIGURE 12.19
Block diagram of a typical 4 × 4 ROM.

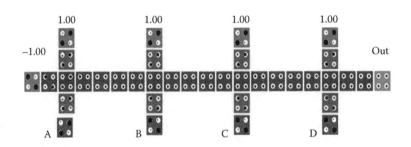

FIGURE 12.20
QCA serial OR.

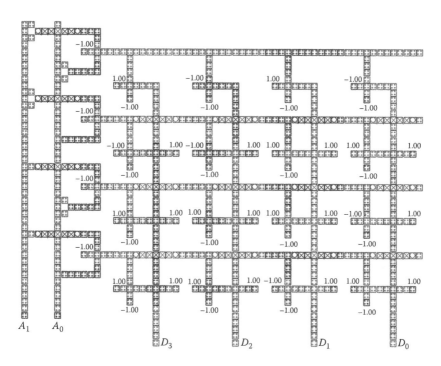

FIGURE 12.21
4 × 4 QDC four-layer ROM layout.

Figure 12.21 depicts the 4 × 4 QCA ROM implemented in four layers. Two via layers have been used to pass data between the main cell layer and cross contact layer. In this ROM layout, the cell spacing is two, while the radius of effect of each cell is 65 nm. It takes one T ($4\varnothing_i$ where each T state consists of four $\varnothing_i$ states) for the decoder to activate the appropriate select of each bit cell, and $1\varnothing_i$ for each bit cell to make the bit available to the output terminal. As a next step, the serial OR makes the bit available at the output data bus. This operation takes $1\varnothing_i$ in its turn. Therefore, it will take $9\varnothing_i$ for any bit in the last column to appear at the data terminal of the ROM.

Figure 12.22 shows the same 4 × 4 ROM implemented in one layer. Taking advantage of the 45° rotated cells, it is possible to cross wires in the same layer. However, in this case, special consideration about the radius of effect and correct cell alignments should be taken into account. The capacity of this ROM can be easily increased via using several layers on top of each other. The data bits are stored in the form of fixed polarity QCA cells. Table 12.5 shows the contents of each ROM in our simulation [32]. Both coherence vector and bistable approximation methods have been exploited in QCADesigner CAD tool to simulate the ROMs. The simulation results have been shown in Figure 12.23. The specifications of these ROMs are in Table 12.6.

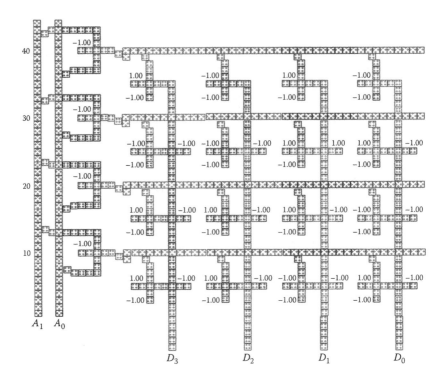

FIGURE 12.22
4 × 4 QDC one-layer ROM layout.

TABLE 12.5

QCA ROM Contents

Address (A1A0)	Data (D3D2D1D0)
00	1010
01	0011
10	1110
11	1101

Due to the operation of the QCA cells in low voltages, a transient noise in clocking scheme may lead to a change in the polarity of data cells in a very short period of time and consequently the read data is wrong. This is more likely to happen in QCA RAMs.

In order to design a more fault-tolerant ROM, we suggest use two quantum-dots in data cells. This reduces the risk of transient data cell polarity change. The challenges of current QCA technology include driving inputs and outputs of the circuit and temperature considerations.

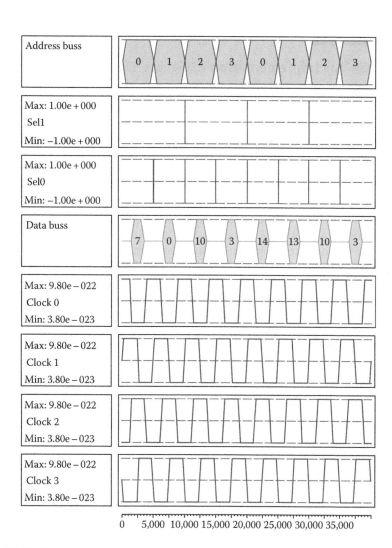

FIGURE 12.23
Simulation of QDC ROM.

TABLE 12.6

QCA ROM Specifications

ROM Type	QCA 4 Layer	QCA 1 Layer
Number of cells	864	692
Access time (T)	9/4	9/4
Area (μm^2)	1.19	1.10
Capacity (Gb/cm^2)	1.8	1.8

12.7 QCA for Implementation of Secure Digital Systems

Power analysis attacks were first introduced by Kocher et al. [33]. In fact, power and EM side channels are the most important ones for implementation of block ciphers. The power consumption as well as the electromagnetic field surrounding a cryptographic module may leak a significant amount of information about the private key. Nowadays, most of digital circuits in cryptographic modules are typically implemented in static CMOS. There is a strong dependency between power consumption of circuits implemented based on this logic style and the data that is processed by the circuit. Due to the difference between parasitic capacitances in the source and drain of a transistor in CMOS technology, when the transistor switches on and off, different amount of current flows through the transistor and leads to different amount of power consumption when the transistor processes a "0" or "1." Consequently, the power consumption as well as the electromagnetic field that is caused by the current flowing in a cryptographic circuit implemented in CMOS leak information about the private key. This current is mainly caused by the charging or discharging of the parasitic capacitances of interconnected wires. More extensive and more general introductions to this logic style and its power consumption characteristics can be found in [34,35]. The power consumption mechanism in QCA) logic is basically different from the CMOS logic.

Columbic interaction causes electrons to move within a cell, not between cells and thus there is no current flow [36]. This avoids many of the heat dissipation and power consumption problems of transistor computing systems. Unlike CMOS, in QCA, the transmission media and logical elements are both comprised of the same basic block—the cell. As such, QCA has been called "processing-in-wire" and is a very suitable alternate technology for cryptographic processors. We have named the cryptographic processors implemented in QCA as the "quantum cryptographic processors" (QCP) [1].

12.7.1 Side-Channel Attacks and Countermeasures

A side channel (for example, the power consumption) of a cryptographic module depends on many parameters. Only one of them is the private key. However, the fact that the side-channel output depends on the private key is often sufficient to reveal it. In order to exploit this dependency between the side-channel output and the private key, an attacker usually builds a model of the side channel. This model is typically not very complex. In fact, attacks conducted in practice have shown that very simple models are often sufficient to reveal the private key [37]. Figure 12.24 depicts the principles of a side-channel attack. On the left side, the figure shows the physical device that is attacked. Its side-channel output is determined by the private key,

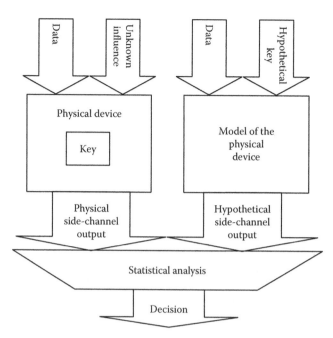

FIGURE 12.24
Principles of side-channel attacks.

the input and the output of the device, and by many other parameters. Some of them are known by the attacker while others are not. The model of the side channel used by the attacker is shown on the right side in Figure 12.24. The model may consider additional parameters besides the key, the input, and the output of the module. However, there is always a certain imperfectness of the model.

In order to determine the private key stored in the attacked module, hypotheses regarding parts of the private key are formulated. Using the model of the side channel, corresponding hypothetical side-channel outputs are calculated. Subsequently, the correlation between the physical side-channel output and the hypothetical ones is determined. If significant stronger correlation occurs for one hypothesis as compared the other ones, this indicates that the hypothesis is correct [38,39]. An important step in power analysis attacks is to derive the hypothetical power consumption values from the intermediate results. Several power models have been proposed so far, namely, the Hamming weight model, Hamming distance model, partitioning power analysis [40], etc. In general, two types of power consumption leakage can be observed. Hamming weight leakage is related to the number of 1's being processed at a time. Transition count leakage is related to the numbers of transitions from 0 to 1 at a given time.

Software implementation of block ciphers often leak information about the Hamming weight of the data they process. The reason for this is that there are processor architectures with pre-charged buses. This means that all wires of the bus are charged to "1" before the key is transferred over the bus. Assuming the parasitic capacitances of all wires are equal, the energy that is needed to switch the bus wires from the pre-charge stage to a part of the key is proportional to the bus width minus the hamming weight of the part of the key. This is a consequence of the power consumption characteristics of CMOS circuits. An attack exploiting the leakage of Hamming weights has been presented in [41]. More about investigation of single leakage and Hamming weight could be found in [38,39,42].

Several countermeasures to power and electromagnetic attacks have been proposed so far. However, each technique may lead to design complexity, more power consumption, size, and speed issues of the entire cryptographic modules. All these strategies can be categorized in two groups: they either try to randomize the intermediate result or take advantage of circuits with data and power consumption independency. These techniques can be implemented in architecture, logic, and algorithm or protocol level. The secure clocked QCA logic [1] we introduce in this section takes advantage of QCA technology with low power consumption and data independency together with complicated clocking scheme that makes it very difficult to make power consumption models for cryptographic processors implemented in SCQCA logic.

12.7.2 SCQCA Logic

We can always get similar functionality of sequential logic from a QCA wire segment spread across several clocking zones, that is, a basic wire implements the master–slave-type data storage, based on neighboring clocking zones acting as flip-flop stages. To make a more secure logic style, we added an additional logic signal "clock." To describe the consequent sequential logic, we introduce a SCQCA D-type flip-flop in this part [1].

D-type latches are the basic elements of synchronous digital logic circuits. Their applications are vast as they are vital elements in any digital circuit in which a memory is present. The function of a D-type latch is similar to a RAM cell, which has been introduced in [36,43]. Although in CMOS digital circuit technology, the D-type latch was introduced before the proposition and fabrication of RAMs, in QCA technology, it was suggested after the introduction of QCARAMs [36]. The structure of a D-type latch and its equivalent circuit in QCA technology has been shown in Figures 12.25 and 12.26. In CMOS technology, D-type flip-flops are implemented using crossing NAND gates. It is also possible to use two D-type latches in master–slave mode in order to make a D-type flip-flop (as shown in Figure 12.27).

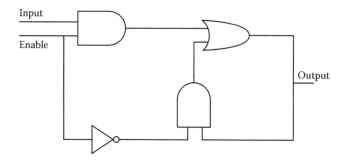

FIGURE 12.25
Structure of a D-latch.

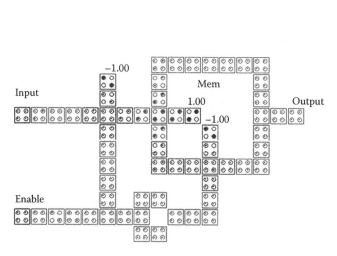

FIGURE 12.26
Structure of a D-latch in QCA.

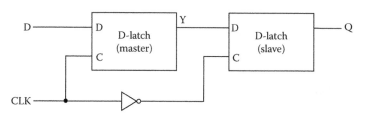

FIGURE 12.27
Structure of the master–slave D-type flip-flop.

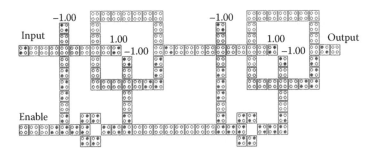

FIGURE 12.28
Structure of a QCA master–slave D-type flip-flop.

Crossing NAND gates are difficult to be implemented in QCA technology. Consequently, the conventional D-type QCA flip-flop is implemented in the master–slave mode.

Figure 12.28 depicts the conventional D-type flip-flop. This flip-flop consists of six majority and two inverter gates. The large area of the circuit and the limitation in the length of QCA wires are main issues when implementing and fabricating circuits in QCA technology [44]. By taking advantage of a level-to-edge converter, it is possible to improve the D-type QCA flip-flop. The level-to-edge converter exploits the intrinsic stages of clocking and zones in QCA. The converter consists of an AND inverter gate. The original signal is transferred into an AND with its inverted delayed copy. The result is a generation of short pulses at the rising edge of the original signal. The QCA edge detector has been shown in Figure 12.29. The D-type flip-flop implemented with this technique has been shown in Figure 12.30. The simulation results obtained with QCA designer verifies the functionality of the proposed flip-flop (Figure 12.31) [1].

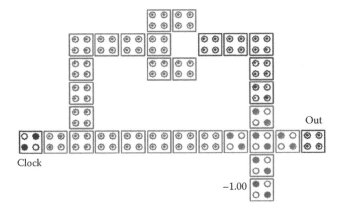

FIGURE 12.29
QCA level to edge converter (edge extractor).

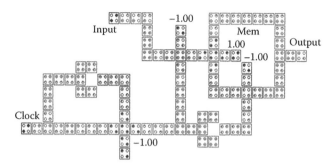

FIGURE 12.30
QCA D-type flip-flop.

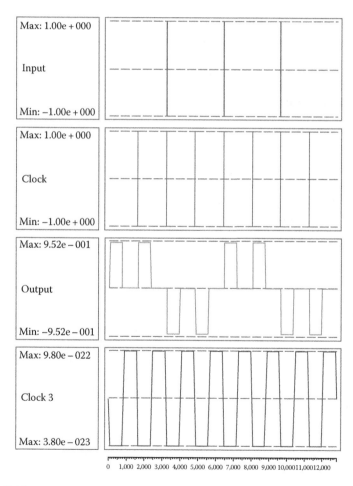

FIGURE 12.31
Simulation of proposed D-type flip-flop.

TABLE 12.7

Specifications of the Flip-Flops

Flip-Flop Type	Conventional DFF	Proposed DFF
Number of cells	123	94
Delay (T)	2	2
Area (μm^2)	0.22	0.14

Table 12.7 shows the comparison of conventional D-type flip-flop with the proposed one from the viewpoint of area, number of cells, and delay. It is very difficult to make a power consumption model of the proposed DFF because of the presence of the quantum mechanical effects and four clocking schemes together with the additional signal "clock" that provides a more secure logic in comparison to CMOS technology.

12.8 Conclusion

It is obvious that with current advances in the digital communication algorithms and the protocols related to the bandwidth-hungry digital data, there is a vital need for advanced devices, circuits, and systems from the point of view of speed, power consumption, and size. Current advances in nanotechnology hold to bring the promising successors of the conventional CMOS technology. It is clear that the Moore's law for scaling down the transistors will not be satisfied forever. QCA is thought to be an attractive technology for future digital telecommunication systems. Moreover, it is also called the "processor in wire" as the signal is processed while it is propagating through the circuit. The chapter provided a comparison of different technologies in the field of digital data storage. It is clear that with the aid of the emerging nanotechnology, the overwhelming digital data can be stored at nanoscale. We reviewed the interesting features of QCA in short and proposed a fault-tolerant QDCROM with capacity of storing 2.5 Gb of information per cm^2 as a paradigm of nanotechnology advances in data-storage mediums.

A new D-type flip-flop structure based on level-to-edge converter was proposed. It is clear that the design of new structures based on QCA which have shorter path, smaller size, and fewer numbers of cells would result in improvements in the functionality and quality of outputs. Side-channel attacks seriously threaten cryptographic modules as they can be implemented with relatively inexpensive equipments. In this chapter, a new approach to implementation of quantum cryptographic modules via QCA technology has been presented. SQCA logic style was introduced through design of a

D-type flip-flop, with additional "clock" signal as a result of nanotechnology advances in developing novel countermeasures and designing more secure modules in cryptography engineering.

References

1. E. Rahimi and S. Mohammad Nejad, Secure clocked QCA logic for implementation of quantum cryptographic processors, *IEEE Applied Electronics Conference*, Czech, pp. 217–220, 2009.
2. W. Porod, Quantum-dot devices and quantum-dot cellular automata, *International Journal of Bifurcation and Chaos*, 7(10), 2199, 1997.
3. A. Vetteth, K. Walus, G. A. Jullien, and V. S. Dimitrov, Quantum-dot cellular automata carry-look-ahead adder and barrel shifter, *IEEE Emerging Telecommunications Technologies Conference*, Dallas, TX, pp. 2–4, 2002.
4. C. S. Lent, P. D. Tougaw, W. Porod, and G. H. Bernstein, Quantum cellular automata, *Nanotechnology*, 4(1), 49–57, January 1993.
5. A. Khurasia and P. Gambhir, Quantum cellular automata, Final Project Report, 2006.
6. A. Fijany, N. Toomarian, K. Modarress, and M. Spotnitz, Bit-serial adder based on quantum-dots, NASA Technical Report, January 2003.
7. W. Wang, K. Walus, and J. Julien, Quantum-dot cellular automata adders, *Third IEEE Conference on Nanotechnology, IEEE-NANO 2003*, San Francisco, CA, pp. 461–464, 2003.
8. M. T. Niemir, Designing digital systems in quantum cellular automata, MSc thesis, Department of Computer Science and Engineering, Notre Dame, IN, 2004.
9. I. Amlani et al., Experimental demonstration of a leadless quantum-dot cellular automata cell, *Applied Physics Letters*, 77(5), 738–740, 2000.
10. A. Orlov et al., Experimental demonstration of clocked single-electron switching in quantum-dot cellular automata, *Applied Physics Letters*, 77(2), 295–297, 2000.
11. A. Orlov et al., Experimental demonstration of a binary wire for quantum-dot cellular automata, *Applied Physics Letters*, 74(19), 2875–2877, 1999.
12. C. S. Lent and P. D. Tougaw, A device architecture for computing with quantum-dots, *Proceedings of the IEEE*, 85(4), 541–557, 1997.
13. S. C. Benjamin and N. F. Johnson, A possible nanometer scale computing device based on an adding cellular automaton, *Applied Physics Letters*, 70, 2321–2323, 1997.
14. C. Single, R. Augke, E. E. Prins, D. A. Wharam, and D. P. Kern, Towards quantum cellular automata operation in silicon: Transport properties of silicon multiple dot structures, *Superlattices and Microstructures*, 28, 429–434, November 2000.
15. C. Single, A. Rugke, and E. E. Prins, Simultaneous operation of two adjacent double dots in silicon, *Applied Physics Letters*, 78, 1421–1423, March 2001.
16. R. P. Cowburn and M. E. Welland, Room temperature magnetic quantum cellular automata, *Science*, 287, 1466–1468, February 2000.

17. G. Csaba and W. Porod, Simulation of field coupled computing architectures based on magnetic dot arrays, *Journal of Computational Electronics*, 1, 87–91, 2002.
18. A. Imre, L. Zhou, A. O. Orlov, G. Csaba, G. H. Bernstein, W. Porod, and V. Metlushko, Application of mesoscopic magnetic rings for logic devices, *4th IEEE Conference on Nanotechnology*, Munich, Germany, pp. 137–139, 2004.
19. C. Lent, Molecular electronics: Bypassing the transistor paradigm, *Science*, 288, 1597–1599, 2000.
20. G. Toth and C. S. Lent, Quasiadiabatic switching for metal-island quantum-dot cellular automata, *Journal of Applied Physics*, 85(5), 2977–2984, 1999.
21. H. Grabert and M. H. Devoret, *Single Charge Tunneling, Coulomb Blockade Phenomena in Nanostructures*, Plenum Press, New York, 1992.
22. C. Wasshuber, *Computational Single-Electronics*, Springer, New York, 2001.
23. M. Liu, Robustness and power dissipation in quantum-dot cellular automata, PhD thesis, 2006, available online at http://etd.nd.edu/ETD-db/theses
24. T. Rui, Z. Fengming, and K. Yong-Bin, Quantum-dot cellular automata SPICE macro model, *15th ACM Great Lakes Symposium on VLSI, GLVLSI'05*, Chicago, IL, pp. 108–111, 2005.
25. T. Rui, Z. Fengming, and K. Yong-Bin, Design metal-dot based QCA circuits using SPICE model, *Microelectronics Journal*, 37(8), 821–827, 2006.
26. SIMON homepage, http://www.lybrary.com/simon/
27. K. Walus, T. J. Dysart, G. A. Julliein, and R. A. Budiman, QCADesigner: A rapid design and simulation tool for quantum-dot cellular automata, *IEEE Transactions on Nanotechnology*, 3(1), 26–31, 2004.
28. G. Toth, Correlation and coherence in quantum-dot cellular automata, PhD thesis, University of Notre Dame, Notre Dame, IN, pp. 56–63, 2000.
29. QCADesigner Documentation, Available online at http://www.qcadesigner.ca
30. J. Timler and C. S. Lent, Power gain and dissipation in quantum-dot cellular automata, *Journal of Applied Physics*, 91(2), 823–831, 2002.
31. E. Rahimi and S. Mohammad Nejad, Quantum-dot cellular ROM: A nano-scale level approach to digital data storage, *IEEE 6th International Symposium on Communication Systems, Networks and Digital Signal Processing*, Graz University of Technology, Graz, Austria, pp. 618–621, July 23, 2008.
32. E. Rahimi and S. Mohammad Nejad, Quantum-dot cellular ROM: The prospective digital data storage, Technical report, Iran University of Science and Technology (IUST), September 2008.
33. P. C. Kocher, J. Jaffe, and B. Jun, Differential power analysis, *Proceedings of Advances in Cryptography, Lecture Notes in Computer Science*, Santa Barbara, CA, 1666, pp. 388–397, 1999, Springer.
34. J. M. Rabaey, *Digital Integrated Circuits*, Prentice-Hall, Upper Saddle River, NJ, 1996. ISBN 0-13-178609.
35. N. H. E. Weste and K. Eshraghian. *Principles of CMOS VLSI Design—A Systems Perspective*, 2nd edn., Addison-Wesley, Reading, MA, 1993. ISBN 0-201-53376-6.
36. S. Frost, A. F. Rodrigues, A. W. Janiszewski, R. T. Raush, and P. M. Kogge, Memory in motion: A study of storage structures in QCA, *First Workshop on Non-Silicon Computing*, Cambridge, MA, 2002.
37. J. Waddle and D. Wagner, Towards efficient second-order power analysis, *CHES 2004*, Cambridge, MA, *Lecture Notes in Computer Science*, 1–15, 2004, Springer.

38. S. Mangard, Securing implementations of block ciphers against side-channel attacks, PhD thesis, Institute for Applied Information Processing and Communications (IAIK), Graz University of Technology, Graz, Austria, 2004.
39. E. Oswald, On side-channel attacks and the application of algorithmic countermeasures, PhD thesis, Institute for Applied Information Processing and Communications (IAIK), Graz University of Technology, Graz, Austria, 2003.
40. T.-H. Le, J. Clediere, C. Canovas, B. Robisson, C. Serviere, and J.-L. Lacoume, A proposition for correlation power analysis enhancement, *CHES 2006*, Cambridge, MA, *Lecture Notes in Computer Science*, 174–176, 2006, Springer.
41. R. Mayer-Sommer, Smartly analyzing the simplicity and the power of simple power analysis on smartcards, *CHES 2000 Proceedings*, Cambridge, MA, *Lecture Notes in Computer Science*, 1965, 78–92, 2000, Springer.
42. F.-X. Standaert, E. Peeters, C. Archambeau, and J.-J. Quisquarter, Towards security limits in side-channel attacks (with an application to block ciphers), *CHES 2006*, Cambridge, MA, *Lecture Notes in Computer Science*, 30–45, 2006, Springer.
43. K. Walus, A. Vetteth, G. A. Jullien, and V. S. Dimitrov, RAM design using quantum-dot cellular automata, *Proceedings of 2003 Nanotechnology Conference*, San Francisco, CA, Vol. 2, pp. 160–163, 2003.
44. M. T. Niemier, A. F. Rodrigues, and P. M. Kogge, A potentially implementable FPGA for quantum-dot cellular automata, *First Workshop on Non-Silicon Computation (NSC-1)*, Boston, MA, 2002.

13

High-Electric-Field-Initiated Information Processing in Nanoelectronic Devices

Vijay K. Arora

CONTENTS

13.1 Introduction

The information propagates in semiconductor devices, circuits, and chips due to charge carriers (electrons or holes) drifting in response to an applied electric field (dc, ac, or electromagnetic). External influences exist in a variety of forms: pressure, chemicals, humidity, stress, light, and electromagnetic fields (Arora, 2009). All of these influences are converted into electrical signals that are processed. In the macroscale device of yesteryears (say 1 cm = 10,000 μm in length), the electric field $\mathcal{E}$ for a processing voltage of 5 V was $\mathcal{E} = V/L = 5$ V/1 cm = 5 V/cm, where V is the voltage (potential difference) applied across the length L of the sample. Most devices on a silicon chip are now of nanometer length (say 100 nm = 0.1 μm). The electric field for this small length is extremely high: $\mathcal{E} = V/L = 5$ V/0.1 μm = 50 V/μm = 50 MV/cm. The linear I–V relation depicted by the familiar Ohm's law becomes nonlinear (Arora, 1985, 2000, 2002) and the current eventually saturates. The analog and digital signal processing is now controlled by the presence of a high electric field exceeding its critical value $\mathcal{E}_c = V_t/\ell_o$. The thermal voltage $V_t = k_B T/q = 0.0259$ V at room temperature ($T = 300$ K) and the typical mean free path is $\ell_o = 100$ nm. Here $k_B = 1.38 \times 10^{-23}$ J/K is the Boltzmann constant and $q = 1.6 \times 10^{-19}$ C is the electronic charge. The critical electric field $\mathcal{E}_c$ is thus 0.259 V/μm. The related critical voltage is $V_c = \mathcal{E}_c L = 2.59$ kV for a $L = 1.0$ cm device and $V_c = 0.259$ V for a μm-length device. Therefore, for an applied voltage that exceeds the critical voltage, Ohm's law and circuits designed based on this law are bound not to make the grade when performance is evaluated.

Another transformation that occurs in signal processing is that the length of the conducting channel is now below 100 nm. As devices are scaled down to an nm-regime, quantum effects due to the quantum nature of the electrons play a predominant role (Arora et al., 1987; Tan et al., 1993). These quantum effects make a conducting channel a waveguide through which electron waves propagate. One distinct feature of the quantum confinement to nanoscale dimensions is that there are standing waves when electrons are confined to dimensions less than the de Broglie wavelength that is typically 10 nm (Fairus and Arora, 2000; Ahmadi et al., 2008, 2009a,b). These quantum effects make devices effectively low-dimensional meaning that analog type characters of a charge carrier exist in less than three dimensions as other dimensions go digital because of the quantum jumps in the energy spectrum. The other factor is that the length of the device may become less than the mean free path. In this case, the electron can avoid collision and may become ballistic (Shur, 2002; Wang and Lundstrom, 2003; Arora, 2009).

In a very high electric field, the drift velocity response to the electric field saturates as all randomly moving vectors in a given device are streamlined and become unidirectional making the intrinsic velocity (extensively discussed by Arora, 2009) the ultimate velocity. This intrinsic velocity is ballistic as it does not depend on traditional scattering processes. For nanoscale devices, both the mobility as well as the saturation velocity are ballistic, which transforms the way we design, evaluate performance, and characterize nanoscale devices (Arora et al., 2007).

13.2 Ohmic (Linear) Transport Defined (Twentieth Century Paradigm)

A bulk resistor in the form of a sheet (Figure 13.1) with dimensions $L_{x,y,z} = L, W$, and T that are much larger than the de Broglie wavelength λ_D is a medium for propagating quantum waves in whatever direction the electric field is applied. The presence of ohmic contacts has its own resistive effect but is normally neglected in the primary stage. As the voltage V is applied to the

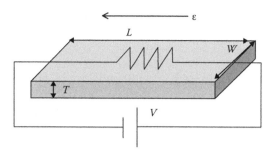

FIGURE 13.1
A sheet resistor with all dimensions
$L_x = L$, $L_y = W$, and $L_z = T$.

homogenous sample across its length L establishing an electric field $\mathcal{E} = V/L$, an electron drift with a velocity v_d results. As a consequence, there is a current $I = n_3 q v_d A_c$ with a charge flow related to the volume carrier concentration $n_3 = N/V$ (N is the total number of electrons in a sample) of the itinerants electrons per unit volume, a drift velocity v_d that is the assumed linear function of the electric field, and an area of cross-section A_c perpendicular to the flow.

The current $I = Q/t$ as the rate of charge flow in a homogeneous sample with no concentration gradients is given by

$$I_n = \frac{Q_n}{t} = \frac{n_3(LWT)(-q)}{t} = -n_3\frac{L}{t}qA_c = -n_3 v_{dn}qA_c \qquad (13.1)$$

$A_c = WT$ is the area of the cross-section perpendicular to the direction of charge flow (current). $v_{dn} = L/t$ is the drift response to the electric field $\mathcal{E}$ modeled by the linear velocity-field relation:

$$v_{dn} = -\mu_{on}\mathcal{E} \qquad (13.2)$$

μ_{on} is the ohmic mobility. This linear response is the origin of Ohm's law that is obtained from Equation 13.1 by the substitution of Equation 13.2 using $\mathcal{E} = V/L$ resulting in

$$I_n = \frac{V}{R_{on}}, \quad R_{on} = \frac{1}{qn_3\mu_{on}}\frac{L}{A_c} = \rho_{3n}\frac{L}{A_c} \qquad (13.3)$$

R_{on} is the ohmic resistance that depends on the material properties of the sample given by the product of resistivity $\rho_3 = 1/qn_3\mu_{on}$ (SI units: $\Omega \cdot m$) and the geometry ratio L/A_c. Designers use the sample properties and geometry ratio to design, characterize, and assess the electric response to an applied stimulation.

As a semiconducting sample contains both electrons and holes, the holes current with each hole charge $+q$ and drift response to the applied electric field as $v_{dp} = +\mu_{op}\mathcal{E}$ is similarly evaluated to give current in the same direction of the electric field.

$$I_p = \frac{Q_p}{t} = \frac{p_3(LWT)(+q)}{t} = +p_3\frac{L}{t}qA_c = +p_3 v_{dp}qA_c \tag{13.4}$$

The total current $I = I_n + I_p$ is unidirectional in the direction of the applied electric field, hence, the electron and hole currents always add, not subtract. The total current I response to the applied voltage is thus given by

$$I = I_n + I_p = \frac{V}{R_o} \quad R_o = \rho\frac{L}{A_c} \quad p_3 = \frac{1}{\sigma} = \frac{1}{(n_3\mu_{on} + p_3\mu_{op})q} \tag{13.5}$$

$\sigma = \sigma_n + \sigma_p = n_3\mu_{on}q + p_3\mu_{op}q$ is the sample conductivity that is the sum of the electron and hole conductivities. Ohmic resistance R_o (in Ω) is the inverse slope of the linear I–V characteristics in the ohmic model.

13.3 Discovery of Sat Law (Twenty-First Century Paradigm)

Ohm's Law enjoyed its superiority in the performance assessment of all conducting materials until it was discovered that the velocity cannot increase indefinitely with the increase of electric fields and eventually saturates to a value $v_{satn(p)}$. The discovery that the drift velocity has a nonlinear response stunned many in the twentieth century. A number of empirical relations to relate drift response to the electric field were put on trial (and are still being tried) to see which one fits the experimental data best. The most prominent (tentatively called Sat law, which emphasizes saturation as well as truth in the Sanskrit language) of these relations is (Greenberg and del Alamo, 1994)

$$v_{Dn(p)} = \frac{-\mu_{on(p)}\mathcal{E}}{[1 + (\mathcal{E}/\mathcal{E}_{cn})^{\gamma_{n(p)}}]^{1/\gamma_{n(p)}}} \quad \text{with } \mathcal{E}_{cn(p)} = \frac{v_{satn(p)}}{\mu_{on(p)}} \tag{13.6}$$

A wide combination of $\mathcal{E}_{cn(p)}$ and $\gamma_n = 1 - 2.8$ values has been utilized to fit the experimental data and can be changed at will in most simulation programs.

The normalized drift response to the normalized electric field is shown in Figure 13.2 for $\gamma = 1$ (normally considered valid for holes), $\gamma = 2$ (normally considered valid for electrons), and a more recent value of $\gamma = 2.8$ (Greenberg and del Alamo, 1994) for electrons in the InGaAs microchannel. Also shown in Figure 13.2 are the extreme linear and saturation limits of the empirical curves. As one can see, a value of γ does not affect the linear

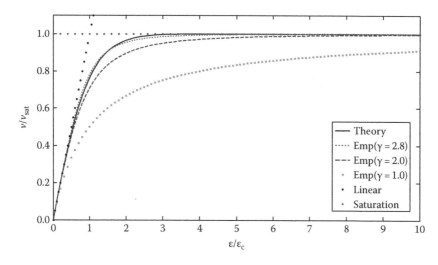

FIGURE 13.2
Drift velocity response to the electric field.

or saturation limits. It only affects the rise toward saturation. A larger value of γ brings the velocity closer to its saturation value faster as the electric field is increased; the mobility and saturation value are unaffected by γ. The theoretical solid curve is an outcome of the theory (Arora, 1985, 2000, 2008; Arora and Das, 1990). According to this theoretical formalism (Arora, 2009), the velocity response to the electric field is given by

$$v_{Dn(p)} = v_{satn(p)} \tanh\left(\frac{\mathcal{E}}{\mathcal{E}_{cn(p)}}\right) \tag{13.7}$$

The critical electric field $\mathcal{E}_c = v_{sat}/\mu_o$, both for electrons and holes, defines the boundary between the Ohm's and Sat laws. When the electric field $\mathcal{E}$ is lower than its critical value $\mathcal{E}_c$ ($\mathcal{E} < \mathcal{E}_c = v_{sat}/\mu_o$ or $\mathcal{E}/\mathcal{E}_c < 1$), it is reasonable to establish the validity of Ohm's law. However, as $\mathcal{E}$ substantially exceeds $\mathcal{E}_c$, the drift velocity saturates to $v_{n(p)} = v_{satn(p)}$ with a typical value in the neighborhood of 10^5 m/s. The typical mobility in a silicon device is 0.1 m^2/V · s. Considering these facts, $\mathcal{E}_c = 10^6$ V/m = 1.0 MV/m = 1.0 V/μm. The applied electric field depends on the length of the device. With a typical voltage of 5 V applied to a cm-length color-coded laboratory resistor, $\mathcal{E} = V/L = 5$ V/0.01 m = 500 V/m is much smaller than $\mathcal{E}_c = 10^6$ V/m. As fabricated microprocessors embrace transistors 1 μm in dimension (Arora and Das, 1990), or even lower as nanometer (nm) scales are encountered, the electric field even with $L = 1$ μm is $\mathcal{E} = V/L = 5$ V/1.0 μm = 5.0 V/μm surpassing its critical value of $\mathcal{E}_c = 1.0$ V/μm. The drift velocity is then closer to its saturation value. The critical voltage $V_c = \mathcal{E}_c L$ at the onset of nonlinear behavior is equal to 10 kV for a cm-long resistor and merely 1.0 V for the μm-long resistor. So any reasonable voltage applied to a microresistor is sure to trigger the breakdown of Ohm's law. It is worth mentioning that true saturation velocity is unobtainable as it requires an infinite electric field. The device will break down at a finite electric field. That is one reason why saturation velocity can only be indirectly determined. The value determined depends on the algorithm or theoretical framework, and hence, a wide variety of values are found in the literature.

It is easier to apply the Sat law for a microresistor as Equation 13.6 is substituted in Equation 13.1 resulting in

$$I_n = \frac{V}{R_{on}} \frac{1}{[1 + (V/V_{cn})^{\gamma_n}]^{1/\gamma_n}} = \begin{cases} \dfrac{V}{R_{on}} & V < V_{cn} \\ I_{sat} = n_3 q v_{sat} A_c & V \gg V_{cn} \end{cases} \tag{13.8}$$

with

$$V_{cn(p)} = \frac{v_{satn(p)}}{\mu_{on(p)}} L \tag{13.9}$$

Equation 13.8 is the empirical relation that fits experimental observations made on a variety of semiconductors, with parameters depending on the sample.

To explain this nonlinear behavior, Arora (1985) predicted that the saturation velocity is limited by the thermal velocity. In recent years, the paradigm developed in 1985 has been extended to include low-dimensional materials by including degenerate statistics (Arora et al., 2007; Arora, 2009). The appropriate saturation velocity for a given dimensionality is then limited by the thermal velocity for samples following nondegenerate statistics and by the Fermi velocity for those following degenerate statistics. The ultimate test of any theory comes from its ability to interpret the experimental data and to be able to make predictions for future designs. It is easier to apply the Sat law for a microresistor of Figure 13.1 as Equation 13.7 is substituted into Equation 13.1 resulting in Arora's law given by

$$I = I_{sat} \tanh\left(\frac{V}{V_c}\right) = \frac{V_c}{R_o} \frac{\tanh(V/V_c)}{V/V_c} \tag{13.10}$$

with

$$R_o = \frac{1}{n_3 q \mu_o} \frac{L}{A_c} = \frac{1}{n_2 q \mu_o} \frac{L}{W} = \frac{1}{n_1 q \mu_o} L \tag{13.11}$$

where

n_3 is the volume carrier concentration (number of carriers per unit volume)

n_2 is the surface carrier concentration (number of carriers per unit area)

n_1 is the linear carrier concentration (number of carriers per unit length)

μ_o is the carrier mobility

L is the length in the direction of current

$A_c = WT$ is the area of the cross-section perpendicular to the charge flow

W is its width

Greenberg and del Alamo (1994) measured the *I–V* characteristics of an InGaAs HFET structure. Their results indicated a direct experimental verification of Equation 13.10. In their measurements, a good fit to Equation 13.8 was obtained for $\mathcal{E}_c = 3.8$ kV/cm and $I_{sat} = 565\ \mu A/\mu m$. $V_c = \mathcal{E}_c L = 0.38$ V for a 1 μm resistor. For a macroresistor of $L = 1$ cm $= 10{,}000\ \mu m$, this value becomes $V_c = \mathcal{E}_c L = 3.8$ kV. Therefore, Ohm's law is valid up to 3.8 kV for a macroresistor of $L = 1.0$ cm. As practical voltage levels are much lower than this extreme value, Ohm's law is generally valid for macroscale resistors. However, with $V_c = 0.38$ V for a 1 μm resistor, the current is close to the saturation value for a reasonable voltage applied. Figure 13.3 is a plot of Equation 13.10 for resistors with a length of $L = 5, 20$, and 80 μm with all other dimensions and material properties remaining the same. The solid curve is the replication of the experimental data for a 5 μm resistor. When plotted on a scale extending to 10 V, 5 μm resistors clearly show a deviation from Ohm's law for relatively low voltages above $V_c = 1.9$ V. For 20 μm resistors, $V_c = 7.6$ V and for $L = 80\ \mu m$, resistor $V_c = 30.4$ V. The 20 μm resistors clearly show a transition from nonohmic to ohmic behavior as the channel length is increased.

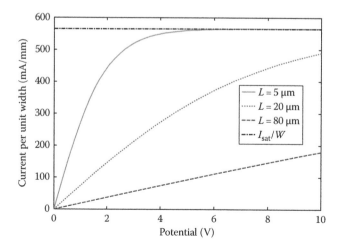

FIGURE 13.3

I–V characteristics of microresistor of length $L = 5, 20$, and 80 μm, approaching the same saturation current.

The 80 μm resistor follows Ohm's law when viewed up to 10 V only since the nonohmic behavior is masked and will become noticeable only when the scale is extended to beyond 30.4 V. This comparison of micro- and macro-resistors shows that $V = V_c$ marks a transition from the ohmic to the nonohmic regime. The current saturation per unit width (I_{sat}/W) depends on the doping density and the saturation velocity, but not on the length of the resistor or scattering-limited mobility, as many earlier works have conjectured. All three curves approach the same saturation value I_{sat} proportional to the width of the channel even if the length and hence the resistance of each resistance is different. The saturation current does not appear to go to the same value when viewed on the 10 V scale. As the scale is extended to the kV regime, the saturation current will be the same for all three resistors.

Normalized *I–V* characteristics with I/I_{sat} plotted as a function of normalized voltage V/V_c follow the same pattern as in Figure 13.2 when the axes are relabeled with I/I_{sat} to replace v/v_{sat} and V/V_c to replace $\mathcal{E}/\mathcal{E}_c$. As *I–V* characteristics become nonlinear, the resistance $R = V/I$ and signal resistance $r = dV/dI$ are given by (Greenberg and del Alamo, 1994)

$$R_{n(p)} = \frac{V}{I_{n(p)}} = R_{on(p)}\left[1 + \left(\frac{V}{V_c}\right)^{\gamma_{n(p)}}\right]^{\frac{1}{\gamma_{n(p)}}}$$

$$= R_{on(p)}\frac{1}{\left[1 - (I_{n(p)}/I_{satn(p)})^{\gamma_{n(p)}}\right]^{\frac{1}{\gamma_{n(p)}}}} \tag{13.12}$$

$$r_{n(p)} = \frac{dV}{dI_{n(p)}} = R_{on(p)}\left[1 + \left(\frac{V}{V_{cn(p)}}\right)^{\gamma}\right]^{1+\frac{1}{\gamma}}$$

$$= R_{on(p)}\frac{1}{\left[1 - (I_{n(p)}/I_{satn(p)})^{\gamma_{n(p)}}\right]^{1+\frac{1}{\gamma}}} = \frac{R_{n(p)}^{\gamma_{n(p)}+1}}{R_{on(p)}^{\gamma_{n(p)}}} \tag{13.13}$$

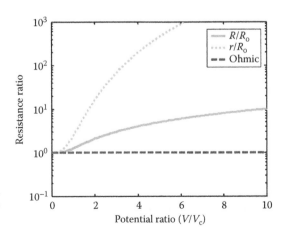

FIGURE 13.4
Resistance blow-up ratio R/R_o and r/R_o as a function of applied voltage ratio V/V_c with $\gamma = 2.8$.

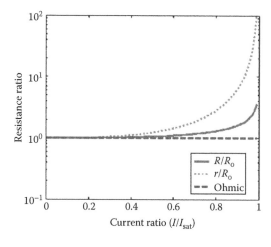

Figures 13.4 and 13.5 show this resistance blow-up effect as a function of voltage and current for $\gamma = 2.8$. The resistance is close to its ohmic value when $V < V_c$, but rises dramatically beyond V_c; incremental (signal) resistance rises much faster than the direct resistance. The current tends to reach its saturation value and as the current saturation is approached, the signal resistance rises sharply and tends to rise to infinity.

13.4 Charge Transport in 2D and 1D Resistors

The charge transport in the two-dimensional (2D) sheet resistor follows the same pattern as in the three-dimensional (3D) sample except now the surface carrier concentration $n_2 = N/LW$ (in m^{-2}) becomes important, so the current is given by

$$I_{n(p)} = \frac{Q_{n(p)}}{t} = \frac{n(p)_2(LW)(\mp q)}{t} = n(p)_2 \frac{L}{t} qW = n(p)_2 v_{Dn(p)} qW \qquad (13.14)$$

In the ohmic domain $v_{n(p)} = \mp \mu_{on(p)} \mathcal{E}$, Equation 13.14 reduces to familiar Ohm's law $I = V/R_o$, but the resistance R_o is now given by

$$R_o = \rho_2 \frac{L}{W} \quad \rho_2 = \frac{1}{(n_2 \mu_{on} + p_2 \mu_{op})q} \qquad (13.15)$$

where ρ_2 is the sheet resistivity (in $\Omega/\square$).

Similarly, for one-dimensional (1D) nanowires, the current and resistance is given by

$$I_{n(p)} = \frac{Q_{n(p)}}{t} = \frac{n(p)_1(L)(\mp q)}{t} = n(p)_1 \frac{L}{t} q = n(p)_1 v_{Dn(p)} q \qquad (13.16)$$

$$R_o = \rho_1 L \quad \rho_1 = \frac{1}{(n_1 \mu_{on} + p_1 \mu_{op})q} \qquad (13.17)$$

where

n_1 is the carrier concentration per unit length (m^{-1})
ρ_1 is the line resistivity (Ω/m)

With modification noted as above, Equations 13.8 and 13.10 are both valid for low-dimensional nanostructures and are in perfect agreement when $\gamma = 2.8$ is considered. The only difference comes in the saturation current $(I_{sat} = n_3 q v_{sat} A_c = n_2 q v_{sat} W = n_1 q v_{sat})$ that is proportional to area A_c in a 3D geometry, width W in a 2D geometry, and is independent of transverse dimensions in a 1D geometry.

13.5 Power Consumption

Velocity saturation limits the current in any resistor. However, it is clear that the outcomes of the theoretical equation (Equation 13.10) agree very well with the empirical equation (Equation 13.8). The current–voltage (*I–V*) characteristic based on Equation 13.10 are given by

$$I_{n(p)} = I_{satn(p)} \tanh\left(\frac{V}{V_{cn(p)}}\right) \qquad (13.18)$$

where $I_{satn(p)} = (V_{cn(p)}/R_{on(p)})$ is the saturation current. The ohmic power consumption law is thus expected to transform when *I–V* characteristics become nonohmic. As an example, let us take $R_{on} = 33.6\ \Omega$ for an InGaAs sheet resistor with $W/L = 20$. As indicated by Equation 13.15, the resistance is the same for all sheet resistors with the same W/L ratio. In a VLSI design, all the dimensions are measured in terms of the technology length (or feature size) λ_s ($L = s_L\lambda_s$, $W = s_W\lambda_s$, $T = s_T\lambda_s$); s_L, s_W, and s_T are scaling parameters. To demonstrate the deviation from established paradigms, two resistors with the same thickness or diffusion depth ($T_1 = T_2$ or same n_2) are considered for performance evaluation. One has the scaling factors $s_{L1} = 5$ and $s_{W1} = 100$ and the other has the scaling factors $s_{L2} = 10$ and $s_{W2} = 200$, assuming the technology feature size $\lambda_s = 1.0\ \mu m$. The relevant dimensions are $W_1 = 100.0\ \mu m$, $L_1 = 5.0\ \mu m$, $W_2 = 200.0\ \mu m$, and $L_2 = 10.0\ \mu m$ giving

$W/L = 20$ for both resistors. The experimentally determined critical electric field is 3.8 kV/cm for the InGaAs diffused layer (Greenberg and de Alamo, 1994) giving $V_{c1} = 1.9$ V and $V_{c2} = 3.8$ V. The power consumed by each of these resistors is given by

$$P = V I_n = V I_{\text{satn}} \tanh\left(\frac{V}{V_{cn}}\right) = \frac{V^2}{R_o} \frac{\tanh\left(\dfrac{V}{V_{cn}}\right)}{\dfrac{V}{V_{cn}}} \tag{13.19}$$

In the ohmic limit $(V < V_c)$, the usual ohmic expression is obtained by

$$P_o = \frac{V^2}{R_o} \tag{13.20}$$

In the saturation limit $(V \gg V_c)$, $P_{\text{sat}} = I_{\text{sat}} V$ is the linear function of the voltage. Thus, the power consumption law changes at the $V = V_c$ transition point with an increase in the voltage. This transition is clearly visible in Figure 13.6. The solid line shows the quadratic behavior for both resistors. However, when current saturation is considered, the one with the lower length (or lower V_c) consumes less power than the one with the higher length (or higher V_c). Arora's law predicts the reduced power consumption in the nonohmic domain $(V > V_c)$ with a linear dependence on voltage, but correctly predicts the quadratic behavior in the ohmic domain $(V < V_c)$.

This is an important information for circuit designers as the power consumption by the devices on the chip and its heat removal is an important

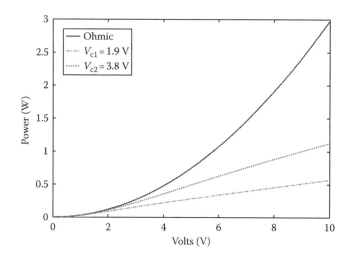

FIGURE 13.6
Power as a function of voltage for two resistors with differing V_c but same resistance R_o.

feature of the VLSI design. Power consumption multiplied by the frequency is a figure of merit in the VLSI design; each component can be traded for optimal design once the framework is implemented.

13.6 Nonohmic Circuit Behavior

As current–voltage characteristics become nonlinear and resistance is no longer a constant, it is natural that familiar voltage divider and current divider rules may not apply. As stated before, the current–voltage relationship is

$$I = \frac{V_c}{R_o} \tanh\left(\frac{V}{V_c}\right) = I_{sat} \tanh\left(\frac{V}{V_c}\right) \tag{13.21}$$

This version of I–V characteristics is more compact than any of the other versions in the literature, and hence is being used in the analysis as follows.

As the length of a resistor plays a predominant role in transforming I–V and resistive behavior, it is worthwhile to see how the voltage will be divided between two resistors having the same ohmic resistance ($R_{o1} = R_{o2}$) but differing dimensions as indicated in Section 13.5. When connected in a series, as in Figure 13.7, an applied voltage V will get divided equally across each resistor according to the voltage division dictated by Ohm's law. However, when Equation 13.21 is used in place of Ohm's law ($I = V/R_o$), the voltage V_1 across R_1 is obtained from

$$V_{c1} \tanh\left(\frac{V_1}{V_{c1}}\right) = V_{c2} \tanh\left(\frac{V - V_1}{V_{c2}}\right) \tag{13.22}$$

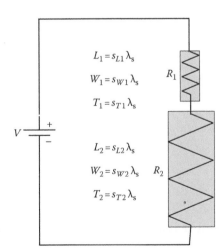

FIGURE 13.7
Voltage divider circuit with two microresistors ($R_{o1} = R_{o2}$).

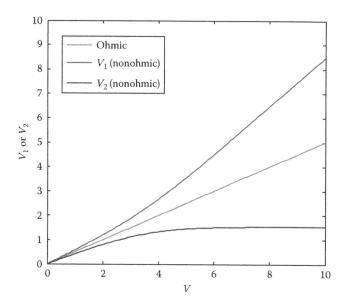

FIGURE 13.8
Voltage division in two microresistors with the same ohmic value ($R_{o1} = R_{o2}$).

with $V_{c1} = 1.9$ V for the 5 μm resistor and $V_{c2} = 3.8$ V for the 10 μm resistor. The output $V_{1,2}$ across the 5 μm resistor is given in Figure 13.8 when the divider circuit is excited by a voltage source of 0–10 V. Also shown are the results expected from Ohm's law. The lower-length resistor ($L = 5$ μm) is more resistive as the voltage across the combination is increased and hence gets more voltage across it. On the other hand, the voltage drop across the 10 μm resistor is less than its ohmic value.

Figure 13.9 shows a current divider circuit where the same two resistors ($R_{o1} = R_{o2} = 33.6$ Ω) are considered in a parallel configuration. With a current

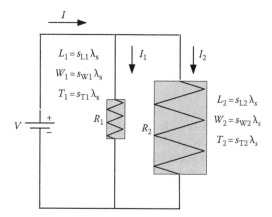

FIGURE 13.9
Current divider circuit with two microresistors ($R_{o1} = R_{o2}$).

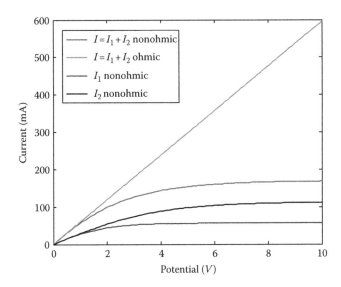

FIGURE 13.10
Ohmic and nonohmic currents in a current divider circuit with two microresistors ($R_{o1} = R_{o2}$).

per unit length of 565 mA/mm, the saturation current for the resistor 1 is $I_{sat1} = 56.5$ mA and for resistor 2 is $I_{sat2} = 113$ mA as the current in the saturation regime is proportional to the width of the resistor. As before, the critical voltages $V_{c1} = 1.9$ V and $V_{c2} = 3.8$ V. As Figure 13.10 shows, the resulting current in each resistor is substantially below its ohmic value. Only when $V < V_c$ can the validity of Ohm's law can be assumed. As V increases beyond V_c, the maximum current that can be drawn from the current source is 170 mA as compared with 595 mA at $V = 10$ V predicted by Ohm's law. When two parallel channels are conducting, more current will pass through the higher length channel even if both resistors have the same ohmic value. Figures 13.8 and 13.10 clearly show that the circuit laws need close scrutiny when micro/nanoresistors are encountered in VLSI circuits.

13.7 CMOS Circuit Design

The basic element of modern digital circuits is the complementary metal-oxide-semiconductor (CMOS) inverter. The CMOS physical structure is shown in Figure 13.11 and its circuit configuration is shown in Figure 13.12. The circuit inverts the logical value of the voltage V_{IN} applied to its input terminal. In a CMOS inverter, the gate of n- and p-MOSFET (NMOS and PMOS) are tied together. A positive gate bias, which turns the NMOS on,

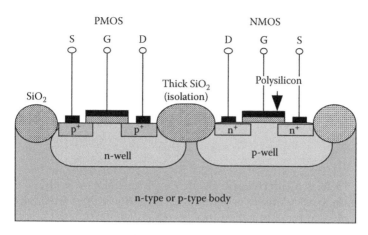

FIGURE 13.11
The physical structure of a CMOS consisting of NMOS and PMOS channels in a twin-well configuration.

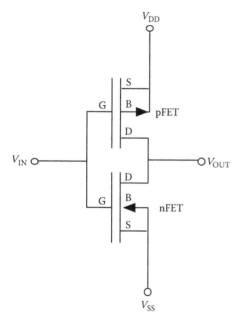

FIGURE 13.12
The CMOS inverter circuit with PMOS as a pull-up transistor and NMOS as a pull-down transistor.

turns off the PMOS and vice versa, and hence, the output is connected to V_{SS} (that is when input is high, output is low). This reverses when the input is low, then, the output is connected to V_{DD} (high). Both stable states correspond to a very low current consumed from the power supply. The switching speed is enhanced when the current $I_{Dn(p)}$ in NMOS and PMOS transistors are the same (Arora and Das, 1990):

$$I_{Dn(p)} = \frac{W_{n(p)} C_{ox} \mu_{on(p)}}{L_{n(p)}} \times \frac{\left(V_{GTn(p)} - \frac{1}{2} V_{DSn(p)} \right) V_{DSn(p)}}{1 + \dfrac{V_{DSn(p)}}{V_{cn(p)}}} \qquad (13.23)$$

with

$$C_{ox} = \frac{\varepsilon_{ox}}{t_{ox}}, \quad V_{GTn(p)} = V_{GSn(p)} - V_{Tn(p)}, \quad V_{cn} = \frac{v_{satn(p)}}{\mu_{on(p)}} L_{n(p)} \qquad (13.24)$$

Equation 13.23 is a direct result of the empirical relation with $\gamma = 1$ that is used for ease in calculations as ohmic and saturation values are unaffected by the value of γ. In the fabrication process, the oxide thickness t_{ox} is the same for both transistors and the threshold voltage can be adjusted so that $V_{Tn} = V_{Tp}$. In a traditional design based on Ohm's law ($V_{DS} < V_c$), the length for both the NMOS and PMOS (minimum feature size) is equal ($L_n = L_p$) and the width is scaled with mobility so the $(W\mu_o)_{n(p)}$ product and hence the channel current in both complementary transistors is equal. To account for the lower mobility of PMOS, the width of the p-channel is scaled so that $W_p = W_n (\mu_{on}/\mu_{op}) \approx 2$ for a typical case of $\mu_{on} = 2\mu_{op}$. However, in a submicron length channel, $V_{cn(p)}$ must also be matched for both channels. Assuming that the saturation velocity v_{sat} is the same for both channels, as is found experimentally, an alternate design for identical current in both channels is

$$\frac{W_p}{W_n} = \frac{v_{satn}}{v_{satn}} \approx 1, \quad \frac{L_n}{L_p} = \frac{\mu_{on}}{\mu_{op}} \approx 2 \qquad (13.25)$$

In the model presented that takes into account the velocity saturation effect, the width is scaled according to the saturation velocity ($I_{sat} = n_2 q v_{sat} W$) as the saturation current is proportional to the product of $v_{sat} W$. This requires that the width be scaled inversely proportional to the saturation velocity. In the linear region ($V < V_{Dsat}$), when the voltage is less than the drain voltage V_{Dsat} at the saturation point, the channel length is scaled according to the mobility to keep V_c identical requiring the length to be scaled proportional to the mobility of the channel.

13.8 Transit Time Delay

As digital signals are processed through micron-length devices, there are two factors that play a predominant role in determining the switching speed and ballisticity of the device. One is the transit-time delay that is the time taken by a carrier (electron or hole) to propagate through the length of the device L. The transit time delay τ_t is then given by

$$
\tau_{tn(p)} = \frac{L}{v_{satn(p)} \tanh\left(\dfrac{V}{V_{cn(p)}}\right)} = \begin{cases} \dfrac{L^2}{\mu_{on} V} & V < V_c \\ \dfrac{L}{v_{satn}} & V \gg V_c \end{cases} \tag{13.26}
$$

The primary reason for making devices small was to cut down this transit time delay. For electrons ($m_n{}^* = 0.057 m_o$) appropriate for the $In_{0.15}Ga_{0.85}As$ channel and surface carrier density of 10^{12} cm^{-2}, $v_{satn} \approx 3.41 \times 10^5$ m/s. Ohm's law overestimates the transit time delay.

The other of the two factors is the RC time constants due to a capacitive load on a circuit (Tan et al., 2007) is discussed in Section 13.9. As resistance blows up in a micro/nano-length channel, so do the RC time constants. This will also influence the analog signal processing as resistance blow-up will affect the impedance of the circuit with an increase in the dc component of the bias.

13.9 RC Time Delay

In digital signal processing, the transit time delay (τ_t) and RC time constants (τ_{RC}) compete in the limiting of the speed of a signal. Considerable progress has been made in reducing the transit-time delay due to the scaling down of the size of the devices that is now in the nano-regime. Efforts are underway to utilize low-resistivity materials and low-k dielectrics to shorten the RC time delay. However, there are intrinsic factors that enhance the RC timing delay due to the resistance blow-up when the step voltage V exceeds the critical voltage V_c for the onset of nonohmic behavior. A prototype RC circuit with a sheet resistor and capacitive load is shown in Figure 13.13.

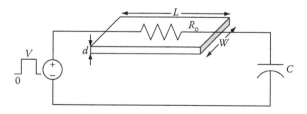

FIGURE 13.13
A prototype RC circuits with the resistor a few nanometer in length.

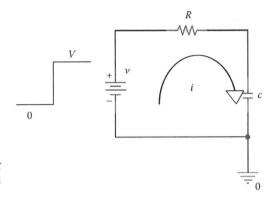

FIGURE 13.14
Equivalent circuit diagram for the *RC*
transient circuit as the input is excited
from 0 V to *V* volts.

Figure 13.14 shows the equivalent circuit where a digital voltage signal
rises from low to high ($v = 0$ to V at time $t = 0$). The rise time of the capacitor
voltage to the ultimate battery voltage will enhance as resistance is enhanced.
The voltage rises slowly because of the resistance blow-up effect. The current
through the resistor responds to the change in the voltage as given by

$$i_R(t) = I_{sat} \tanh\left(\frac{v_R(t)}{V_c}\right) \tag{13.27}$$

with

$$i_R(0) = I_{sat} \tanh\left(\frac{V}{V_c}\right) \tag{13.28}$$

The potential V is divided across the resistor and capacitor

$$V = v_R(t) + v_C(t) = V_c \tanh^{-1}\left(\frac{i(t)}{I_{sat}}\right) + \frac{q(t)}{C} \tag{13.29}$$

Differentiating Equation 13.29 with respect to t gives

$$0 = V_c \frac{1}{1 - \left(\dfrac{i}{I_{sat}}\right)^2} \cdot \frac{1}{I_{sat}} \frac{di}{dt} + \frac{i}{C} \tag{13.30}$$

Considering that $R_o = V_c/I_{sat}$, the equation can be written in the differential
format

$$\frac{1}{1 - \left(\dfrac{i}{I_{sat}}\right)^2} \frac{di}{dt} + \frac{i}{R_o C} = 0 \tag{13.31}$$

Separating the variable $i(t)$ and t on each side of the equation and substituting $\tau_o = R_o C$ yields

$$\frac{1}{i}\frac{di}{1-\left(\dfrac{i}{I_{sat}}\right)^2} = -\frac{1}{\tau_o}dt \tag{13.32}$$

Separating Equation 13.32 into partial fractions and integrating with the constant of integration $\ln K$ yields

$$\int di\left[\frac{1}{i}+\frac{1}{2}\frac{1}{I_{sat}-i}-\frac{1}{2}\frac{1}{I_{sat}+i}\right] = -\frac{t}{\tau_o}+\ln K \tag{13.33}$$

Integration yields

$$\ln\ i - \ln\left(I_{sat}-i\right)^{\frac{1}{2}}-\ln\left(I_{sat}+i\right)^{\frac{1}{2}} = -\frac{t}{\tau_o}+\ln\ K \tag{13.34}$$

Equation 13.34 can be further condensed to

$$\ln\left[\frac{i}{K(I_{sat}-i)^{\frac{1}{2}}(I_{sat}+i)^{\frac{1}{2}}}\right] = -\frac{t}{\tau_o} \tag{13.35}$$

or

$$\frac{i}{\left(I_{sat}^2-i^2\right)^{\frac{1}{2}}} = K\,e^{-\frac{t}{\tau_o}} \tag{13.36}$$

The constant K can be evaluated by invoking the initial condition of Equation 13.28. This gives

$$K = \frac{\tanh\left(\dfrac{V}{V_c}\right)}{\left(1-\tanh^2\dfrac{V}{V_c}\right)^{\frac{1}{2}}} \tag{13.37}$$

Utilizing $\operatorname{sech}^2(x)=1-\tanh^2(x)$ identity yields

$$K = \sinh\left(\frac{V}{V_c}\right) \tag{13.38}$$

Therefore,

$$\frac{i}{\left(I_{sat}^2-i^2\right)^{\frac{1}{2}}} = \sinh\left(\frac{V}{V_c}\right)e^{-\frac{t}{\tau_o}} \tag{13.39}$$

In the ohmic domain ($i \ll I_{sat}$ and $V \ll V_c$), Equation 13.39 gives

$$\frac{i}{I_{sat}} = \frac{V}{V_c} e^{-\frac{t}{\tau_0}} \Rightarrow i(t) = \frac{V}{R_0} e^{-\frac{t}{\tau_0}} \tag{13.40}$$

Equation 13.39 can be organized to solve for $i(t)$:

$$i^2 = \left(I_{sat}^2 - i^2\right) \sinh^2 \frac{V}{V_c} e^{-2\frac{t}{\tau_0}} \tag{13.41}$$

or

$$i^2 \left(1 + \sinh^2 \frac{V}{V_c} e^{-2\frac{t}{\tau_0}}\right) = I_{sat}^2 \sinh^2 \frac{V}{V_c} e^{-2\frac{t}{\tau_0}} \tag{13.42}$$

$$i^2 = \frac{I_{sat}^2 \sinh^2 \frac{V}{V_c} e^{-2\frac{t}{\tau_0}}}{1 + \sinh^2 \frac{V}{V_c} e^{-2\frac{t}{\tau_0}}} \Rightarrow i(t) = I_{sat} \frac{\sinh \frac{V}{V_c} e^{-\frac{t}{\tau_0}}}{\left(1 + \sinh^2 \frac{V}{V_c} e^{-2\frac{t}{\tau_0}}\right)^{\frac{1}{2}}} \tag{13.43}$$

Equation 13.43 can be checked for initial current $i(0)$ to correctly give

$$i(0) = I_{sat} \frac{\sinh \frac{V}{V_c}}{\left(\cosh^2 \frac{V}{V_c}\right)^{\frac{1}{2}}} = I_{sat} \tanh \frac{V}{V_c} \tag{13.44}$$

Equation 13.43 also gives $i(\infty) = 0$ as expected. In the limit $L \to \infty$, Equation 13.43 reduces to the ohmic expression

$$i(t) = \frac{V}{R_0} e^{-\frac{t}{\tau_0}} \tag{13.45}$$

Hence, the solution of Equation 13.43 satisfies all boundary conditions correctly.

Equations 13.43 and 13.45 show that the initial current will be substantially higher in the ohmic model as compared with that obtained from the non-ohmic current-saturation-limited model. In the ohmic model, the initial current $i_O(0) = V/R_0$. In the nonohmic model, it will be $i_{NO}(0) = (V_c/R_0) \tanh (V/V_c)$. The ratio of the nonohmic to ohmic initial current and related power consumption ratio P/P_o is given by

$$\frac{i_{NO}(0)}{i_O(0)} = \frac{P}{P_o} = \frac{\tanh(V/V_c)}{(V/V_c)} \tag{13.46}$$

This ratio approaches 1 in the ohmic regime $V < V_c$ as expected. However, in the regime $V \gg V_c$, the ratio decreases as $(V/V_c)^{-1}$. This drop in the initial current

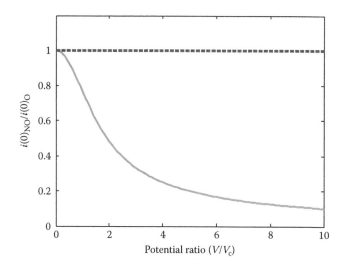

FIGURE 13.15
The ratio of initial charging current in the nonohmic model to that in the ohmic model. The dashed line represents ohmic current. The relative power ratio follows the same pattern.

as the capacitor starts charging is shown in Figure 13.15. The power consumption $P = VV_c/R_o$ in the nonohmic regime not only is smaller but is also a linear function of the applied step voltage as compared with the quadratic behavior in the ohmic regime ($V < V_c$). This transformed behavior affects the figure of merit with a trade-off between the frequency and the power.

Figure 13.16 indicates the charging response of the capacitor of $C = 1$ pF connected in the series with a resistor of $W = 100$ μm and $R_o = 16.8$ Ω. The approach toward the full potential of the capacitor is very slow, especially for the shortest resistor as compared with the ohmic response indicative of a considerable enhancement of the RC timing delay. This observation is consistent with the linear resistance rise with applied voltage when $V > V_c$. For an ac signal, the differential resistance rise is even larger.

For comparison, $t = \tau_{RC}$ is defined as the time at which the capacitor potential is $(1 - e^{-1})$ of the higher-logic potential V. The ratio of this transformed time delay to its ohmic value is obtained as

$$\frac{\tau_{RC}}{\tau_o} = \ln\left[\frac{\sinh(V/V_c)}{\sinh(V/eV_c)}\right] \tag{13.47}$$

In the regime where $V < V_c$ ($L \to \infty$), this ratio reaches unity. In the other extreme ($L \to 0$), the ratio is

$$\frac{\tau_{RC}}{\tau_o}(L \to 0) \approx (1 - e^{-1})\frac{V}{V_c} \tag{13.48}$$

This ratio rises linearly with potential in the nonohmic regime.

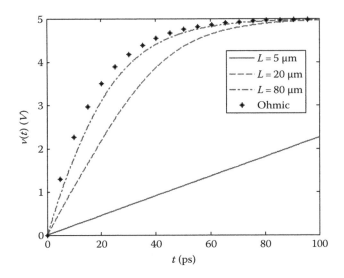

FIGURE 13.16
The response of capacitor in a nano-*RC* circuit as voltage is increased from low to high.

Figure 13.17 shows the comparison of two time delays for channels with a length of $L = 1$, 5, and 20 μm. In the long-channel limit ($L \to \infty$), the transit time delay is $L^2/\mu_o V$. Here, μ_o is the ohmic mobility and $\mathcal{E} = V/L$ is the applied electric field. In the short-channel limit ($L \to 0$), it is L/v_{sat} independent of the applied voltage.

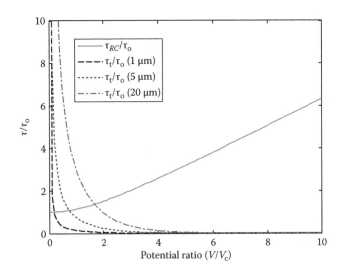

FIGURE 13.17
The normalized *RC* time delay and transit time delay as a function of normalized applied voltage.

The response of a resistive circuit to a digital signal is slow due to the enhancement of the RC time delay. However, power consumption is less. Therefore, a trade-off in the circuit design exists. This trade-off can be usefully utilized in the extraction of parasitic elements in the contact regions.

13.10 Conclusion

The theory of high-field-initiated transport has a rich and interesting history. Since World War II, starting with the work of Shockley (1951) and Ryder and Shockley (1951) who observed deviations from Ohm's law, which led to current and drift velocity becoming sublinear as the electric field (or voltage) across the device was increased. It was the availability of supercomputers in the 1970s that created intense activity in the high-field transport. Monte Carlo simulations were tried and are still in vogue by inputting a number of scattering processes and adjusting parameters until the desired outcome is obtained. The consensus emerging out of Monte Carlo experiments are that the saturation velocity is scattering-limited that may be controlled by the optical phonon energy. However, no clear connection of the saturation velocity to the ohmic low-field mobility emerged from these studies. In 1985, a high-field distribution was proposed by Arora (1985) that for the first time indicated the ballistic nature of the saturation velocity whose magnitude was comparable with the carrier thermal velocity in a nondegenerate semiconductor. While dealing with the nanoscale channels with lengths smaller than the mean free path, the possibility of ballistic mobility had been raised (Saad et al., 2008, 2009a,b; Natori, 1994) that is now confirmed by the experimental results of Lusakowski et al. (2005). Thus, both the mobility and saturation velocity may become ballistic. In fact, a number of authors (Mugnaini and Innaccone, 2005a,b) tried to invoke the ballistic nature of transport based on the earlier work of Buttiker (1986). The higher mobility leading to higher saturation velocity was conjectured, but has not proven to be true. One may wonder what is the limit of the critical voltage $V_c = (V_t/\ell_o)L$. This question is explored by Saad et al. (2009a,b). As mean free path is limited by the device length ($L \approx \ell_o$ or smaller), it is clear that $V_c \approx V_t$. In this case, Ohm's law may not be obeyed only for miniscule voltages that are smaller than the thermal voltage V_t. Given the thermal noise in most devices, it is not practical to apply voltages smaller than V_t. Hence, all future circuit design must take into account the failure of Ohm's law in circuit design, performance evaluation, and characterization. The possibility of a multivalley transfer in a high-field-initiated transport was raised by Sharma and Arora (2005). Similarly, switching rise and fall times of digital signals that were explored earlier by Samudra et al. (1994) are consistent with the paradigm presented.

The new paradigm includes the impact of the failure of Ohm's law. The ultimate current (saturation current) is shown to be limited by the saturation velocity that may be limited by the appropriate average of the thermal velocity for nondegenerate samples and the Fermi velocity for the degenerate samples (Arora, 2009). The saturation velocity may be lower than the ultimate velocity due to the onset of a quantum emission that depends on a particular experimental setup (Fairus and Arora, 2000). When applied to the voltage division and current division circuits, the lower-length resistors are found to have higher resistance as compared with the higher-length resistor even if their ohmic values are the same. When applied to the CMOS circuit design, a new scaling law that keeps the width the same and scales the length of the transistors is found to give equal currents in the two transistors. The velocity overshoot leading to the ultimate saturation velocity has recently been presented by Tan et al. (2009). The patterns emerging from these studies can be easily extended to nanowire circuits (Ahmadi, 2009).

Acknowledgments

This work is supported by distinguished visiting Professor Grant, grant number 77514 of the Universiti Teknologi Malaysia—the venue for this research. The excellent hospitality of the Faculty of Electrical Engineering in providing a conducive environment for this research is gratefully acknowledged. The author thanks Michael L. P. Tan and Desmond Chek for assistance in formatting this manuscript and offering several valuable suggestions. Brain Gain fellowship of the Academy of Sciences Malaysia funded by the Ministry of Science, Technology, and Innovation (MOSTI) is gratefully acknowledged.

References

Ahmadi, M. T., Ismail, R., and Arora, V. K. 2008. The ultimate ballistic drift velocity in a carbon nanotubes. *Journal of Nanomaterials*. Article ID number 769250, 8pp, doi: 10.1155/2008/769250.

Ahmadi, M. T., Tan, M. L. P., Ismail, R., and Arora, V. K. 2009a. The high-field drift velocity in degenerately-doped silicon nanowires. *International Journal of Nanotechnology*, 6: 601–617.

Ahmadi, M. T., Lau, H. H., Ismail, R. et al. 2009b. Current–voltage characteristics of a silicon nanowire transistor. *Microelectronics Journal* 40: 547–549.

Arora, V. K. 1985. High field distribution and mobility in semiconductors. *Japanese Journal of Applied Physics* 24: 537.

Arora, V. K. 2000. Quantum engineering of nanoelectronic devices: The role of quantum emission in limiting drift velocity and diffusion coefficient. *Microelectronics Journal* 31(11–12): 853–859.

Arora, V. K. 2002. Drift-diffusion and Einstein relation for electrons in silicon subjected to a high electric field. *Applied Physics Letters* 80(20): 3763–3765.

Arora, V. K. 2009. Theory of scattering-limited and ballistic mobility and saturation velocity in low-dimensional nanostructures. *Current Nanoscience* (CNANO) 5: 227–231.

Arora, V. K. and Das, M. B. 1990. Effect of electric-field-induced mobility degradation on the velocity distribution in a submicron-length channel of InGaAs/AlGaAs heterojunction MODFET. *Semiconductor Science and Technology* 5: 967.

Arora, V. K., David, M. S. L., and Morkoc, H. 1987. Mobility degradation in a quantum-well heterostructure of GaAs/AlGaAs Prototype. *Applied Physics Letters* 50: 1080.

Arora, V. K., Tan, M. L. P., Saad, I., and Razali, I. 2007. Ballistic quantum transport in a nanoscale metal-oxide-semiconductor field effect transistor. *Applied Physics Letters* 91: 103510.

Buttiker, M. 1986. Role of quantum coherence in series resistors. *Physical Review B* 33: 3020.

Fairus, A. T. M. and Arora, V. K. 2000. Quantum engineering of nanoelectronic devices: The role of quantum confinement in mobility degradation. *Microelectronics Journal* 32: 679–686.

Greenberg, D. R. and del Alamo, J. A. 1994. Velocity saturation in the extrinsic device: A fundamental limit in HFET's. *IEEE Transactions on Electron Devices* 41: 1334–1339.

Łusakowski, J., Knap, W., Meziani, Y. et al. 2005. Ballistic and pocket limitations of mobility in nanometer Si metal-oxide semiconductor field-effect transistors. *Applied Physics Letters* 87: 053507.

Mugnaini, G. and Iannaccone, G. 2005a. Physics-based compact model of nanoscale MOSFETs—Part I: Transition from drift-diffusion to ballistic transport. *IEEE Transactions on Electronic Devices* 52: 1795–1801.

Mugnaini, G. and Iannaccone, G. 2005b. Physics-based compact model of nanoscale MOSFETs—Part II: Effects of degeneracy on transport. *IEEE Transactions on Electronic Devices* 52: 1802–1806.

Natori, K. 1994. Ballistic metal-oxide-semiconductor field effect transistor. *Journal of Applied Physics* 76: 4879–4890.

Ryder, E. J. and Shockley, W. 1951. Mobilities of electrons in high electric fields. *Physical Review B* 81: 139–140.

Saad, I., Ismail, R., and Arora, V. K. 2008. Investigation on the effects of oblique rotating ion implantation (ORI) method for nanoscale vertical double gate NMOSFET. *Solid State Science and Technology Letters* 15(2), 69–76.

Saad, I., Tan, M. L. P., Aaron, C. E. L. et al. 2009a. Scattering-limited and ballistic transport in a nano-CMOS circuit. *Microelectronics Journal* 40, 581–583.

Saad, I., Tan, M. L. P., Hii, I. H. et al. 2009b. Ballistic mobility and saturation velocity in low-dimensional nanostructure. *Microelectronics Journal* 40, 540–542.

Samudra, G., Yong, A. K. F., Lee, T. K. et al. 1994. The role of velocity saturation on switching delay of RC-loaded inverter. *Semiconductors Science and Technology* 9: 1108–1116.

Sharma, A. and Arora, V. K. 2005. Velocity-field characteristics in the multivalley model of gallium arsenide. *Journal of Applied Physics* 97: 093704–093705.

Shockley, W. 1951. Hot electrons in germanium and Ohm's law. *Bell System Technical Journal* 30: 990–1034.

Shur, M. 2002. Low ballistic mobility in submicron HEMT. *IEEE Electron Device Letters* 23: 511–513.

Tan, L. S., Chua, S. J., and Arora, V. K. 1993. Velocity-field characteristics in selectively-doped $GaAs/Al_xGa_{1-x}As$ quantum well heterostructures. *Physical Review B* 47(20): 13868–13871.

Tan, M. L. P., Saad, I., Ismail, R. et al. 2007. Enhancement of nano-RC switching delay due to the resistance blow-up in InGaAs. *NANO* 2(4): 233–237.

Tan, M. L. P., Arora, V. K., Saad, I. et al. 2009. The drain velocity overshoot in an 80-nm metal-oxide-semiconductor field-effect-transistor. *Journal of Applied Physics* 105: 074503–(1–7).

Wang, J. and Lundstrom, M. 2003. Ballistic transport in high electron mobility transistors. *IEEE Transaction on Electronic Devices* 50: 1604–1609.

14

Packaging and Assembly of Microelectronic Devices and Systems

S. Manian Ramkumar

CONTENTS

14.1 Introduction

Ever since its invention in 1949, the transistor has served as the fundamental building block shaping the electronics industry's growth. The transistor's full potential was realized when several hundred to several thousands to several hundred millions of them were integrated into a single silicon device (silicon chip) or integrated circuit (IC) using the photolithography process. As predicted by Gordon Moore, the cofounder of Intel Corporation, the number of transistors built on a single silicon chip had doubled every 18 months and currently continues to double on an annual basis. Silicon chips today dissipate more power, require more inputs and outputs, and operate at ever-increasing speeds. This level of transistor integration into a single chip, without increasing its size, results in packaging issues that include material selection, manufacturing, input/output (I/O) management, and thermal management. The packaging of silicon chips is required to spread the closely spaced I/Os on the chip (typically 100–200 μm pitch) to a more widely spaced component lead arrangement, leadless termination arrangement, or solder ball arrangement (typically greater than 300 μm pitch) for ease of handling and assembly onto printed circuit boards (PCBs), as shown in Figure 14.1.

Packaging is a multifaceted, highly interdisciplinary technological field facing significant challenges due to miniaturization and multifunctional integration. Finding the best combination of size and weight reduction with

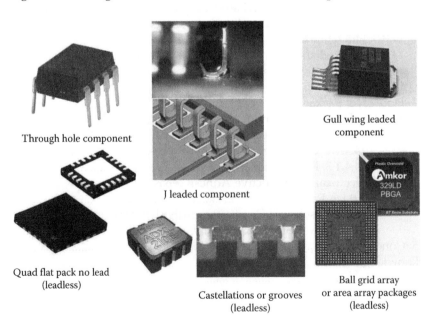

FIGURE 14.1
Leaded, leadless and area array component packages.

functional integration is essential at the silicon chip level packaging (Level 1) and also the PCB assembly level packaging (Level 2). Silicon chip packaging is in the midst of transitioning from single chip systems to higher density multi-chip systems within a single package. Also, most of the packaging has been reduced to the scale of the silicon chip, thereby improving the efficiency of the packaged device. Improved package efficiency enhances its performance and response to signals. During the course of the silicon chip development, the packaging has also evolved from being a leaded through-hole package to a bumped chip level area array package. As the trend toward higher I/O, higher performance, and higher PCB manufacturing yield continues, packages containing an array of solder balls on the bottom side of the package, such as ball grid array (BGA), chip-scale packages (CSP), flip chip (FC) packages, and wafer level chip-scale packages (WL-CSP), find more usage in the electronics industry than leaded packages. Vertically stacked three-dimensional (3D) package design, with area array solder ball interconnection between the stacked packages, is gaining widespread acceptance in mobile phone technologies and other telecommunication applications today.

The electronics package miniaturization has spawned a revolution in miniaturized sensors and micromechanical devices, which find a wide array of applications in consumer electronics, telecommunication, complex medical systems, global positioning and tracking, guidance and navigation, etc. Micromechanical sensors for pressure and acceleration have used semiconductor manufacturing technology for several decades, but recent innovations allow further miniaturization, greater flexibility, and compatibility with microelectronics. Biosensors and chemical sensors that can sense gases and chemicals are being perfected using IC manufacturing techniques and will be used in medical, food processing, and chemical processing applications. Even though miniaturized microelectronics devices will find usage in a variety of applications, the scale of the physical products that make use of these devices is still considerably larger and requires further packaging to ease the assembly of these devices into the product (Figure 14.2).

The proliferation of electronics into portable devices, consumer products, automotive, military, aerospace, medical, and industrial applications is forcing more compact and ruggedized packaging. This chapter will provide a basic description of the Level 0, Level 1 and Level 2 packaging of electronics devices (Figure 14.3).

14.2 Classification of Electronic Devices (Level 0)

Electronic devices can be broadly classified as passive and active. This classification is based on the fact that passive components do not require power to operate, whereas active components require power to operate and the power

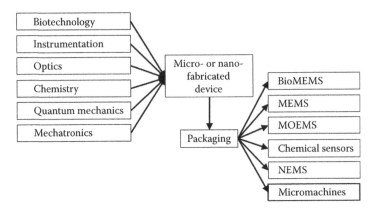

FIGURE 14.2
Packaging for technology integration into products.

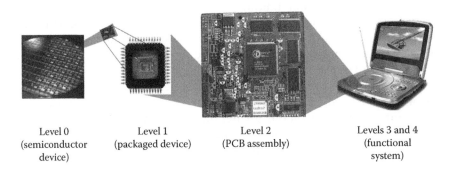

| Level 0 (semiconductor device) | Level 1 (packaged device) | Level 2 (PCB assembly) | Levels 3 and 4 (functional system) |

FIGURE 14.3
Electronics packaging levels.

applied has the ability to control electron flow. The signal passing through the passive components get altered but do not undergo any power gain, which is an increase in the power or amplitude of the signal. Most common types of passive components include resistors, capacitors, inductors, connectors, switches, etc. On the other hand, active devices have the ability to alter signals by amplifying, switching, or rectifying the signals that are passed through the devices. This unique ability to alter the signal comes from the use of semiconductors (silicon or germanium), materials that can be doped appropriately to produce more free electrons in them (N-type) or more holes in them (P-type), which can receive the electrons (Figure 14.4). Combining the P-type and N-type semiconductors provides the most common type of active components, the diodes and transistors, also referred to as discrete actives (Figure 14.5).

Many thousands to many millions of transistors integrated into a single device, at the micrometer scale, are referred to as ICs or microelectronic devices.

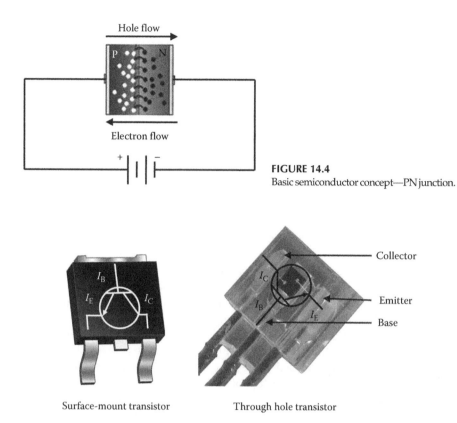

FIGURE 14.4
Basic semiconductor concept—PN junction.

Surface-mount transistor   Through hole transistor

FIGURE 14.5
Transistor.

These devices can be integrated with optical functions, sensing, mechanical actuation, biological or chemical sensing, and communication functions to form microsystems. Microsystems today incorporate nanomaterials and nanostructures and are produced using nanomanufacturing techniques to include sensing and actuation at the nanoscale. This forms the basis for nanoelectronics.

14.2.1 Microelectronic Devices

Microelectronic devices are silicon-based digital logic elements that have several millions of transistors integrated to form logic circuit elements on a micron scale. These logic elements are the building blocks of the processor technology and have evolved from vacuum tubes to transistors to very large-scale integration of transistor circuits on a single silicon chip (typically 10×10 mm size). The logic elements are formed by combining P-type and N-type semiconductors (NPN- or PNP-type transistors) (Figure 14.6). Complementary metal-oxide semiconductor (CMOS) technology has been the

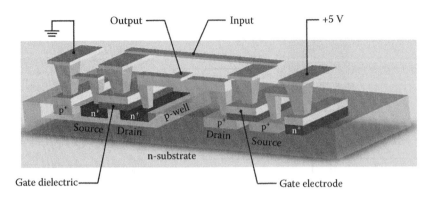

FIGURE 14.6
IC structure/transistor integration on silicon.

backbone for the development of very large-scale and giga-scale integrated logic circuits. Communication among several of these ICs is enabled only by their assembly onto a substrate (PCB, ceramic, or silicon) containing multiple layers of circuit traces and interconnect vias.

The single transistor element of a microelectronic device is formed on the silicon substrate using the photolithography process. The process starts with the formation of an N-type or P-type base silicon substrate with a layer of silicon dioxide (SiO_2) grown on top of it. This process is followed by the application of photoresist and masking the resist with a positive image of the pattern, to be etched on the SiO_2. This pattern will correspond to the regions where P-type or N-type semiconducting elements are required on the base substrate. Exposing the photoresist material to an ultraviolet (UV) light source hardens the exposed regions of the resist leaving the unexposed section soft enough to be washed off. Once the soft photoresist is washed off, the exposed SiO_2 is chemically etched and the exposed base substrate is diffused with the required semiconducting element. The above process is repeated to form another semiconducting element on to the previous element, thus creating a bipolar junction transistor (multilayer structure). Once the many thousands or millions of transistor junctions are developed and the entire logic circuit is built on the silicon chip, further aluminum metallization is added to integrate many such logic circuits. The aluminum metallizations or pads that are added enable the next level of packaging and interconnection.

Many of the features on the microelectronic devices are moving into the nanometer scale. This will further shrink the size of the fundamental transistor element, thereby enabling an increase in density, higher speeds, and low power consumption. Nanotechnology will be capable of encompassing a trillion molecular-scale transistor devices within a square centimeter (Figure 14.7). The main challenge is not in making these nanoscale devices but in interconnecting them in appropriate circuit architectures and in managing power dissipation.

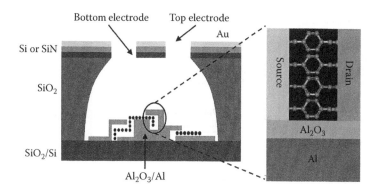

FIGURE 14.7
Molecular transistor.

14.2.2 Optoelectronic Devices

Optoelectronic devices, as the name indicates, use a combination of opto (light and optics) and electronics in one device. Some of the common smaller scale optoelectronic devices used in our daily life include light-emitting diodes (LEDs), the read-write head of CD and DVD players, document scanners and copiers, barcode readers, laser pointers, etc. (Figure 14.8).

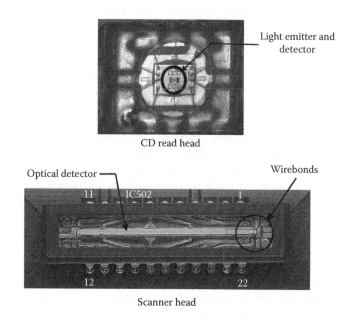

FIGURE 14.8
CD read head and scanner head.

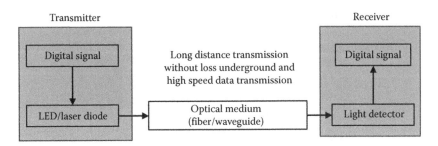

FIGURE 14.9
Optoelectronics communication schematic.

On a much larger scale, when light is used as a method of communication, it increases the ability to transmit large amounts of information (high bandwidth) at very fast rates over long distances and without much loss.

Optoelectronic devices have the ability to convert digital signals into optical information (light) or vice versa. The digital signal converted at the source into light is transferred to the desired destination by the use of glass fiber or embedded optical waveguides in the PCB (Figure 14.9). These optical transmission paths provide electrical isolation and the elimination of ground loops, noise, crosstalk, and shielding from electromagnetic induction (EMI) and radio frequency (RF) interference. This improves signal transmission efficiency and signal integrity tremendously. The conversion of electrical energy into light energy is referred to as electroluminescence and the conversion of light energy into electrical energy is referred to as photodetection.

Electroluminescence devices include LEDs and laser diodes. These are usually made of compound semiconductors that are neither silicon nor germanium. Compound semiconductors are a combination of Group IIIA elements and Group VA elements from the periodic table, which have one less electron and one additional electron in the valence shell, respectively. This combination forms the PN junction where electrons from the N-type semiconductor combine with the holes in the P-type semiconductor (Figure 14.4). During this combination, the excess energy of the electron is released as photons (light). LEDs are typically made of gallium arsenide (GaAs), gallium phosphide (GaP), aluminum gallium arsenide (AlGaAs), induim gallium nitride (InGaN), etc. The element combination along with the applied voltage provides the color for the LED and the wavelength. This basic technology finds a wide variety of applications in displays.

Laser diodes, as the name indicates, are LED-type devices capable of producing a spatially coherent, narrow, low divergent beam of light using stimulated emission on a semiconductor scale. Stimulated emission is effected by providing a gain medium within the laser diodes. The gain medium is a material with properties that allows it to amplify light by repeated reflection of photons until they gain the energy required to leave the material. Typically, a thin layer of AlGaAs sandwiched between P-type and N-type AlGaAs layers acts as the gain medium. The P-type and N-type AlGaAs layers are in turn sandwiched between a P-type and N-type GaAs to form the most commonly used laser diode, referred to as the injection laser diode or the quantum well separate confinement heterostructure (SCH) laser diode. The term injection refers to the electrons that are pumped into the device through the P-type material. The quantum wells within the gain medium represent the nanostructure within this optoelectronic device. The refractive index of the GaAs layers is lower than the inner layers thereby containing the light effectively within this region, which allows for effective stimulation. Also, the ends of the heterostructure device are polished to act as mirrors to reflect the photons back into the gain medium for more recombination or stimulation. Typically, one of the mirrored surfaces is partially transparent and the output laser beam is emitted through this surface on the side (Figure 14.10).

Other laser diodes referred to as vertical cavity surface emitting lasers (VCSELs) provide laser output perpendicular to the active region of the device rather than from the sides, thereby allowing tens of thousands of these laser diodes to be processed simultaneously on a 3 in. GaAs wafer (Figure 14.11). VCSELs utilize optical grating on the top and bottom side of the active layer to perform the stimulated emission. The gratings are made from alternating high and low refractive index multilayers capable of reflecting the photons back into the active region until they gain enough energy to leave the grating. In common VCSELs, the upper and lower mirrors are doped as P-type and N-type materials.

Photodetection can be accomplished using photoemission, photoconduction, or the photovoltaic phenomenon. Photoemission is the emission of electrons when light energy strikes a photosensitive material acting as a cathode within the device. The electrons emitted from the surface of the photosensitive materials are referred to as photoelectrons and are drawn

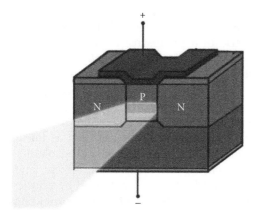

FIGURE 14.10
Side emitting laser.

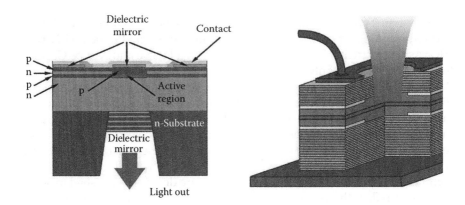

FIGURE 14.11
VCSEL.

toward the anode. The amount of light and its intensity, incident upon the cathode determines the current flow and the voltage drop across the load resistance. Current flow can also be controlled by adjusting the potential applied to the anode. Generally, alkali metals such as lithium (Li), sodium (Na), potassium (K), etc. are highly photosensitive elements. Hence, a coating of alkali metals is provided on the photosensitive materials to improve photosensitivity. These are electron producing materials. The most common solid–state photoemitter is the charge-coupled device (CCD) used in cameras.

Photoconduction uses the change in resistance of a photosensitive material to detect the incident light energy. This technology uses a photoconductive material tied to a load resistance and a power source. The circuit is physically closed through the photoconductive material, but not electrically. When light rays hit the photoconductive material, the energy of the incident light dislodges the valence electrons, thereby decreasing the resistance of the photoconductive material. This in turn increases the flow of current through the load and will be proportional to the intensity of the incident light.

The photovoltaic phenomenon is the method of producing voltage or electromotive force (EMF) based upon the incident light. This is predominantly used in solar cells. It makes use of a photosensitive material sandwiched between a thin filter metal and a dissimilar metal. When light energy is incident on the photosensitive material, electrons are released causing flow to the dissimilar metal and hence current flow into the circuit. In the case of photovoltaics, there is no external power source connected to the terminals of the cell.

There are other kinds of active integrated optical devices such as linear displays, liquid crystal displays, photodiodes and transistors, opto-isolators, etc. that require proper packaging to ensure the mechanical, optical, and electrical integrity of the system. Mechanical integrity is achieved by physically sealing the entire component, optical integrity is achieved by providing proper glass windows and filters, and electrical integrity is achieved by properly routing the interconnections on the solid–state devices to the next level of packaging.

14.2.3 Microsystems

Microsystems, which include the micro-electro-mechanical systems (MEMS) and micro-opto-electro-mechanical systems (MOEMS), involve the integration of sensing and actuation elements along with the electronics, on a common silicon substrate, through micro-fabrication technology. The sensing and actuation can be mechanical, optical, biological, or chemical. Microsystems can also incorporate microfluidic pathways to enable sensing of biological and chemical matter and also the flow of matter in ink-jet printing type applications. While the electronics are fabricated using the IC manufacturing process sequence (CMOS process), the microsystem components are fabricated using compatible "micromachining" processes that selectively etch away parts of the silicon wafer or add new structural layers to form the microsystem devices. Prominent examples of microsystems in mass markets include ink-jet printer heads, tire pressure sensors, accelerometers for crash sensing and airbag deployment, gyroscopes for sensing orientation, etc. (Figure 14.12).

The capability to sense, analyze, compute, and control, all within a single chip of the microsystem, provides new and powerful products. Microsystems

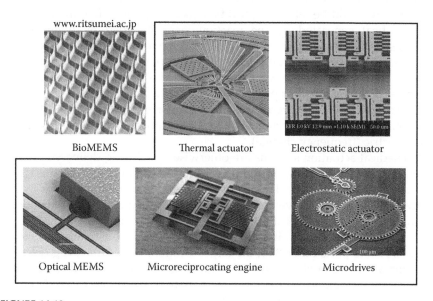

FIGURE 14.12
Microsystems sensors and actuators. (Courtesy of Scandia National Laboratories, Albuquerque, NM.)

have become an enabler fundamental to the convergence of technology (mechanics, electronics, and optics) and science (chemistry, physics, biology, and medicine). As a breakthrough technology, by allowing unparalleled synergy between previously unrelated fields such as biology and microelectronics, many new microsystems and nanotechnology applications have emerged expanding beyond that which is currently identified or known.

Two aspects that have contributed significantly to the growth in microsystems are the miniaturized scale of these systems and low cost of fabrication using existing techniques. In many instances, small systems can perform actions that large systems cannot, simply because of their small size. This can be due to the applications needing limited space or access (minimally invasive surgery) or the advantages of lightweight (aerospace applications) and energy-efficient applications (mobile applications). Miniature systems can use physical effects at small scales, such as more efficient chemical reactions or negligible influence of gravity. The second aspect is the manufacturing cost of microsystems, which provides unique possibilities. Material cost is negligible and materials that would otherwise be cost prohibitive can be used. Technology derived from IC-fabrication processes allows the production of miniature components in large volumes and hence low cost.

One of the prime areas of growth for nanotechnology and microsystems is biological or chemical sensing devices useful in biomedical electronics applications. These biomedical sensing systems are special kinds of devices that manipulate the biological signals and produce an equivalent electrical signal so that the condition of the biological entity could be monitored and necessary actions could be taken. Typical biological responses that are measured include heat generated by the reaction, heat absorbed by the reaction, changes in the distribution of charges, movement of electrons produced in a reduction oxidation reaction, light output during the reaction or a difference in light absorbance between the reactants and products or effects due to the mass of the reactants or products. Based on the type of response recorded by these devices, biosensors are classified as calorimetric (heat), potentiometric (ionic potential), amperometric (current), optical (light absorption), or piezoelectric (frequency change).

Biomedical actuation systems are otherwise called drug delivery systems. These devices sense the surrounding conditions and deliver drug dosage at the appropriate location. This is more effective because the drug when delivered using the conventional method (tablet) undergoes tremendous changes before it reaches the target location. The moment a drug is consumed in the tablet form, it starts mixing with the bloodstream and hence poses a risk to parts other than the target location. Hence, the drug delivery systems are more effective and efficient. These devices constitute both the sensing and mechanical actuation system, both of which are controlled by a processor. The actuation parts consist of a chamber that stores the drug and releases it only when it is activated by an electrical signal. The interconnections have to be reliable because the component should perform the intended function under all environmental conditions. These devices sense the basic form of response, much like an ON or OFF condition, and actuate the drug delivery system.

14.3 Device Packaging (Level 1)—Component

Device packaging or Level 1 packaging involves the assembly and interconnection of the microelectronic, optoelectronic, or microsystems device onto a lead frame arrangement or substrate. The lead frame is the framework that provides mechanical support to the die. The electrical connections between the die and the leads are established by means of a wire bond. The lead frame consists of a die paddle, to which the die is attached, and leads, which serve as the means for an external electrical connection to the outside world (Figure 14.13). Lead frames are constructed from a flat sheet of metal either by stamping or etching processes. The use of substrates is becoming more commonplace when compared with lead frames, as the industry migrates more and more toward leadless and area array packages.

The primary functions of Level 1 packaging are to provide electrical connections from the device I/O to the lead frame, spread the closely spaced I/Os on the device to a more wider spacing on the lead frame, protect the device from environmental or handling damage, provide heat dissipation for effective thermal management, and provide the mechanical strength for the device and interconnect structure. This section will introduce the various methods for device level packaging including a brief discussion on the various types of substrates.

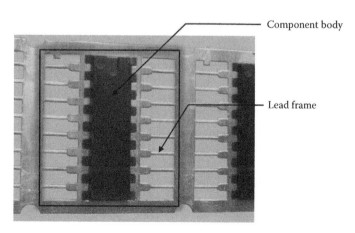

FIGURE 14.13
Lead frame structure.

14.3.1 Substrates

Substrates provide mechanical support to one or more devices and also provide an electrical interconnection between them. They are also responsible for redistributing the closely spaced I/Os of the device to a coarser pitch to facilitate the next level of packaging. In Level 1 packaging, substrates are used primarily in BGA and CSP style packages, which are replacing leaded through hole components and some leaded surface-mount components. These leaded components tend to be large pitch and do not lend themselves easily to miniaturization. A substrate can either be rigid or flexible with either single or multiple conductive layers that are sandwiched between insulating layers. Typically, the term substrate is used to represent interconnect structures made of any material. The classification of the substrate is shown in Figure 14.14.

14.3.1.1 Rigid Organic Substrates

Organic substrates, as the name indicates, are made of resin-impregnated base material. These are typically referred to as PCB or printed wiring boards (PWBs) and are the most commonly used substrates for a variety of applications. These substrates are primarily made up of three materials, which include the fabricated laminate material (resin and reinforcement combination), resin-impregnated glass fiber cloth—B staged (prepreg), and treated copper foil (Figure 14.15). Prepreg is a term used to describe "pre-impregnated" glass fibers with the impregnated resin partially cured. Prepreg has a unidirectional weave structure that adds strength to the laminate. Many layers of prepreg form a laminate of required thickness. The prepreg will cure completely with the application of heat and pressure during the lamination process.

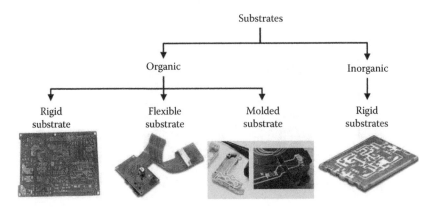

FIGURE 14.14
Classification of substrates.

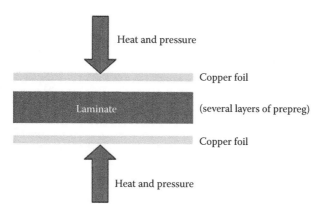

FIGURE 14.15
Basic building unit of a rigid substrate.

Organic substrates provide several advantages such as a low dielectric constant and hence higher signal speed, low cost, volume production capability, good thermal expansion match with plastic IC packages, and are easily reworkable. The major disadvantages of organic substrates include moisture absorption, substrate warpage, and multilayer separation (delamination). Multilayer rigid PCB fabrication involves laminating several pre-etched double-sided PCBs (which form the inner layers) with prepreg sheets between them and a sheet of copper on the top and bottom sides to form the outer layers (Figure 14.16). The outer layer copper traces are protected by a solder mask and the exposed copper pad is protected using a metallic or nonmetallic surface finish to prevent oxidation of the copper.

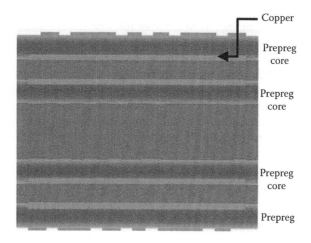

FIGURE 14.16
Multilayer rigid organic substrate stackup.

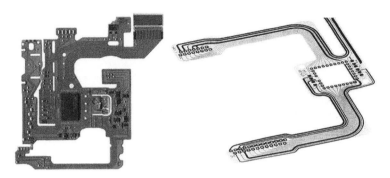

FIGURE 14.17
Flexible substrates.

14.3.1.2 Flexible Organic Substrates

A flexible substrate consists of a conductive foil and a flexible dielectric base. The ability of the substrate to bend allows them to be used as interconnects. Flexible substrates are primarily designed to replace bundles of wires and cables that connect different PCBs (Figure 14.17). A film cover layer, which is usually made up of polyimide or polyester film, is used to protect the conductors on the flexible substrate. Stiffeners are usually attached to the substrate to provide mechanical strength. Adhesives are used instead of prepreg to bond more layers to form a multilayer flex substrate. Multilayer flexible substrates are not used commonly, as the flexibility of the substrate decreases with an increase in layer count.

14.3.1.3 Inorganic Substrates

These substrates are made up of nonmetallic materials. Some widely used inorganic substrate materials are 99% pure alumina (ceramic), 96% alumina and beryllia (BeO), aluminum nitride (AlN), and silicon carbide (SiC). Beryllia, AlN, and SiC have high thermal conductivity. The coefficient of thermal expansion (CTE) of inorganic substrates closely match the CTE of silicon. These substrates provide excellent resistance to moisture when compared with organic substrates and therefore do not cause any substrate warpage. Component packages containing inorganic substrates can be hermetically sealed for use in harsh and critical application environments. The primary disadvantage of inorganic substrate is its higher dielectric constant, which leads to a loss in signal power.

14.3.2 Wire-Bonded Device Attach

The wire bonded device attach method utilizes thin gold wires to connect the aluminum I/O terminations on the silicon chip to the gold-coated

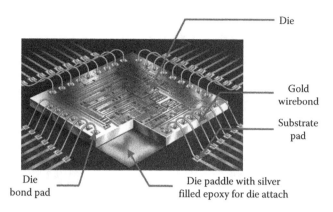

Die

Gold
wirebond

Substrate
pad

Die
bond pad

Die paddle with silver
filled epoxy for die attach

FIGURE 14.18
Gold wire-bonded chip attach.

pads on the substrate to which the chip is attached (Figure 14.18). Each pad on the substrate, where the gold wire terminates, links the chip I/O to a lead frame structure or a via hole that terminates in a solder ball on the bottom side. This lead or solder ball provides the interconnection to the next level of packaging. The gold wires are typically 25 μm (0.001" or 1 mil) in diameter or higher.

The process starts with the attachment of the silicon chip, face up, to the substrate using silver-filled epoxy (die attach adhesive). The silver-filled epoxy consists of flakes of silver dispersed in an epoxy medium. The volume fraction of the silver flakes in the epoxy medium (typically >20%) allows it to conduct in all directions (isotropic conductivity). After the silver-filled epoxy is dispensed onto the substrate, the inactive side of the silicon chip is placed on the epoxy and it is cured. The curing process allows the epoxy to shrink thereby bringing the silver particles closer together and in contact with each other, allowing for both electrical and thermal conductivity through the cured epoxy medium. The die-attach epoxy serves as a ground connection or provides a good thermal dissipation medium. The silicon chip is typically assembled onto a large gold-coated copper pad for effective thermal dissipation. This pad is also referred to as the die attach paddle. After assembling the die onto the paddle, the assembly is subjected to plasma cleaning to enable gold wire bonding. The bonding of the wire is performed using the thermosonic (thermo-compression + ultrasonic [US]) bonding technique.

The thermo-compression technique forms a ball bond on the chip I/O and the ultrasonic technique forms a wedge bond on the substrate pad. The wire bonding process is sequential as it forms the connection one pad at a time. The process starts at the chip end by using a high current arc that melts the gold wire, which is then pressed against the I/O termination on the chip to form the ball and the wire is then drawn to form a loop and the other end of the loop is welded to the gold-coated pad on the substrate (Figure 14.19).

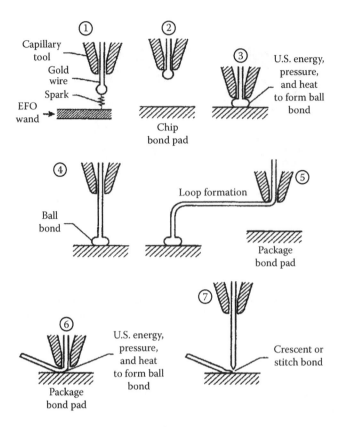

FIGURE 14.19
Wire bond process. (From National Bureau of Standards, Roadmaps of Packaging Technology.)

This process is repeated for each I/O on the silicon chip. When the molten gold comes into contact with the aluminum (Al) termination on the silicon chip, it forms the intermetallic that enables the metallurgical bonding and the interconnection. The aluminum and gold have a tendency to diffuse, over a period of time, resulting in voids in the interconnection referred to as Kirkendall voids. Once the wire bonding is completed, the wires can be protected using plastic overmold or by using glob top encapsulation (Figure 14.20).

14.3.3 Flip Chip Device Attach

In the case of FC attach, gold balls or high melt solder balls are attached to the I/O points on the silicon chip and the chip is flipped to match the pads on the substrate for assembly, hence the name FC technology (Figure 14.21). When gold balls are used for the attach, the gold is attached directly onto the aluminum I/O terminations on the silicon chip. This is accomplished by adding the gold balls using the wire bond technique and completing the

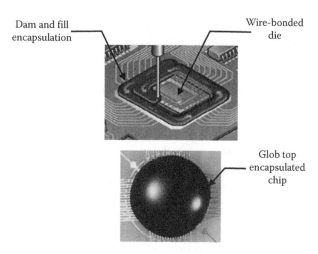

FIGURE 14.20
Glob top encapsulation.

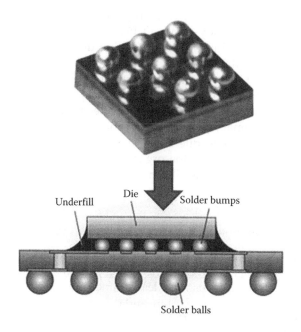

FIGURE 14.21
FC attach.

process without forming the wire loop. On the other hand, when solder balls are used, specialized metallization is required between the aluminum I/O termination on the chip and the solder ball. This metallization is referred to as the under bump metallization (UBM). The primary purpose of the UBM is to enable the interconnection between the aluminum on one end and the tin (Sn) in the solder on the other end. Solder is typically a mixture of tin, which is the primary metal, and other secondary metals such as lead (Pb), copper (Cu), silver (Ag), bismuth (Bi), antimony (Sb), etc. Unlike gold and aluminum, tin and aluminum do not form the necessary intermetallic to form the joint. The most common UBM metallization is chromium-copper-gold (Cr-Cu-Au), although other metallizations, such as titanium-tungsten-gold (Ti-W-Au), nickel-gold (Ni-Au), etc., are also used. The solder ball is formed by printing solder containing paste onto the UBM.

When using gold bumps on the FC, the chip is attached to the substrate pads by using the thermo-compression bonding technique, which involves the use of heat and pressure and a mild scrubbing action to allow the gold bump to form the interconnection with aluminum. When using solder bumps for device packaging (Level 1), typically high melt solder alloy is used (10%Sn-90%Pb) that melts at >300°C. The primary reason for this is to avoid remelting of the solder within the packaged device when it is assembled onto the PCB with a low melt solder. If FC attach is used for PCB level assembly, without device level packaging, the solder balls will be made of low temperature solder alloy—tin-lead (63%Sn-37%Pb) and tin-lead-silver (62%Sn-32%Pb-2%Ag), which melts at 183°C or tin-silver-copper (Sn-4%Ag-0.5%Cu), which melts at 217°C.

The assembly of FC on organic substrates requires the use of underfill. Underfill is silica-filled epoxy that is dispensed in between the silicon chip and the substrate after the FC component has been soldered to the substrate (Figure 14.22). The primary purpose of the underfill is to minimize the effect of strain on the solder joint, due to the mismatch in the CTE between the silicon chip (4–6 ppm/°C) and the organic substrate (15–18 ppm/°C). This mismatch is minimal when the silicon chip is assembled on inorganic substrates (4–8 ppm/°C).

14.3.4 I/O Redistribution—Chip Level and Wafer Level

The two device packaging techniques discussed thus far involve the redistribution of the closely spaced I/O on the silicon chip to a coarse pitch outside the area of the silicon chip using a lead frame or solder ball array. These techniques take up valuable real estate and do not lend themselves easily for miniaturization. The newer technique currently employed in device level packaging redistributes the closely spaced I/O within the area of the silicon chip itself. This redistribution can be performed on the individual devices or on the entire wafer containing many devices. When accomplished at the

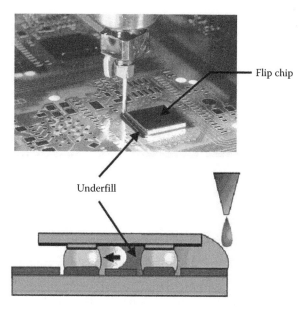

FIGURE 14.22
FC with underfill. (Courtesy of Asymtek, Carlsbad, CA.)

wafer level, it allows for the processing of many devices in a single process. This I/O redistribution technique reduces the real estate required for spreading the I/O and contains the redistribution within the silicon chip area, thereby allowing the rapid miniaturization and effective use of the device and substrate real estate. The redistribution is accomplished by placing an interposer (rigid or flexible substrate) between the silicon chip and the solder bumps. The interposer provides the redistribution metallization (traces and pads) that connect the ultrafine pitch I/O on the chip to the coarse pitch of the solder bump array (Figure 14.23).

The I/O redistribution at the individual device level uses interposers while the wafer level redistribution uses the photolithography process and metallization process to build the dielectric layer and cover layer with a metallic interconnect layer interposed between them. The small form factor of these packages provides the advantages of FC without the concerns of handling damage, as the devices are packaged and can be tested for functionality.

14.3.5 Single-Chip Packaging

Single-chip packages are components with a lead frame or solder ball containing a single silicon chip inside. This chip can either be wire bonded or solder bump attached to the substrate. Single-chip packaging is predominantly used even today. Many of the ICs are packaged as single chip packages

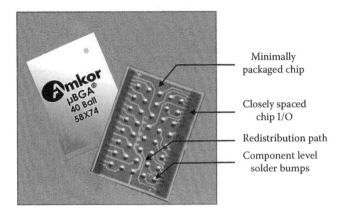

Minimally
packaged chip

Closely spaced
chip I/O

Redistribution path
Component level
solder bumps

FIGURE 14.23
CSP.

to be able to handle heat dissipation requirements and also to keep the cost low. The necessary passive components are assembled separately to support the single-chip package on the PCB.

14.3.6 Multi-Chip Packaging

Multi-chip packages are components with many silicon chips interconnected on a substrate and attached to the PCB with lead frame arrangement or solder balls. These are complex packages that are multifunctional. In many instances, the passive components required to support these packages are also assembled within these packages, and in some of the high-density multi-chip packages, the passives are also embedded within the multilayers of the substrate. Multi-chip packages pose severe design challenges and are also cost prohibitive in many instances.

Multi-chip packages come in one of many forms (Figure 14.24). These include silicon stacking with wire-bonded chip attach or a combination of wire bonding and solder bumped FC attach, many silicon chips attached within the multi-chip package substrate with integrated or embedded passives, and many WL-CSPs assembled onto the component substrate with passives. Multi-chip packages provide considerable performance benefits and also size and mass reduction. In many instances, it provides greater reliability for the assembly, as it reduces the number of defects per million opportunities (DPMO) during the assembly process. Many of the disadvantages of multi-chip package include design constraints, a lack of design tools, the need for high-density interconnection substrates, I/O management, cost, and thermal management.

Package stacking

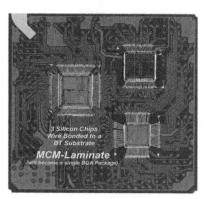

Multiple silicon chips in
one component package

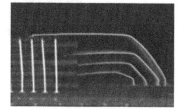

Die stacking

FIGURE 14.24
MCM and stacked die package.

14.4 PCB Level Packaging (Level 2)

This level of packaging involves the assembly of packaged active and passive components onto a PCB. There are three approaches that are used, which include through hole assembly, surface-mount assembly, and mixed assembly (Figure 14.25). The most prevalent technology is the surface-mount assembly and mixed assembly. Mixed assembly uses a combination of both through hole and surface mount. The main reasons for this is the unavailability of certain high power devices in surface-mount format due to the requirement for high thermal dissipation and the use of components such as connectors that require constant application of force during cable insertion. Apart from these it is also very expensive to switch completely from through hole designs to surface-mount designs and a complete migration to pure surface-mount assembly requires considerable investment in equipment and infrastructure.

14.4.1 Solder-Based Component Attach Process

The most common method for component attach is using solder. This has been in use for a long time and the process is reasonably well understood. The use of solder provides a long working life for the assembly and also provides excellent electrical performance and mechanical strength. Solder does not get affected by moisture and provides mass assembly and curing. This section will discuss the surface-mount assembly process (Figure 14.26) in more detail.

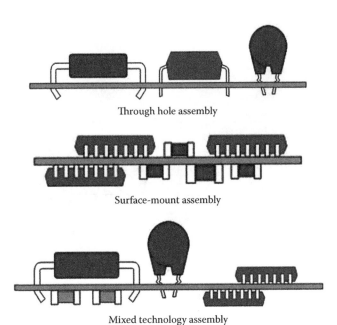

FIGURE 14.25
Through hole, surface mount, and mixed technology assembly.

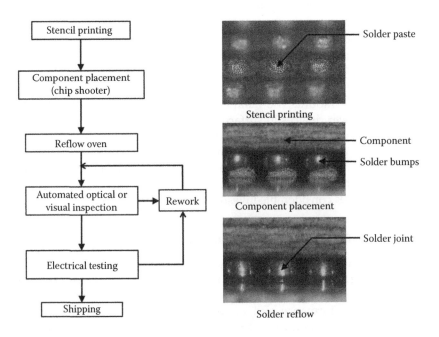

FIGURE 14.26
Surface-mount assembly process.

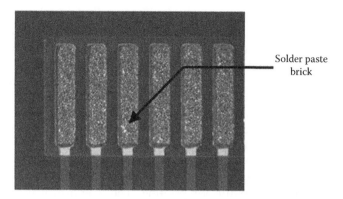

Solder paste
brick

FIGURE 14.27
Solder paste brick on a pad.

14.4.1.1 Solder Paste Print Process

The solder paste print process is used for printing a required volume of solder paste onto the pads of the PCB, using matching apertures on a stencil foil (Figure 14.27). A squeegee is used to drag the paste across the stencil by applying a certain amount of pressure and at a certain speed to enable the print process. The PCB is also adequately supported during the print process to avoid any bowing or flexing and it makes contact with the stencil. The solder paste consists of minute spherical solder particles dispersed in flux. The solder particles are responsible for forming the solder joint and for attaching the component leads or the solder balls onto the pads of the PCB. The flux is responsible for suspending the particles in the paste, protecting the particles from getting oxidized in the paste, cleaning the surface of the pads and the leads before soldering, protecting the leads and pads from oxidation during the soldering process, and forming the flux residue by collecting the byproducts of the cleaning action.

The solder particles can either be a tin-lead alloy or a lead-free alloy. The solder particle size is determined based upon the smallest aperture dimension on the stencil foil to be used in the process. The use of nanosized solder particles is also being considered in order to reduce the melting temperature of the alloys. The flux is made up of resin or rosin (also referred to as the gelling agents), which acts as a glue to hold the components in place and also provides the body for the paste. The gelling agent is also responsible for protecting the surface of the pads and leads during the soldering process and for forming the residues. The mount of gelling agents in flux determines the amount of residue formation and hence the required cleaning after assembly. No clean flux predominantly used in electronics assembly today produces minimum residue that requires no cleaning. Also, the rosin-based gelling agents used in no clean flux forms a thick residue shell entrapping all the corrosive byproducts released during the soldering process. Apart from the gelling agents, the other important ingredients in the flux include the activators, which are chlorine-containing acids and salts used for providing the cleaning action. During the cleaning process, the chlorine (ions) are separated and left behind within the residue. If these ionic contaminants are not entrapped within the residue and are exposed to moisture and bias (flow of current) in the field, it will eventually result in dendrite growths that will short circuit adjacent conductors and affect the assembly functionality. Other ingredients of the flux include thixotropic agents that control the viscosity behavior of the solder paste during the print process, special additives if necessary, and solvents to enable the mixing of constituents of the paste to form a homogeneous mixture for printing. The length of exposure of the solder paste to the ambient conditions determines the extent of solvent evaporation and eventual drying of the solder paste and aperture clogging during the print process. The thixotropic rheology of solder paste allows the paste to become thinner during the print process, for effective deposition through the apertures, and to regain its viscosity after the paste has been deposited onto the pads of the PCB.

The stencil foil used in the print process provides the aperture openings for depositing a predetermined volume of solder paste onto the pads of the PCB. The foil is typically attached to an extruded aluminum frame using a flexible mask (Figure 14.28). The foil is predominantly made of stainless steel and is manufactured using the chemical etching process or the laser machining process. When using nickel for the foil material, it is manufactured using the electroforming process. The chemical etching process and the laser machining process are material removal processes that produce rough aperture wall finish requiring secondary operations such as electropolishing and

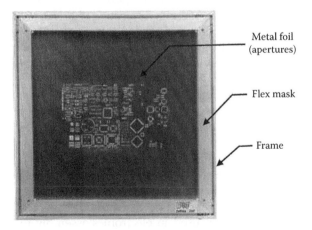

Metal foil
(apertures)

Flex mask

Frame

FIGURE 14.28
Flex mask stencil.

nickel flashing. Nickel flashing provides a coating of nickel on the rough surfaces of the aperture thereby smoothing it. The electroforming process on the other hand is a material build-up process at the atomic level, which provides a very smooth surface in the as manufactured condition itself without requiring any secondary operations. The proper aperture dimensions (length, width, and foil thickness or the diameter and foil thickness) and the finish on the aperture walls will determine the paste transfer efficiency or the amount of paste that gets deposited onto the pads of the PCB.

The two most important process parameters that affect the print process are the squeegee pressure and speed. The squeegee used predominantly in the print process is a 250 μm (10 mil) thick spring steel blade (Figure 14.29). Both the squeegee pressure and the speed affect the thixotropic rheology of the paste or its viscosity. Excessive pressure and higher speeds during the print process tend to reduce the viscosity considerably and also result in flux separation. A proper balance of these two parameters will allow adequate thinning of the paste for effective printing and also the regaining of viscosity after print. The squeegee pressure ranges from 1 to 2 pounds per linear inch of squeegee length and the speed ranges from 0.5 to 3 in./s. Other important aspects during the print process include an adequate amount of solder paste to roll in front of the squeegee blades and also proper board support and alignment from the stencil to the pads. Board and stencil alignment is

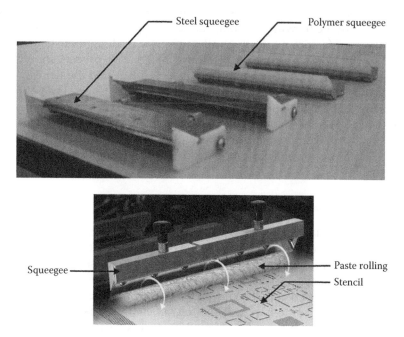

FIGURE 14.29
Polyurethane and metal squeegee blades.

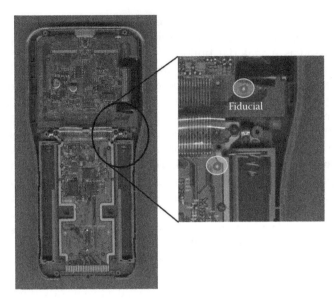

FIGURE 14.30
Fiducials on a PCB.

accomplished either manually or by using a vision system to match the fiducial on the PCBs with matching fiducials on the bottom of the stencil. Fiducials are reference marks, typically a circular or doughnut-shaped disk, etched on the corners of the PCB (Figure 14.30). The process should also ensure no filming of paste between the squeegee blade and the stencil, which could lead to the eventual clogging of the fine pitch apertures. The volume of solder paste deposited for fine pitch leaded surface-mount components is more critical than the solder ball array components because the solder balls in the area array component melt during reflow along with the solder ball particles in the paste and accommodate for any insufficient print volume and coplanarity error.

14.4.1.2 Component Placement Process

The component placement process is the most flexible but expensive process step in surface-mount assembly. The flexibility comes from the different types and sizes of components that can be assembled using the placement machine, and the expense comes from the cost of placement equipment, the number of machines required to have effective line balancing, and the number of feeders required to handle the variety of components. Each machine is capable of handling in the range of 50–150 feeders. Feeders are required to present the components at the proper location and in the proper orientation, consistently, for automated assembly.

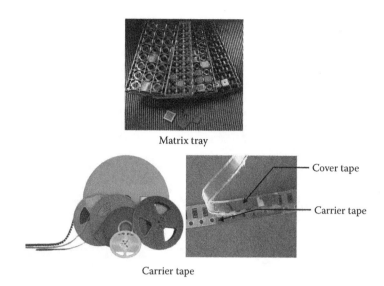

Matrix tray

Carrier tape

FIGURE 14.31
Tape and reel feeder and matrix tray feeder.

The components are typically packaged in tape and reel or matrix trays for effective feeding. Tape and reel packaging is the most common as it provides the versatility and flexibility in handling a wide variety of components, except fine and ultrafine pitch-leaded components, which are packaged in matrix trays (Figure 14.31). Also, the tape and reel accommodates a large number of components in one reel depending on the size of the component. The tape consists of a punched cardboard or thermoformed plastic carrier with a cover to protect the components, which is attached using a heat-sensitive adhesive or a pressure-sensitive adhesive. The carrier and cover can be antistatic or static dissipative in nature to prevent component damage due to electrostatic discharge (ESD). Fine and ultrafine pitch components are packaged in matrix trays to prevent lead damage.

Apart from the feeders, the placement machines also rely heavily on the vision systems to provide corrections for minor component misalignment when picked up from the feeders. Component pickup from the feeders is done using vacuum nozzles. Fiducial recognition is also used in component placement machines for board alignment. The component placement machines utilize turret heads with multiple nozzles for high-speed placement of small passives, discrete actives and low I/O IC components (Figure 14.32a), and precision heads with a single nozzle for fine and ultrafine pitch components (Figure 14.32b). The component specifications defined within the component database of the placement machine controls the Z height movement of the nozzle to place components of varying geometries and sizes without damaging the component or the PCB.

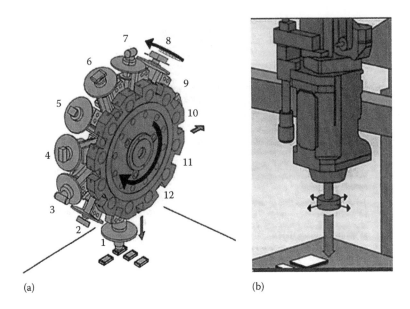

(a)   (b)

FIGURE 14.32
(a) Turret head and (b) single head for placement. (Courtesy of Electronics Assembly Systems Divisions-Siemens Logistics and Assembly Systems.)

14.4.1.3 Reflow Soldering Process

The reflow soldering process starts when the populated PCB enters the reflow oven and extends until the board exits the oven (Figure 14.33). After component placement, the populated PCB enters the reflow soldering oven,

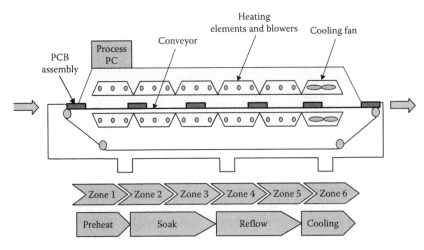

FIGURE 14.33
Reflow soldering oven.

typically 6–10 ft long separated into multiple heating zones and one or two cooling zones at the end. The assembled boards travel through the oven either on a mesh conveyor or on edge conveyors. The heat is distributed throughout the oven by recirculating air or nitrogen. The reflow process involves the mass heating of the solder alloy particles in the paste to form the solder joint interconnection.

The identification of an optimal reflow oven recipe for a given assembly, tin-lead or lead-free, is very challenging. Typically, the reflow oven zones and conveyor speed are set to certain values, based on the process engineer's knowledge and experience, to obtain an initial profile. Once the initial profile is obtained, the set points are adjusted to obtain the near optimum profile, either manually or with the help of a manual prediction tool or automated oven recipe search engines. Even though this process is iterative, the automated process optimization software tools, with a good understanding of the reflow profile process and requirements, significantly reduce the number of iterations and the time to set up a reflow oven (Figure 14.34).

The soldering process parameters along with the solder paste constituents control the inter diffusion of two or more compatible metals to form the intermetallic compound (IMC) and the bulk solder matrix microstructure of the joint (Figure 14.35). The IMC forms the actual bond between the component termination and the solder and the solder and the pads on the PCB. Tin (Sn) is one metal that forms IMCs with many other metals used in electronics such as copper (Cu), silver (Ag), gold (Au), nickel (Ni), and palladium (Pd). Therefore, Sn is a very important element in any solder including lead-free.

The various stages of the reflow process are preheat, soak, reflow, and cool down, shown as a time–temperature graph and also referred to as the reflow

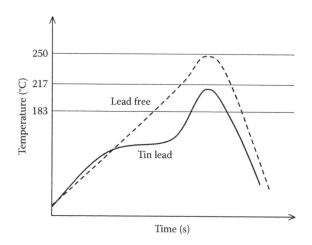

FIGURE 14.34
Typical tin-lead and lead-free reflow profiles.

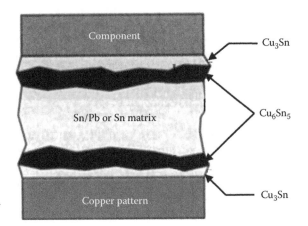

FIGURE 14.35
IMC and bulk matrix for tin-
lead and lead-free solder joint.

profile. The reflow process window defines the limits of these reflow parameters. The preheat stage represents the initial stage of heating when the
temperature of the PCB with stenciled solder paste and mounted components is raised from room temperature to a temperature close to the activation
of flux. The temperature to which the assembly is raised is determined by the
properties of the flux and the solder alloy. It is a general practice to heat to
a temperature slightly below the flux activation temperature. The slope of
the heating curve during this stage is fairly linear. If the rate of preheat is high
or low, it can cause a number of undesirable results. Furthermore, if the
temperature of preheat exceeds flux activation or decomposition temperatures, the solder pastes tends to dry out before melting, resulting in solder-
ability issues.

The second stage of the reflow profile, referred to as soak, is primarily used
to bring all components along the length and breadth of the board to a
uniform temperature to achieve a minimal thermal gradient across the
board. In this stage, an extremely low rate of heating ($\approx 0.5°C/s$) is used;
essentially preparing the board to go through the reflow in the next stage.
The primary purposes of this stage is to bring the entire board, components,
and solder paste to a uniform temperature, activate the flux so that it dissolves
any oxides on the pad and the component lead metallurgies, dissolve any
oxides on the solder powder particles, expose clean surfaces for metallurgical
bond formation, and facilitate the escape of volatile components of the flux.
The volatiles when present in the solder after the melting point of the alloy
is reached may lead to voids and solder splattering. The two important
parameters that determine the effectiveness of this stage are soak temperature
and time.

After the soak stage, the board and the components enter the reflow stage
where the temperature is raised above the solder alloy's melting temperature. In a molten state, tin from the solder reacts with the pad and component termination metallurgies to form metallurgical bonds. The two

important parameters that determine the nature and thickness of IMC formed are peak temperature, which is the highest temperature that the solder alloy experiences and time above liquidus (TAL), defined as the time for which the alloy is maintained above its melting point. It is a commonly accepted practice to set peak temperatures 30°C–40°C higher than the alloy melting point and an average TAL of 30–120 s, depending on the board structure and the component mix. Elevated temperatures enable better melting of the solder joint constituents while sufficient time above the melting point is necessary to form joints of uniform composition (solder joint matrix). Enough time should be provided for tin from the solder to react with other metals to form the IMC. Either of the parameters in excess will deteriorate the joint quality due to the increased thickness of brittle intermetallic phases. Adequate care has to be taken not to exceed the temperatures that the components and board materials are qualified for, based on the moisture sensitivity level (MSL) specifications. This constraint is even more pronounced in the lead-free reflow process, due to the fact that the most components are still qualified for a maximum temperature of 245°C assuming a tin-lead soldering environment, which has a peak temperature of 220°C.

This cool down stage in the profile occurs at the ramp down after reaching the peak temperature. A part of this stage occurs when the solder is in the liquid state and hence the rate of cooling in this region is critical to the integrity and long-term reliability of the interconnections formed. It is a commonly adopted practice to use a cooling rate of less than 4°C/s for second level interconnects reflow. Faster cooling results in fine grains of matrix while slow cooling (lower ramp down) results in a coarse grain structure. The size and structure of the grains formed determine the reliability of the solder joint. A finer structure has a better fatigue resistance but would fail under creep loading while a coarser structure may have a relatively better creep strength. The cooling rates should not be very high, in order to avoid any thermal shock on the board. Very high cooling rates may lead to nonuniform grain structure formation and hence affect the grain growth with time.

14.4.2 Electrically Conductive Adhesive-Based Component Attach Process

Electrically conductive adhesives (ECAs) consist of a mixture of adhesive matrix and conductive fillers. The typical adhesive matrix by itself is non-conductive (has very high resistivity) although there are some new intrinsically conductive adhesives under development. The adhesive matrix provides the adhesion, mechanical strength, and protection of the metallic contacts, while the fillers within the matrix provide the electrical conductivity between the component termination and the pad. The fillers are typically flakes or particles made of metal or metal-coated polymers. The adhesive matrix is either a thermoplastic or a thermoset material. Thermoplastic materials become soft when reheated after curing and become hard when cooled. Repeated melting does not change their basic properties. Thermoset materials, on the other hand, become permanently hard once cured, due to the catalytic action and cross-linking of the polymer chains. Thermosetting adhesives (epoxies) are the most common, the most prevalent, and also the most diverse. Typical ECAs use epoxy resin as their matrix, which has superior properties over other polymers, such as high adhesion and low dielectric constant. Epoxy adhesives can be either a one-part system where the curing agent is already mixed in or a two-part system where the curing agent is mixed just before use. Two-part systems have a long shelf-life but a short working life. Two-part systems could also have special additives added to them to enhance their cure rate, flexibility, peel resistance, impact resistance, etc.

The fillers used in conductive adhesives are metal particles that are made of gold (Au), silver (Ag), nickel (Ni), indium (In), copper (Cu), chromium (Cr), or lead-free alloy (Sb-Bi). The filler particles are available as spheres, flakes, fibers, and granules but the optimum geometry is that which provides both the best contact with neighboring filler particles and adhesion to the polymer. Flakes, due to their high aspect ratio, provide more particle-to-particle contact, greater conductivity, and consistency of product performance. Silver is the widely used filler material due to its excellent electrical conductivity and low contact resistance between particles. Copper oxidizes fast and turns nonconductive. Although nickel oxidizes slowly, it is less malleable and hence cannot be formed into particles of a desired size and shape. Gold and indium are expensive. Nanoconductive particles are being embedded into the ECAs to enhance the conductivity of the ECA joints.

The electrical and mechanical performance of an ECA is dependent on the ratio of the polymer matrix to filler in the composition. Electrical conduction is provided by the filler content, and high conductivity requires high filler content. On the other hand, achieving high mechanical adhesion and adhesive strength requires a low filler content. These contradicting constraints therefore require the appropriate volume fraction of fillers in the ECA, with allowance for manufacturing tolerances. The percolation theory plays a very important role in determining the volume fraction of filler needed. It identifies the critical volume fraction where the classical insulating to conducting transition takes place. It increases as the filler concentration increases and the transition occurs at a fixed fraction called the threshold of percolation. Based upon the physical characteristics and the filler volume fraction, ECAs can be classified into two categories: isotropic conductive adhesives (ICAs) and anisotropic conductive adhesives (ACAs), which are discussed in detail in the following sections (Figure 14.36).

The assembly process using ECAs is similar to that of the solder paste process. The ECA is applied on the PCB using either a stencil printing process or a needle dispensing process, followed by component placement and cure. Since adhesives do not offer self-alignment capability like solders, component

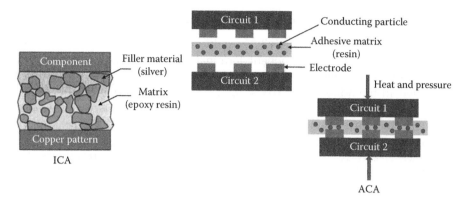

FIGURE 14.36
ICA and ACA comparison.

placement should be highly accurate. After placement, the adhesive is cured by heating in an oven with or without the application of pressure or by using UV radiation in the presence of a catalyst. Typically, a cure profile similar to a reflow profile is used but the cure temperature is in the range of 100°C–160°C, which is very low when compared with lead-free solder technology.

14.4.2.1 Isotropic Conductive Adhesive Attach

ICAs offer electrical conductivity in all directions. The volume fraction of fillers is in the range of 20%–30%, determined based on the percolation theory. The particle size of the fillers in ICAs is in the range of 1–20 μm. The most popular filler material used in ICAs are silver flakes and are available in heat or room temperature cured one- or two-part systems. Typical ICAs require a cure temperature of 120°C–150°C and a cure cycle of greater than or less than 10 min, respectively.

ICAs typically have low electrical conductivity before cure and it increases due to the shrinkage of the epoxy matrix during the cure process. Adhesive matrix shrinkage is shown to have a strong correlation with electrical conductivity. ICAs with higher shrinkage show lower bulk resistance or better conductivity. ICAs have been used extensively in applications requiring electrical and/or thermal conduction. ICAs were considered as possible replacements for solder, but were not applicable for fine and ultrafine pitch packaging.

Due to its conductivity in all directions, precise printing of the ICA material onto fine pitch circuit board attachment pads is required and is a huge challenge. Therefore, ICAs find use in several applications that do not require the precise application of the adhesive such as die attach, heat sink attach, etc.

14.4.2.2 Anisotropic Conductive Adhesive Attach

ACAs offer electrical conductivity in only one direction, i.e., in the Z-axis direction, between the component termination and the pad. A very low fraction of conductive fillers, typically in the range of 5%–20% by volume, is loaded in the adhesive matrix so that conductivity is only in the Z-axis direction. Insufficiency of fillers results in the gap between particles, which prevents interparticle contact. This eliminates conductivity in the X-axis and Y-axis directions.

Particle size in ACAs is in the range of 3–15 μm. During the curing process, the ACA needs the application of pressure in order to capture a monolayer of particles between the mating pads. Fine pitch components are mounted using ACAs, since the ACAs in paste form can be stencil printed over the entire footprint area without individual pad openings. Thus, ACAs can be used for extremely fine pitch technologies. ACAs are also available as films. New formulations of ACAs are available that consist of conducting filler coated with a low melting point and nontoxic metal, which can be fused to achieve metallurgical bonding between adjacent particles, as well as to the substrate. The FC bonding technology using ACAs has been used for decades in the area of glass substrate and for assembling FC on rigid and flexible substrates.

Several new ACA variants have been introduced in the market either by changing the filler material or by adding additional nonconductive fillers to enhance the CTE. The only drawback with all of these is the assembly method. All of these ACA variants still require a curing process that includes the application of heat and pressure, which requires specialized tooling and also sequential assembly. The application of pressure has been identified as a major factor in the performance of these ACAs. Any surface variations in the component, pad, or the extent of particle deformation influence the performance and the long-term reliability of these assemblies.

14.5 Conclusion

Packaging plays a very important role in the usability of the semiconductor chip device in a product. Packaging at various levels provides a myriad of functions, including mechanical support, electrical interconnections, environmental protection, and thermal management. With the continued miniaturization and the use of nanomaterials and nanostructures, packaging provides the scale for these devices to be assembled using the standard surface-mount assembly process. Component level packaging (Level 0) is migrating more and more to area array solder ball packages at the scale of the chip itself in order to maintain the size advantage and also provide the robustness for handling and assembly. This chip-scale packaging incorporates the redistribution of the closely spaced I/Os on the chip to a more coarsely spaced pads within the area of the chip itself. This packaging, when done at a wafer level, provides economies of scale apart from the above-mentioned advantages. Packaging the devices for the SMT assembly process ensures high yield and keeps assembly costs low.

Acknowledgments

The author acknowledges all of the graduate research assistants, specifically Harish Gadepalli in the Center for Electronics Manufacturing and Assembly (CEMA) at the Rochester Institute of Technology, for providing valuable insights into the manuscript content and its organization and for creating and compiling the images. The author also acknowledges the continued support of Speedline, Asymtek, Siemens, Heller, KIC, Malcom, OKI International, PACE, SMARTSonic, Indium, Cookson, Glenbrook, and ACE Production and their unwavering partnership with CEMA through equipment and material donations.

Bibliography

1. Goddard III, W. A., Brenner, D. W., Lyshevski, S. E., and Lafrate, G. J. (2007). *Handbook of Nanoscience, Engineering and Technology*. Boca Raton, FL: CRC Press.
2. Lee, J.-O., Lientschnig, G., Wiertz, F., Struijk, M., Janssen, R. A., Egberink, R. et al. (2003, May 1). Keywest. Retrieved February 23, 2009, from Keywest, http://qt.tn.tudelft.nl/publi/2003/keywest2002/keywest2002.html
3. Mongillo, J. (2007). *Nanotechnology 101*. Westport, CT: Greenwood Press.
4. Morris, J. E. (2008). *Nanopackaging Nanotechnologies and Electronics Packaging*. New York: Springer.
5. Seippel, R. G. (1989). *Optoelectronics for Technology & Engineering*. Englewood Cliffs, NJ: Prentice Hall.
6. Ulrich, R. K. and Brown, W. D. (2006). *Advanced Electronic Packaging*. New York: John Wiley & Sons, Inc.

15

An Architectural Perspective on Digital Quantum Switching

I-Ming Tsai and Sy-Yen Kuo

CONTENTS

15.1 Introduction to Classical Switching Technologies

Since Alexander Graham invented the telephone in the 1870s, classical telecommunication technologies have been dominating the way people exchange information for more than 100 years. Basically, a telecom network allows end-users, each with a terminal (telephone), to establish a connection with any other party (or parties) in the network. In order to avoid a fully meshed architecture, switching units are required to build a realistic network. Traditionally, there are two paradigms in doing digital switching: circuit switching and packet switching. The classical telephone network is a circuit switching network, while the Internet is a packet switching network.

In circuit switching, a path is decided before the connection is established. The connection is established at the time of call setup and released when the call is torn down. During the whole session, the route is dedicated and exclusive. After the path is selected, the function of the switching module is to transfer the data from the input port to the corresponding output port. Figure 15.1 shows an example topology of telecom networks. Assuming the path $S–K–D$ (dash line) is selected at call setup, the function of node K is to receive the data from link $S–K$ and send the data to link $K–D$. Modern switching equipments employ the *time-sharing* concept to process all the input links. Assuming the traffic described above is assigned to time slot S_0

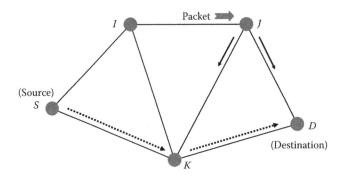

FIGURE 15.1
A network diagram describing circuit switching and packet switching.

of link *S–K* and time slot S_2 of link *K–D*, then the function of the switching module at node *K* is to transfer the data (i.e., a classical 0 or 1) from S_0 of link *S–K* to S_2 of link *K–D*. This operation is executed periodically until the session is terminated.

In packet switching, no predetermined path is allocated beforehand. A packet usually has a header carrying the information such as destination addresses. When a packet arrives at a node, it is forwarded to the desired node according to its header. Taking Figure 15.1 as an example, a packet from node *S* might reach node *J* on its way to node *D*. The function of node *J* is to receive the packet, read its destination (node *D* in this case), and send the data to a link according to a *routing table*. This (usually dynamically changed) table shows where the packet should be forwarded. As a result, the physical path can be either *S–I–J–D* or *S–I–J–K–D*, depending on the network condition (e.g., congestion or link failure). Unlike circuit switching, there is no call setup and release. Each node forwards the packet according to the routing table. The time-sharing concept still applies to a packet switching network. Modern packet switching networks take packets that share the same transmission line as input and each packet is transferred to its destination individually.

The primary difference of these two switching paradigms is *data dependency*. The operation of a circuit switch is independent of the contents of the input traffic. It performs the operation periodically with no need to check the data. However, a packet switch performs the switching function according to the header of the input packets. The control module checks the header and refers to a table, which maps the destination address to an output port. Another important characteristic of packet switching is the problem of *output contention*. Because packets from different input ports may have the same destination at the same time, packets need to be buffered for scheduling. This is one of the reasons why a packet gets delayed or even dropped under heavy traffic load. As a contrast, because a physical path or time slot is reserved, output contention is not a problem in circuit switching once the path is allocated.

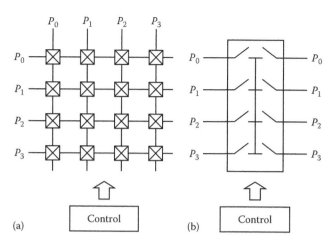

FIGURE 15.2
Examples of switching in the (a) space domain and (b) time domain.

Although significant differences exist between circuit switching and packet switching, the main function of each node is to move data from an input port to an output port. In the field of classical digital switching, various techniques have been designed to perform this function. For example, data can be switched in the space domain, the time domain, the wavelength domain, etc. The crossbar switch, shown in Figure 15.2a, is an example of a space domain switch. In this architecture, a rectangular array of cross-points serves as a simple space switch. Every output port can be reached by every input port in a nonblocking way by closing a single cross-point. A connection is established by closing proper cross-points to select a path from the inlet to the outlet.

A device that switches the data in the time domain is called a time division switch. Taking Figure 15.2b as an example, the connection from the inlet P_1 to the outlet P_3 is established by closing their corresponding switch. This process is executed for each of the connections in a cyclic way to achieve the switching. Primarily owing to the low cost of semiconductor devices, the implementation of a time division switch is usually done by using digital memory. Data received over an incoming port is written into the memory. The switching is accomplished by reading out the individual bits in the desired time slot, which is equivalent to connecting the inlet to the outlet for data transfer.

In addition to space or time switching, wavelength switching has received increasing attention in recent years. In a wavelength switch, an input wavelength division multiplexing signal is split and led to individual tunable wavelength filters, each of which extracts a specific wavelength signal from the input signal. The output optical signal is then applied to a wavelength converter, which results in a wavelength interchange.

15.2 Quantum Bits and Quantum Circuits

Compared with classical digital switching technology, quantum information science (or quantum computing) is a relatively new field of study. Quantum computers were first discussed in the early 1980s (Benioff 1980, Feynman 1982, Deutsch 1985). Since then, a great deal of research has been focused on this topic. Remarkable progress has been made due to the discovery of secure key distribution (Bennett and Brassard 1984), polynomial time prime factorization (Shor 1994), and fast database search algorithm (Grover 1996). These results have made quantum information science the most rapidly expanding research field.

The basic information carrier in quantum information science is a two-level quantum bit—qubit. In a two-level quantum system, each bit can be represented using a basis consisting of two eigenstates, denoted by $|0\rangle$ and $|1\rangle$. These states can be either spin states of a particle ($|0\rangle$ for spin-up and $|1\rangle$ for spin-down) or energy levels in an atom ($|0\rangle$ for ground state and $|1\rangle$ for excited state). These two states can be used to simulate the classical binary logic. A classical binary logic value must be either ON (1) or OFF (0), but not both at the same time. However, a bit in a quantum system can be any linear combination of these two states. The state $|\phi\rangle$ of a quantum bit can be written as

$$|\phi\rangle = c_0|0\rangle + c_1|1\rangle,$$

where $c_0, c_1 \in C$, and $|c_0|^2 + |c_1|^2 = 1$. The state shown above exhibits a unique phenomenon in quantum mechanics called *superposition*. When a particle is in such a state, it has a part corresponding to $|0\rangle$ and a part corresponding to $|1\rangle$ at the same time. When you measure the particle, the system is projected to one of its basis (i.e., either $|0\rangle$ or $|1\rangle$). The overall probability for each state is given by the absolute square of its amplitude. Taking the state $|\phi\rangle = c_0|0\rangle + c_1|1\rangle$ as an example, the coefficient $|c_0|^2$ and $|c_1|^2$ represents the probability of obtaining $|0\rangle$ and $|1\rangle$, respectively. Two or more qubits can jointly form a quantum system. A two-qubit system is spanned by the basis of the tensor product of the two individual spaces. Hence, the joint state of qubit A and qubit B is spanned by $|00\rangle_{AB}$, $|01\rangle_{AB}$, $|10\rangle_{AB}$, and $|11\rangle_{AB}$, i.e.,

$$|\phi\rangle_{AB} = c_0|00\rangle + c_1|01\rangle + c_2|10\rangle + c_3|11\rangle,$$

where $c_0, c_1, c_2, c_3 \in C$ and $|c_0|^2 + |c_1|^2 + |c_2|^2 + |c_3|^2 = 1$. The notations described above can be generalized to multiple-qubit systems. For example, in a three-qubit system, the space is spanned by a basis consisting of eight elements ($|000\rangle, |001\rangle, \ldots, |111\rangle$).

A quantum system can be manipulated in many different ways, called *quantum gates*. A quantum gate can be represented in the form of a linear operation. For the purpose of illustration, we will describe some of the quantum gates in the following paragraphs.

1. NOT gate

   A quantum *Not* (N) gate applied on a single qubit changes the two components in the basis as $|0\rangle \rightarrow |1\rangle$ and $|1\rangle \rightarrow |0\rangle$. Due to linearity, the gate transfers a superposition state $|\phi\rangle = c_0|0\rangle + c_1|1\rangle$ into the state $|\phi'\rangle = c_0|1\rangle + c_1|0\rangle$. The symbol of an N gate is shown in Figure 15.3a. Note that the horizontal line connecting the input and the output is not a physical wire as in classical circuits, it represents a qubit under time evolution.

2. Control-Not gate

   The *Control-Not* (CN) gate is a two-qubit gate, which does the following transformation: $|00\rangle \rightarrow |00\rangle$, $|01\rangle \rightarrow |01\rangle$, $|10\rangle \rightarrow |11\rangle$, and $|11\rangle \rightarrow |10\rangle$. The symbol of a CN gate is shown in Figure 15.3b. It consists of one *control* (upper) qubit, which does not change its value, and a *target* (lower) qubit, which changes its value only if the control qubit is $|1\rangle$. Assuming the control qubit is x and the target qubit is y, the gate can be written as $CN|x, y\rangle \rightarrow |x, x \oplus y\rangle$, where the "$\oplus$" denotes exclusive-or.

3. Hadamard and control-Hadamard gate

   A quantum *Hadamard* (H) gate applied on a single qubit changes the two components in the basis as $|0\rangle \rightarrow (|0\rangle + |1\rangle)/\sqrt{2}$ and $|1\rangle \rightarrow (|0\rangle + |1\rangle)/\sqrt{2}$. Due to linearity, the gate transfers a superposition state $|\phi\rangle = c_0|0\rangle + c_1|1\rangle$ into the state $|\phi'\rangle = ((c_0 + c_1)|0\rangle + (c_0 - c_1)|1\rangle)/\sqrt{2}$. Similar to the CN gate, a Control-Hadamard gate can be built. It does the Hadamard transform on the target qubit whenever the control qubit is $|1\rangle$. The symbol of a Hadamard gate

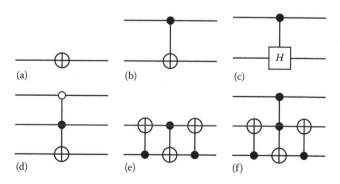

(a)   (b)   (c)

(d)   (e)   (f)

FIGURE 15.3

Symbols of (a) Not, (b) CN, (c) C-Hadamard, (d) CCN, (e) Swap, and (f) C-Swap gate.

is usually denoted by H in a box and the symbol of a Control-Hadamard gate is shown in Figure 15.3c.

4. Control-Control-Not gate

   The *Control-Control-Not* (*CCN*) gate is a three-bit gate transfer $|110\rangle \rightarrow |111\rangle$ and $|111\rangle \rightarrow |110\rangle$, other components in the basis remain unchanged. Assuming the first two qubits are the control qubits, the gate can be written as $CCN|x,y,z\rangle \rightarrow |x,y,(x \cdot y) \oplus z\rangle$. The symbol of a *CCN* gate is shown in Figure 15.3d. Note that if the "dot" on the control qubit is a "circle," then the function is activated only when the qubit is $|0\rangle$. The same notation applies on other gates.

5. Swap gate

   It is interesting that if we put three *CN* gates together, with the middle one upside down, the circuits will perform a *Swap* operation. More specifically, the two qubits will exchange their quantum states, just like they are physically replaced by each other. The symbol of a *Swap* gate is shown in Figure 15.3e.

6. Control-swap gate

   Similar to the CN gate, a swap gate can be controlled by a control qubit, as shown in Figure 15.3f. If the control qubit is in $|1\rangle$, the two target qubits are swapped (exchanged), otherwise their states remain unchanged. Note that if the control qubit is in a superposition state, each qubit will have a state that is a mix of both qubits.

Further generalizations can be made on the quantum gates described above. They involve *rotation* and *phase shift* that control the phase difference and relative contributions of the eigenstates to the whole state. For more information on quantum operation and quantum gates, see the article by Nielsen and Chuang (2000).

15.3 Quantum Networks and Quantum Switching

Many of the applications in quantum information science involve sending or receiving photons between two (or even multi) parties. For example, teleportation (Bennett et al. 1993) and dense coding require a quantum channel to exchange qubits. Moreover, quantum key distribution (Bennett and Brassard 1984) also involves sending qubits over the quantum channel. Although these applications have been demonstrated via either optical fiber or free space, a *switch-based quantum network* is needed to make these applications available to all end-users in a network. In this section, we present several architectures so that quantum information carriers (qubits) can be switched in the quantum domain.

15.3.1 Architecture 1—Cycle Implementation

The first architecture is based on *permutation* and *cycle* implementation using quantum circuits (Tsai and Kuo 2002). The idea is developed from the fact that any switching configuration is actually a permutation and can be implemented using a series of swap gates. As described before, the function of a switching module is to move the data from an input port to the corresponding output port. This can be viewed as a qubit permutation. A permutation can also be expressed as disjoint cycles. A cycle is basically an ordered list, which can be represented as $C = (e_1, e_2, \ldots, e_{n-1}, e_n)$. The order of the elements describes the operation. In the cycle described above, it takes $e_1 \to e_2$, $e_2 \to e_3, \ldots, e_{(n-1)} \to e_n$, and finally $e_n \to e_1$. The number of elements in a cycle is called *length*. A cycle of length 1 is called a *trivial* cycle, which can be ignored as it does not change anything. A cycle of length 2 is called a *transposition*.

Quantum switching is equivalent to reshuffling the quantum states for each of the qubits. Since a permutation can be decomposed into disjoint cycles, the implementation actually consists of executing cycles of various lengths in parallel. Because a cycle of length 1 does not permute anything, no circuit is required for a trivial cycle. For a cycle of length 2, the transposition can be done by three *CN* gates, as shown in Figure 15.3e. It is interesting that, for any length n ($n \geq 3$), a cycle can be done by six layers of *CN* gates without ancillary qubits. Figure 15.4 shows a cycle of length 4 and its decomposition. If we define the following nonoverlapping qubit transpositions $X = (0, 1)$ $(2, 3)$ and $Y = (0, 2)$, the cycle can be implemented using $C = YX$. The circuit implementation of these two functions is shown in Figure 15.5.

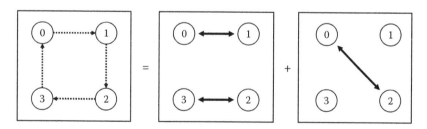

FIGURE 15.4
A diagram implementing cycles by disjoint transpositions.

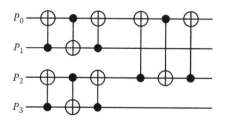

FIGURE 15.5
Cycle implementation with six layers of quantum gates.

Note that both X and Y consist of disjoint transpositions and can be executed in parallel using three layers of CN gates. As a result, each cycle can be performed using six layers of CN gates, as shown in Figure 15.5. This achieves the constant time complexity of a qubit permutation. Although the control module can be implemented in classical circuits, a variation has been proposed and discussed to implement the control module in quantum circuits (Sue 2009).

The architecture described above is strict-sense nonblocking. It can be applied to perform either circuit switching or packet switching. Compared with a traditional space or time switch, the proposed switching mechanism is more scalable. Assuming an $n \times n$ quantum switch, the space consumption grows linearly, i.e., $O(n)$, while the time complexity is only $O(1)$. Based on these advantages, a high throughput switching device can be built simply by increasing the number of I/O ports.

15.3.2 Architecture 2—Quantum Superposition

The second architecture is based on sending quantum superposition over a classical self-routing network (Shukla et al. 2004). This architecture takes advantage of the quantum superposition and can be applied to any classical switch with contention, such as Banyan networks. Figure 15.6 shows a typical 8×8 Banyan network. In this self-routing switching network, the destination address in the header is used to decide which output port the path should be sent to. If the ith bit in the address field is 0, then in the ith stage the packet will be sent to the upper output port. Similarly, if the

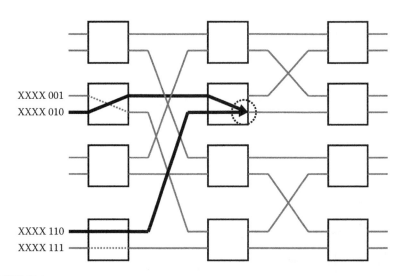

FIGURE 15.6
A diagram showing a Banyan network and a case of contention.

address bit is 1, the packet will be sent to the lower output port. A disadvantage of this network is that there will be contention for an output port if their destination bits are the same.

The idea of this design is to send the two packets contending for the same port using quantum superposition. A simple scenario of this idea for a 2×2 switch is depicted in Figure 15.7. Two qubit streams (P_0, P_1), each with its own destination address bits (A_0, A_1), are fed into the switch. The address qubits that determine the output ports are processed as follows:

1. If $A_0 = |0\rangle$ and $A_1 = |1\rangle$, the switch gate is set in the *through* state. Packet P_0 goes to port 0 and packet P_1 goes to port 1.
2. Similarly, if $A_0 = |1\rangle$ and $A_1 = |0\rangle$, then the switch gate is set in the *cross* state. Packet P_0 goes to port 1 and packet P_1 goes to port 0.
3. However, if $A_0 = |0\rangle$ and $A_1 = |0\rangle$, the control qubit of the switch would be a superposition of $|0\rangle$ and $|1\rangle$. This causes the switch gate to create an equally weighted superposition of the two packets at *both* output ports.
4. Similarly, if $A_0 = |1\rangle$ and $A_1 = |1\rangle$, the switch gate also creates an equally weighted superposition of the two packets at both output ports.

As described above, this design creates a superposition of the contending packets at the desired output. However, as a side effect, a complementary, undesirable, superposition is also created at the other output port. There are two problems on these unwanted packets. First, some output ports might receive packets that are not addressed to them. Second, the incorrectly routed packets will interfere with the routing of correctly routed packets, since their address bits will be used by later stages of the network. This further increases

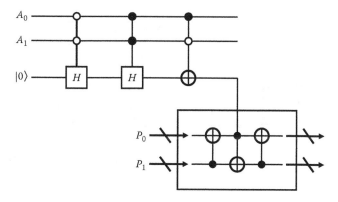

FIGURE 15.7
An idea of sending packets through a contending port with superposition.

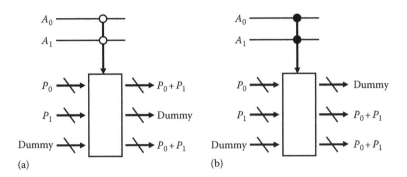

FIGURE 15.8
(a) Dummy packets are sent to port 2 and (b) Dummy packets are sent to port 1.

errors in the routing. The problem can be solved by introducing dummy ports and dummy packets, as shown in Figure 15.8a and b. The idea is that the undesired packets can be directed to the dummy port, and a dummy packet can be sent to the output port to avoid duplicated packets.

By introducing dummy packets and dummy ports, all packets are routed through the network without undergoing any interference and blocking. A simple measurement destroys the superposition and gives only a particular output sub-permutation, which is equivalent to classical routing through a Banyan network with random packet drops in case of contentions.

15.3.3 Architecture 3—Destination Address Sorting

The third way of building a quantum switch is to sort the destination address and the data packets using a sorting algorithm (Cheng and Wang 2006).

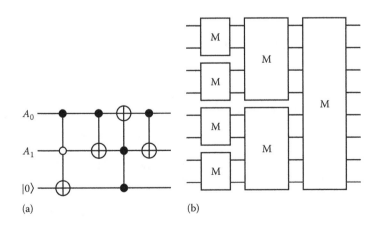

FIGURE 15.9
(a) The idea of a quantum comparator and (b) block diagram of the merge sort.

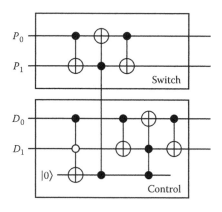

P_0

P_1

Switch

D_0

D_1

$|0\rangle$

Control

FIGURE 15.10
An idea of sorting the address qubits and the corresponding packet.

This design follows the classical merge sort algorithm, which sorts a list of length n ($n > 1$) as follows. First, divide the unsorted list into two sublists of about half the size. Then, sort each sublist recursively by reapplying merge sort and then merge the two sublists back into one sorted list. An important building block in merge sort is to compare two numbers and exchange their order. This can be done using a quantum *control-exchange* gate, as shown in Figure 15.9a. Figure 15.9b shows a circuit block diagram of quantum merge sort. Each module in the diagram (M) is a quantum bitonic merge sort module.

With the quantum merge sort circuits, a quantum switch can be easily implemented. The quantum switching consists of two parts: a switch module and a control module, as shown in Figure 15.10. The switch module is responsible for permuting all the input qubits into their corresponding output ports. The data permutation in the switch module is controlled by setting up destination addresses in the control registers.

Taking Figure 15.10 as an example, two input packets are fed into port P_0 and P_1 and their destination addresses are setup at D_0 and D_1. To perform a correct switching, we can then do a sorting on the destination addresses. If the address qubits are swapped, we swap the data packets too. After the sorting, all input data would be switched to their corresponding destination port correctly. In other words, quantum switching can be implemented by a quantum sorting algorithm that sorts all of the input qubits into an increasing order according to their destination port number. Once the switching is completed, the address qubits can be recycled for the next packet.

15.4 Conclusion

Telecommunication technology is one of the most rapidly growing fields in the engineering community. Due to the emergence of quantum networks, various techniques have been proposed to perform the efficient switching of quantum information. In this chapter, we have reviewed some of the most important techniques for carrying quantum bits through a quantum information network. As we have seen, to support quantum traffic, a wide range of technologies are available. We hope some of these promising architectures can be used to deploy a high throughput switching network so the ever increasing end-user demand can be met.

References

Benioff, P. 1980. The computer as a physical system: A microscopic quantum mechanical Hamiltonian model of computers as represented by Turing machines. *J. Stat. Phys.*, 22(5), 563–591.

Bennett, C. and Brassard, G. 1984. Quantum cryptography: Public key distribution and coin tossing. *Proceedings of the IEEE International Conference on Computers Systems and Signal Processing*, Cambridge, MA, pp. 175–179.

Bennett, C., Brassard, G., Crépeau, C., Jozsa, R., Peres, A., and Wootters, W. 1993. Teleporting an unknown quantum state via dual classical and Einstein-Podolsky-Rosen channels. *Phys. Rev. Lett.*, 70(13), 1895–1899.

Cheng, S. T. and Wang, C. Y. 2006. Quantum switching and quantum merge sorting. *IEEE Trans. Circ. Syst. I*, 53(2), 316–325.

Deutsch, D. 1985. Quantum theory, the Church-Turing principle and the universal quantum computer. *Proc. Roy. Soc. Lond. A*, 400, 97–117.

Feynman, R. 1982. Simulating physics with computers. *Int. J. Theoret. Phys.*, 21, 467–488.

Grover, L. 1996. A fast quantum mechanical algorithm for database search. *Proceedings of the 28th Annual ACM Symposium on the Theory of Computing*, Philadelphia, PA, pp. 212–219.

Nielsen, M. A. and Chuang, I. L. 2000. *Quantum Computation and Quantum Information*. Cambridge University Press, Cambridge, U.K.

Shor, P. 1994. Algorithms for quantum computation: Discrete logarithms and factoring. *Proceedings of the 35th Annual IEEE Symposium on the Foundations of Computer Science*, Santa Fe, NM, pp. 124–134.

Shukla, M. K., Ratan, R., and Oruc, A. Y. 2004. A quantum self-routing packet switch. *Proceedings of the 38th Annual Conference on Information Sciences and Systems CISS'04*, March 2004, Princeton, NJ, pp. 484–489.

Sue, C. C. 2009. An enhanced universal $N \times N$ fully nonblocking quantum switch. *IEEE Trans. Comput.*, 58(2), 238–250.

Tsai, I. M. and Kuo, S. Y. 2002. Digital switching in the quantum domain. *IEEE Trans. Nanotechnol.*, 1(3), 154–164.

16

Powering the Future with Nanotechnology

Harry Efstathiadis

CONTENTS

16.1 Introduction

Nanotechnology is the study of the control of matter on an atomic and molecular scale. Generally, nanotechnology deals with structures that are 100 nm or smaller. Nanotechnology is viewed as having huge potential in areas as diverse as biosciences, integrated circuit fabrication and electronics, power generation, and energy storage. Governments, academia, and businesses across the world have started to invest substantially in their development. Recently, Lux Research predicted that nanotechnology will impact most manufactured goods and represent $2.6 trillion in global manufacturing output by 2014.[1] The United States, like the rest of the world, faces not only rising energy costs but is concerned about environmental impact and worries about economic competition. Thus, there is a worldwide need to identify ways to meet the growing energy demand with clean, renewable sources and to create new jobs and new business opportunities. The U.S. government recently announced plans to promote clean, renewable energy, and reduce the country's electricity consumption by 15% by 2015 to reduce energy dependence on foreign oil, address global climate change, and create new jobs.

The world energy need is estimated to grow from its current 13 terawatt-year (TW-year) to up to 30 TW-year by 2050. As the world is trying to meet this energy demand, it is struggling to address environmental and climate challenges due to greenhouse gases while facing a steady decline of fossil fuels. There is no single or simple answer to this energy demand and environmental issues. A synergy of several technology-based solutions could possibly address these issues, such as energy efficiency, renewable energy, nonpolluting transportation fuels, the separation and sequestration of CO_2, and next-generation nuclear energy technologies. Nuclear power, for example, is a primary candidate for reduced carbon emission electricity production. However, several issues, such as the grade and purity of uranium ores, nuclear fuel reserves, waste disposal, and public concern, ensure that it is not a very attractive solution. Hydroelectric power generation is another technology currently used in energy production but it is limited by supply expansion. On the other hand, wind power is gaining a lot of attention lately and is considered to be another important electricity production source. However, it is estimated that on its own it might not be able to address the global demand in energy due to the limited amount of energy production it can provide.

Since approximately 125 TW of solar power reaches the earth at any time, solar energy might be the most promising energy source to meet the global energy demand in the future. Solar energy offers the promise of increased energy security, a cleaner environment, and significant economic benefits. Solar power is particularly valuable for reducing the load in the United States' electric grid and for lowering the risk of major blackouts. This semiconductor-based technology converts sunlight directly into electricity with no moving parts, consuming no fuel, and creating no pollution. It is a distributed energy resource that can be deployed throughout the United States, improve grid reliability, lower distribution and transmission costs, and be sited at the point of use with minimal or no environmental impact.

Among several photovoltaic (PV) technologies, bulk c-Si accounts for ~80% of the PV market today. This is due to its relatively high cell efficiency along with a well-established infrastructure. But c-Si technology suffers from the high cost of silicon, with the wafer being the most expensive part of the manufacturing process. Gartner Inc. has issued its first forecast of the solar cell market, estimating that it will grow at a 17% compound annual growth rate between 2008 and 2013 to reach $34 billion in revenue in 2013.[2] It is estimated by Linx Consulting that the PV material market sector is expected to grow from $2.3 billion to ~$15 billion by 2015.[3] Growing worldwide demand for "green" energy sources aligns with shrinking PV module prices based on improved manufacturing technologies and less-expensive raw materials, in which nanotechnology advancements are expected to play an important role.

Government support and incentives of solar energy are essential in PV market growth because solar energy remains dependent on subsidies to be cost-competitive with the conventional forms of electric power generation. During the last two decades, European governments have collaborated closely with research institutions and industries aiming to reduce their dependence on hydrocarbon-based electricity sources. The United States is expected to follow a similar approach in its alternative-energy strategy. With government support, the cost is expected to be reduced at a value at which PV-generated electricity is economically viable without external support.

On the technology side, crystalline silicon (c-Si) is expected to remain the dominant PV technology during the next 10–15 years, despite technological advancements and fast growth in thin-film PV sales. Gartner Inc. forecasted that the PV market will grow to reach 23.4 GW by 2013. The global solar PV market grew to at least 5.5 GW of electrical power converted using PV methods in 2008 compared with 2.4 GW in 2007.

In general, solar cell technologies can be classified into three generations. First-generation solar cells are the large area, Si-based, PV cells that, despite their high manufacturing costs, have dominated the solar panel market mainly due to their high efficiency. It was also expected that the "first-generation" silicon wafer-based solar cells would be replaced by a "second generation" of thin-film technology using much less Si and probably involving something other than Si semiconductor material.[5] Technologies based on materials such as amorphous silicon, copper indium diselenide, cadmium telluride, and now thin-film silicon have been regarded as viable thin-film candidates to replace first-generation solar cells. The technology that will dominate the PV market is the one that produces a high power output per unit area and has the lowest possible future manufacturing costs in a large area. However, due to the dropping prices of polysilicon, the use of thin-film technology has not exploded as was expected.[5] Such an analysis makes it likely that PV, in its most mature form, will probably evolve to a "third generation" of high-efficiency energy conversion, thin-film technology as originally developed by Martin Green of the University of New South Wales.[6] Achieving competitive cost/efficiency ratios will require breakthroughs even in mainstream technologies. The range of efficiency values targeted is almost triple the currently targeted range of 15%–20%, which could lead close to the thermodynamic limit (Carnot limit) of conversion power efficiency of 93%. In tandem cell configurations, power efficiency can be increased by stacking cells of different bandgap absorbing layers such as amorphous silicon/amorphous silicon germanium at the expense of increased device complexity due to material integration. However, as opposed to this "serial" approach, better-integrated "parallel" approaches are possible that offer similar efficiency to tandem cells. These alternatives could become feasible with the likely evolution of materials technology over the next 10–15 years. There would be an enormous impact on economics if these new concepts could be implemented in thin-film form, making PVs one of the cheapest known options for future energy production.

In addition to tandem cells, a number of integrated "parallel" conversion approaches have also been suggested with the potential to achieve efficiency levels similar to tandem cells.[4] One category that includes tandem cells is based on having multiple energy threshold processes available in the one device. Another example of this approach is "the impurity photovoltaic effect," during which excitation and recombination are allowed between the valence and the conduction band, as in standard solar cells and also between these bands and a third impurity band. Other quantum multiplication approaches are to create two lower energy photons from a single high energy photon, or an electron plus a lower energy photon.

Another approach is based on using sunlight to heat an absorber, then the energy extracted from the heated absorber can be converted to electricity. A technique of this type is the solar thermal electric approach, where heat from the absorber drives a heat engine or thermoelectric devices; there are devices based on quantum well structures with a layer thickness of approximately 10 nm. Another thermal approach is thermo-PVs, where the absorber reradiates the absorbed energy. A solar cell is then used to convert the light emitted by this radiator.

Another category of third-generation solar devices is organic-based photovoltaics (OPVs). The development of materials from organic cells can be leveraged by the increasing research and development (R&D) of organic materials for displays, actuators, and transistors. The key advantages of organic-based solar cells are that polymers and small organic molecules are inexpensive, they posses a high absorption coefficient, and they can be fabricated at low process temperatures with high-speed, high throughput techniques such as roll-to-roll or screen printing on flexible plastic substrates. If achieved efficiencies are comparable or even lower than other technologies of solar cells, the OPVs low fabrication cost may be a convincing argument favoring them. A key issue that needs to be addressed is the poor stability of organic materials leading to device degradation due to a loss of interfacial adhesion, changes in film morphology, and the interdiffusion of elements and components.

A large variety of ways to fabricate efficient OPVs have been investigated. These approaches include organic–organic and organic–inorganic composites, dye-sensitized nanostructured oxide cells, and multilayer devices of organic semiconductor layers similar to tandem cells. Up to 5%–6% efficiency values have been claimed from OPV devices, which approach the efficiency values reported by amorphous silicon-based solar cells. However, the values reported have not been confirmed or certified by an independent national laboratory such as the National Renewable Energy Laboratory in Golden, Colorado.

Thus, developing techniques for fabricating nanostructures inexpensively in large areas and volume, i.e., manufacturing them, is an area that requires substantial effort; nanoscience and nanotechnology will not be fully successful until they have provided the base for manufacturing technologies that are economically viable. It is probable that methods developed for fabrication of silicon grains, for example, in the >100 nm size range, will not work in the 20–10 nm size range. Thus, it will probably be necessary to develop an entire new suite of manufacturing methods for nanostructures. Two examples in this area will be described in Section 16.2.

16.2 Main Focus

The efficiency of silicon and gallium arsenide (GaAs)-based solar cells has increased significantly during the past few decades. Efficiencies as high as 24% in silicon and more than 40% in multi-junction GaAs cells have been achieved in small size devices. Sanyo Electric Co. Ltd. announced in May 2009 that it has achieved the highest energy conversion efficiency of 23% at the research level for a c-Si solar cell in a practical size (≥ 100 cm^2) using Sanyo's proprietary heterojunction with an intrinsic thin-layer PV cell.[7] However, due to the high cost of materials and processing, the cost of producing electrical power from these high efficiency cells is still around \$3.00/W for silicon and much higher for the multi-junction cells. Currently, the cost of fossil fuel–generated electricity is on average around \$0.50/W. This large price gap must be reduced before PV can become a viable energy option without the need for large subsidies. Second-generation thin-film processes may reduce the cost of \$/W but these approaches have not yet made it to the marketplace. One method of lowering the cost of the PV material is to use a concentrator to collect light over a large area and focus it to a smaller PV cell area. This approach can substantially reduce the PV cell area; however, if large concentration ratios (concentrator area/PV cell area) are used, the PV cell design must be capable of handling the corresponding higher power and heat dissipation levels. These factors increase material and processing costs. In addition, for optimum effectiveness, the concentrator should track the sun in its movement over daily and annular cycles. The tracking system further adds to the overall cost and makes the system less effective in utilizing diffuse illumination.

Many nanotechnology achievements have contributed to the advancements of PV technology. Most of these advancements have been demonstrated in small device areas, such as less than 1 in^2. The expectation of the research teams working in solar cell technologies is to demonstrate high efficiency devices and to scale up the device and pass it from the research stage to the R&D stage. However, most of the time, this is not a straightforward or an easy case. Two examples are demonstrated below where advancements in nanotechnology were demonstrated in areas as large as ~ 1 m^2. The first is in scaling plasma-enhanced chemical vapor deposition (PECVD) for large areas of nanocrystalline silicon (nc-Si) and the second is related to large areas of nanostructured polymer-based solar cells.

Radio-frequency (RF) PECVD from silane has long been the dominant method for depositing hydrogenated amorphous silicon (a-Si:H) films on glass substrates to manufacture PV modules or thin-film transistor arrays for displays. Most of this work utilizes plasma sources at the standard frequency of 13.56 MHz.[8] A limitation of this approach is that increasing the deposition rate leads to poorer stability of the material and reduces the film material density. For this reason, deposition rates in practical large area depositions are generally limited to less than 0.2 nm/s.

Recently, there has been exploration of the use of very high frequency in the range of 30–300 MHz for PECVD of nc-Si films.[9,10] It was found that it yields much faster deposition rates at the same power density, and there are suggestions that the films may be of good quality due to a high generation rate of atomic hydrogen. Furthermore, because of its greater activity, VHF appears to lead in a more natural way to the growth of micro- and nc-Si:H films. However, the majority of the VHF work to date has been conducted in small-area systems of ~100 cm^2, and depositions attempted over large areas were found to be exceedingly nonuniform.[11]

The use of a combination of junctions of a-Si/nc-Si or a-Si/a-SiGe would be highly desirable for the manufacturing of PV modules since it could allow thin-film Si PV conversion efficiencies to be increased by about 50%, translating to an increase in module level efficiencies from 6%–7% to 10%–11%. Such an efficiency increase would enable low-cost thin-film silicon to compete with polycrystalline silicon.

The targeted PV devices require a thickness of the nc-Si component of 2.0–2.5 μm in order to ensure adequate absorption of light in the wavelength range of 750–1000 nm. This is 5× thicker than the a-Si used in a-Si PV modules, and this dictates a corresponding increase in the deposition rate. However, conventional 13.56 MHz PECVD yields a rate of only 0.1–0.2 nm/s for nc-Si rendering the conventional approach uneconomical. Another issue in the fabrication of p-i-n devices concerns the extreme sensitivity of the un-doped i-layer to low levels of unintended dopant cross contamination. Another serious drawback of the conventional 13.56 MHz PECVD approach is the narrow process window that is found for the satisfactory growth of nc-Si. Often, device-quality nc-Si is obtained only in a restricted area of the substrate and not over its entirety. Furthermore, film thickness nonuniformity issues on large area substrates using VHF in conventional reactor designs represents a barrier to the development of VHF.

There are several technical barriers to extend PECVD to the relatively unexplored VHF domain. These are the lack of understanding of gas phase and surface chemistry and the lack of useful models for predicting the thickness uniformity distribution over the deposited area. These technical barriers have so far prevented the commercial production of advanced hybrid thin-Si PV modules and slowed the exploitation of thin-film technology generally.

A possible solution to these problems, i.e., growth rate, texture uniformity, process window, economics, composition, and thickness uniformity, is to accomplish VHF growth of nc-Si over large areas with excellent thickness, composition, and grain size uniformity and accelerate these efforts via computer modeling.

Plasma excitation by VHF instead of RF power enjoys some specific advantages for processing nc-Si-type thin films. The following is a list of key processing issues that could be addressed by higher excitation frequency:

1. *Higher deposition rate.* For a given power density (W/cm^2) applied to the plasma zone, higher deposition rates can be readily obtained due to a more efficient generation of hydrogen (as well as SiH_x) radicals by VHF vs. RF-glow discharge (GD), allowing nc-Si to be formed with lower hydrogen dilution, thus increasing the flux of growth species to the substrate surface. Higher film growth rates (without deterioration of required film properties) is one of the most critical considerations in the commercial viability of the nc-Si PV technology, as a thick ($\sim$2 μm) nc-Si absorber layer is needed for adequate optical absorption. Using conventional RF-GD at 0.1–0.2 nm/s, this thickness corresponds to several hours of deposition. But with VHF excitation, the limitation on throughput is expected to be substantially reduced. In general, a broader processing window exists for the nucleation and growth of nc-Si thin films at various rates by using VHF rather than RF, partly due to the reduction in high-energy ion bombardment on the growth surface.

2. *Thickness and texture uniformity.* This is the most critical issue in nc-Si process scaling-up for module fabrication. Uniformity refers not only to the thickness, but more importantly to the "phase" or degree of crystallinity of the nc-Si films. VHF excitation can help to improve the deposition uniformity of nc-Si layers compared with what can be achieved with RF power for large-area deposition. Lower hydrogen dilution can be used for nc-Si deposition. For a given substrate area and gas flow rate, the percentage depletion of SiH_4 across the substrate surface is reduced with VHF vs. RF-GD. This helps to establish a more uniform growth profile, which is also very important in maintaining a more uniform "phase" of nc-Si, as the optimal nc-Si film is grown near the edge of the transition to the amorphous phase. In addition, the processing window for high quality nc-Si by RF-GD is very narrow. The best nc-Si for solar cell application is deposited under conditions very close to the borderline of a-Si to nc-Si transition. Inferior solar cells are obtained when the nc-Si:H absorbers become microcrystalline. Unlike poly-Si, large grain sizes are not preferred for nc-Si PV devices, as further increased optical absorption is more than offset by reduced open circuit voltage and fill factor. The narrow processing window makes nc-Si solar cells difficult to scale-up and reproduce for large area module processing. While it is extremely difficult to maintain plasma conditions near the "edge" over large areas using RF-GD, which requires very high H_2/SiH_4 dilution ratio, the much lower H_2/SiH_4 dilution needed in VHF-GD makes it a much easier case.

3. *Faster nucleation with a higher density of nuclei.* These two features are particularly beneficial for high-efficiency nc-Si device fabrication.

The fast, relatively easy nucleation of Si:H on a foreign or amorphous substrate, such as glass, is highly desirable for forming a thin but effective p-type "seed" or nucleation layer. Otherwise, a thicker p-type seed layer is not only time consuming, but also detrimental to solar cell performance due to short-wavelength optical loss in the narrow-gap seed layer. Furthermore, it is known that the presence of boron deters nucleation, above and beyond the difficulty of nucleating pure Si:H without any dopant. Furthermore, the texturing of the superstrate (e.g., SnO_2 or ZnO) needed for light trapping makes nucleation difficult. Thus, with RF-GD, it has been proven to be a challenging task to quickly generate a thin, transparent, boron-doped nucleation layer by the RF-GD with the necessary optical transparency, degree of crystallinity, and doping level. The seeding step is a most critical element in fabricating high-performance nc-Si-based PV devices. There has been an observed loss of transparency of SnO_2 due to the initiation of the intense H-rich RF plasma (seeding) in the absence of a thick ($\sim$100 Å) p-type a-SiC:H protective layer (yet its presence also results in optical loss). When the duration of the nucleation plasma is shortened (or when lower power is applied) by the use of VHF-GD, a much thinner protective layer can be used to boost the overall transparency of layers in front of the i-layer.

4. *Wider bandgap nc-Si:H alloys.* For reduced optical loss through doped layers, it is highly desirable to have wider bandgap alloys of nc-Si, such as $Si_{1-x}O_x$:H or $Si_{1-x}C_x$:H ($x < 0.1$), by adding small amounts of alloying feedstock into the plasma feed gas mixture. This is especially helpful in multi-junction devices such as p-i-n p-i-n cells. Alloying elements greatly impede the seeding and growth of nc-Si films. With VHF excitation producing a high density of hydrogen radicals, the growth of such wider bandgap nc-Si alloy films should be much more feasible.

In summary, nc-Si development efforts are focussed on tight control of material properties over large deposition areas, i.e. in the order of 1.0 m^2. The film properties targeted are thickness nonuniformity less than 15% over the entire deposition area, growth rate of greater than 0.2 nm/s, efficiency of single-junction (p-i-n) nc-Si:H solar cell of 7%, efficiency of two-junction (p-i-n/p-i-n) hybrid a-Si = nc-Si solar cell of 11%.

Another example of nanotechnology is to develop superior, low-cost PV cells with high conversion efficiencies by incorporating and coupling both semiconductor nanocrystals and single wall carbon nanotubes (SWNTs) into conjugated polymers, perform first-principles modeling and simulation of the optical and electrical properties of the active layer, and engineer layer surface morphology and interfaces in the cell.

Both semiconductor nanocrystals and SWNTs can be synthesized to have high electron affinity and therefore have the ability to dissociate the excitons that are generated in photoactive polymers upon optical absorption. Semiconductor nanocrystals make it possible to tailor the absorption spectrum of the material to best match the incident solar spectrum through the adjustment of size, shape, and composition. SWNTs primarily serve the purpose of increasing mobility by improving the conduction process. When these constituents are combined—since the polymer, nanocrystals, and SWNTs may each absorb in a different spectral region—it may be possible to produce a series of junctions in a polymeric solar cell that would be analogous to a conventional triple-junction solar cell. And apart from the potential advantages of the mixture of nanocrystals and SWNTs in the solar cell, the actual coupling of the nanocrystals to the SWNTs is likely to provide the most efficient combination of exciton dissociation and carrier transport leading to efficiencies approaching theoretical numbers of 60%.

Successful nanoengineering efforts can achieve up to 30% of the theoretical performance targets, while the anticipated cost of doped conjugated cells could reach the commercially proposed target of about \$0.33/W. This could provide a huge incentive to consumers to adopt PV cells as a clean energy source.

The ultimate goal of current engineering efforts is the development of a low-cost technology for PV devices based upon the incorporation of nanomaterials into conjugated polymers. Such nanoengineering projects are focused on demonstrating that high purity carbon nanotubes can be produced in large volume exceeding 100 g/h with controllable sizes optimized for incorporation into a polymer matrix for PV devices.[12] It is important to demonstrate the reproducibly of producing large quantities of fully controllable semiconductor nanocrystals with quantum yields over 40% for the optimum absorption spectrum for use in solar cells. Current nanotechnology issues in this area are establishing correlations between the composition, morphology, and polymerization mechanism of active layers and their electric conductivity; identifying reactions occurring at various interfaces, as well as chemical changes within the polymer blend induced by electrical stress and exposure to ultraviolet light; demonstrating that semiconductor nanocrystal and SWNT integrated complexes can be synthesized with optoelectronic properties such that higher power conversion efficiencies can be obtained; and demonstrating PV films made from combinations of carbon nanomaterials and semiconductor quantum crystals that have film efficiencies exceeding 20%. Achieving over 10% power conversion efficiency in large commercial-sized OPV modules made in commercial quantities that are reproducible and highly reliable as well as dependable in outdoor performance over extended periods of time are critical issues that could be overcome with the help of nanotechnology advancements. In addition, ensuring that cost targets of fabricating OPV systems can reach $10/m^2$ with efficiencies over 5% resulting in a targeted module manufacturing cost of about $0.2/W (significantly below the commercial cost goal of $0.33/W) is very important for the advancement of OPVs.

After the discovery in 1991 that the transfer of photoexcited electrons from conjugated polymers to fullerenes is very efficient, it took 10 more years until organic solar cells reaching 2.5% efficiency were reported.[13,14] Research efforts devoted to the development of PV devices based on organic materials for the conversion of sunlight to electricity have produced promising results in recent years. Although less efficient than corresponding inorganic (semiconductor-based) devices, organic devices show promise as a cost-effective means of providing a widespread application of the PV energy conversion. Polymer-based PV devices are processed from solution thus avoiding the costly processing steps required to produce semiconductor solar cells (using lithography under high temperatures and in a vacuum). The economic gain of reduced cost promises to outweigh the liability of reduced efficiency.

The efficiency of inorganic solar cells is typically around 10% while that for organic devices is on the order of 4%–5% or less. Recent advances, however, have led to efficiencies as high as 6%. Low carrier mobility is one of the major deficiencies of organic solar cells. Charge carriers are produced by the absorption of photons to produce electron–hole pairs. Oppositely charged carriers (i.e., electrons and holes) move in opposite directions in the device due to gradients in the chemical potentials.

The mobility determines the likelihood that a photogenerated charge carrier will reach an electrical contact before recombining. The recombination of electrons and holes is a stochastic process that is governed by a certain lifetime. Carriers with low mobility spend a longer time in the device than carriers with higher mobility and are therefore more likely to recombine. The short-circuit current is determined by the difference in the rate at which carriers are created (determined optical properties of the material) and the rate at which they combine.

It has been shown that mobilities can be increased through the introduction of second-phase conducting materials into the polymer matrix and the introduction of SWNTs increased cell performance.[15] The nanotubes behave as electron acceptors providing an increased-conductivity path for electron conduction. Similar performance gains were obtained upon introduction of semiconductor nanorods into the polymer matrix.[16] In this study, the nanorods not only lead to increased mobility but also provide greater flexibility to tailor the absorption spectrum of the material by controlling the size, shape, and chemical composition of the semiconductor nanorods.

Further improvements have also been obtained through efforts to reduce barriers to charge transport. The introduction of a thin layer of LiF on Al electrodes resulted in an increase in short-circuit current, open circuit voltage, and fill factor.[17] Two mechanisms explaining these improvements are (1) lowering the effective Al work function and (2) the formation of a dipole layer between the Al electrode and the organic material, resulting in a vacuum level offset between these two materials. In other words, the LiF/Au contact is more selective in that holes are blocked while electrons are more easily transported across the contact.

Nanotechnology faces several challenges in this arena. Some of the challenges are that the role of the organic/electrode interfaces for the charge collection in solar cells has not been investigated in detail. In addition, the ITO–polymer interface is not well understood and controlled. It has been proposed by many research groups that the actual organic/metal interface is characterized by the presence of interface dipoles generating a potential near the electrodes. According to the direction of the dipoles, the photogenerated charges approaching the electrode vicinity may either be repelled or attracted toward the electrode. Initial studies indicated that the organic/electrode interfaces affect the PV device characteristics, which may be due to a variety of phenomena occurring at the organic/metal interface.

Utilizing appropriately designed nanocrystal and nanotube complexes to achieve the optimization of an ideal additive for polymeric solar cells that exhibit high electron affinity and high electrical conductivity could improve the device efficiency.

CdSe core/shell nanocrystals with absorption onset optimized for PV conversion efficiency to achieve efficient charge carrier transport to a conductive polymer is a possible solution. Another possibility is to employ carbon nanotube materials. The chemical vapor deposition (CVD) process has been shown to produce carbon nanotubes at low cost and high quality. More importantly, this process is able to produce high-purity nanotubes by the in situ etching of amorphous carbon during the growth process. Traditional purification processes are not only tedious but also result in severe damage to the carbon nanotube structure.

The synthesis of both SWNTs and multiwalled nanotubes (MWNTs) has been already developed and demonstrated in both patterned growth on substrates as well as bulk. A recently developed floating catalyst CVD method can generate approximately 0.5 g/h of SWNTs, and this process is being scaled up. It was discovered to be a powerful technique where MWNTs can be selectively grown on silica patterns on Si/SiO_2 and other substrates. The patterns are produced by photolithography and arrays of nanotubes grown on the machined silica areas perfectly vertically aligned to exactly the same heights. Using catalysts such as Pd during CVD synthesis can alter the basic tubular structure of the nanofibers and structures of fibers can be completely altered to, for example, stacked cones. The successful control of the positioning and alignment of nanotube arrays to a large extent, by making use of their substrate selective growth, has been achieved.

1. *Controlling the bandgap of nanotubes.* The multiple possibilities of carbon nanotube structures defined by the tube diameter and chirality offer great promise for creating next-generation functional electronic/photonic devices and microsensors. Intensive efforts have been placed on creating miniaturized field effect transistors and chemical/biological sensors based on carbon nanotubes. Most of the above methods investigated the tunable electronic behavior of nanotubes when surrounded by different gas/liquid media. Apart from electronic properties, there has been a substantial interest in the optical behavior of carbon nanotubes, including the bandgap absorption and photo- or electroluminescence for their application in fields such as PV. The ability to tailor the nanotube surface by treating them with various organic solvents and couple with nanocrystals, which in turn modifies the electronic and optical properties of these materials, will also play an important role in enhancing the optoelectronic properties.

   The bandgap optical absorption of SWNTs weakens after exposure to the air. This happens because the adsorption of foreign gases could cause a charge transfer between gas molecules and nanotubes diminishing their bandgap absorption. This can severely hamper the use of SWNTs for various applications. It has recently been shown that this bandgap absorption can be gradually recovered by treating nanotubes in organic solvents.

2. *Metallic nanostructures.* A technique has already been demonstrated to synthesize the aligned metallic nanorods to explore their effect on cell efficiency by embedding them into the polymer matrix.[18] Coupling these with semiconductor nanocrystals could improve cell efficiency. Important parameters to control are the length and diameter of the nanorods since their one-dimensional (1D) character makes them suitable as charge transporters. Anopore membranes were employed along with electroplating as a cost-effective nonvacuum technique to produce aligned metallic nanorods. Copper and silver nanorod arrays have already been fabricated by the electrodeposition of metal into anopore membranes with pore sizes of 20, 100, and 200 nm.

3. *Integration of carbon nanotubes and semiconductor nanocrystals for superior solar cells.* The most widely investigated nanomaterials for polymeric solar cells have been semiconducting nanocrystals, fullerenes, and SWNTs.[19] In particular, CdSe quantum dots (QDs) and nanorods have all shown viability as successful polymer solar cell additives. Under AM1.5 illumination, 90% w/w CdSe nanorods in P_3HT and 86% w/w CdSe tetrapods in OC_1C_{10}-PPV have been reported to produce a power conversion efficiency of 1.7% and 1.8%, respectively.[20] Fullerenes have also been a popular additive

for polymeric solar cells with multiple variations in the chemical structure shown to promote high electron acceptor properties. Recent results have demonstrated a power conversion efficiency of 2.5% under AM1.5 illumination for a device with an active material of methanofullerenes incorporated into a modified-PPV polymer with LiF/metal electrodes. The latest nanomaterial to show success as an additive in polymeric solar cells is the SWNT. Incorporation of SWNTs into P$_3$OT has demonstrated high open-circuit voltages, nearly 1.0 V, although the power conversion efficiencies are currently below 2%.

The high electron affinity of QDs and fullerenes provide them with the ability to dissociate the polymeric excitons, leading to improvement in these types of devices. Also, since the electrical conductivity of the photoactive polymers routinely employed is low, the QD or fullerene additive is primarily responsible for electron transport to the negative electrode. However, these materials have a low aspect ratio (high percolation threshold in the polymer), and have therefore necessitated very high doping levels (consistently >75% w/w). In comparison, the low percolation threshold of SWNTs coupled with their extraordinary conductivity allow for the significant enhancement of electron transport at even very low doping levels (<1% w/w). Therefore, if a material that encompasses both high electron affinity and high electrical conductivity can be utilized as an additive, a further enhancement in the conversion efficiency of polymeric solar cells is expected. In addition, a review of the optical absorption energies and carrier transport for an optimal nanomaterial-polymer solar cell may suggest a suitable route for additive material synthesis.

The ideal cascade of energy transitions for an optimal nanomaterial-polymer solar cell would include photon absorption by the components over the entire air mass zero (AM0) spectrum and dissociation of the excitons by the highest electron affinity material nearest the exciton.

In addition to the absorption by the polymer and QDs, it has been shown that semiconducting SWNTs can also absorb light and create electron–hole pairs that would contribute to the photoconductivity in these types of devices. Ultimately, the holes are transported by the polymer to the positive electrode and the dominant electron path is through the percolating SWNTs to the negative electrode. Due to the fact that the polymer, QDs, and SWNTs may each absorb in a different spectral region, the possibility exists that these nanomaterials could be combined in such a way as to produce a series of junctions in a polymeric solar cell that would be analogous to a conventional triple-junction solar cell. Although a mixture of QDs and SWNTs in a polymer may prove worthwhile, coupling of the QDs to SWNTs would presumably provide the most efficient combination of exciton dissociation and carrier transport.

16.3 University-Based Research and Development Activities

The increasingly more complex research requirements of emerging generations of high technology products are causing a radical transformation in the scope and magnitude of university-based R&D activities. In most emerging scientific fields, individual research projects managed by a single principal investigator have become a thing of the past. They have been replaced with multidisciplinary, multi-investigator, vertically and horizontally integrated R&D enterprises that require a state-of-the-art, multidimensional, infrastructure to ensure timely delivery. Corresponding educational programs and associated instructional tools must mirror this radical transformation.

Energy and Environmental Technology Applications Center (E2TAC): The E2TAC was established in June 1998 by New York State to formulate the science and technology base required to develop and implement new, environmentally friendly alternative energy sources. The emphasis is on system-on-a-chip (SOC) technologies and nanosystem devices for advanced energy and environmental sensor applications. It represents a statewide partnership that involves major research universities, federal R&D organizations, and private industries. Examples include the University at Stony Brook-SUNY, Brookhaven National Laboratory, and Benet Laboratory (Watervliet Arsenal).

NanoFab 200: NanoFab 200 is the only fully-integrated 200 mm wafer nanofabrication facility at a university in the United States. With a net asset value in excess of $75 million, NanoFab 200 is designed to provide a one-stop shop for the design, processing, integration, and demonstration of next-generation nanosystems on a 200 mm wafer platform. Its primary function is to serve as a technically aggressive, economically competitive environment for the prototyping/integration of SOC, including emerging micro- and nanosystems, biological devices and components, and sensors for telecommunications, energy, and the environment.

Also, UAlbany was recently named the New York State Center of Excellence in Nanoelectronics. A primary mission of the Center of Excellence is to establish state-of-the-art instructional facilities and associated advanced workforce development programs to meet the highly skilled workforce needs of the nanoelectronics-based industries. In this respect, CNSE's curriculum is intended to "provide distinguished graduate and professional programs which reflect the distinctive strengths" of the UAlbany faculty, and "which are competitive regionally, nationally and internationally for students of exceptional academic ability."

By ensuring the proper dissemination of fundamental knowledge concepts and new frontier scientific innovations in nanoscience and nanoengineering, it provides an excellent vehicle for the university to establish a national and international academic recognition for the university as one of the best educational, training, research, and economic outreach resources in these rapidly expanding fields. The excellent synergy and tight coupling between CNSE and E2TAC, the energy-related part of CNSE, is critical to the successful achievement of these goals.

Among the many strengths of the CNSE, one of the most visible has been in the area of microelectronics. Microelectronics and nanotechnology involve the tiny engines and brains that drive millions of products and services. Computers, digital cameras, and cell phones all work due to the manipulation and control of materials at the nanometer level—and the CNSE offers a unique environment for research and education involving these advanced device structures. As device dimensions continue to shrink, the demands placed on the students going into this field will be significant, although the potential benefits are undeniable.

Other challenges and opportunities abound. Clean efficient energy is a scientific challenge and global goal to help reduce mankind's footprint on the environment. Many technologies have the potential to meet this need, such as new generations of advanced PVs or solar cells based on thin-film nanocrystalline materials that offer higher efficiencies. Advanced fuel cells and advanced hydrogen storage technology require R&D in nanomaterials and manufacturing.

16.4 Conclusion

PVs is an emerging industry with remarkable similarities to the semiconductor industry in terms of materials, process, process integration, equipment, yield (manufacturing), and innovation leading to next-generation devices in which nanotechnology is playing an important role. In a potential trillion dollar market, where a "solar PV" Moore's law really is not necessary, declining $/kWh solar cell electricity competes with a rising $/kWh incumbent.

References

1. www.semiconductrnet.com
2. http://www.semiconductor.net/article
3. http://www.semiconductor.net/article/CA6656931.html?industryid = 47534
4. M. A. Green, *Third Generation Photovoltaics: Advanced Solar Energy Conversion*, Springer, New York (2006).
5. A. Hand, Electronic media, PV Society, May (2009), www.pvsociety.com.
6. M. A. Green, *Physica E* 14(1), 65–70, 2002.

7. http://sanyo.com/
8. L. Li, Y-M. Li, J. A. Selvan, A. E. Delahoy, and R. A. Levy, *Thin Solid Films* 451–452, 269–273 (2004).
9. O. Vetterl, F. Finger, R. Carius, P. Hapke, L. Houben, O. Kluth, A. Lambertz, A. Mück, B. Rech, and H. Wagner, *Sol. Energy Mater. Sol. Cells* 62, 97–108 (2000).
10. U. Kroll, J. Meier, P. Torres, J. Pohl, and A. Shah, *J. Non-Cryst. Solids* 227–230, 68–72 (1998).
11. P. Delli Veneri, L.V. Mercaldo, C. Minarini, and C. Privato, *Thin Solid Films* 451–452, 269–273 (2004).
12. M. W. Rowell, M. A. Topinka, M. D. McGehee, H.-J. Prall, G. Dennler, N. S. Sariciftci, L. Hu, and G. Gruner, *Appl. Phys. Lett.* 88, 233506 (2006).
13. N. S. Sariciftci et al., *Science* 258, 1474 (1991).
14. S. E. Shaheen et al., *Appl. Phys. Lett.* 78(6), 841 (2001).
15. E. Kymakis and G. A. J. Amaratunga, *Appl. Phys. Lett.* 80, 112 (2001).
16. W. U. Huynh, J. J. Dittmer, and A. P. Alivisatos, *Science* 295, 2425 (2002).
17. C. J. Brabec, S. E. Shaheen, C. Winder, N. S. Sariciftci, and P. Denk, *Appl. Phys. Lett.* 80, 1288 (2002).
18. S. Huang, H. Efstathiadis, P. Haldar, H. G. Lee, B. Landi, and R. Raffaelle, *Mater. Res. Soc. Symp. Proc.*, vol. 836, pp. 49–53, Boston, MA (2005).
19. B. J. Landi, P. Denno, R. DiLeo, W. VanDerveer, R. Raffaelle, H. Efstathiadis, and P. Haldar, *19th Space Photovoltaics Research and Technology (SPRAT) Conference Proceedings*, Cleveland, OH (2005).
20. D. H. Son, S. M. Hughes, Y. Yin, and A. P. Alivisatos, *Science* 306, 5698, 1009 (2004).

17

Commercialization of Nanotechnology: Opportunities and Issues

Sohail Anwar

CONTENTS

17.1 Introduction

Nanotechnology being a platform technology feeds its output into numerous industries, which use these inputs to improve their products. Consequently, any effort to commercialize this technology has to be supported by scientific and engineering research in conjunction with an innovative well-funded product development and marketing program involving all downstream industries that are going to utilize nanotechnology products (Dhillon et al. 2008). There is no doubt about the potential of nanotechnology to impact numerous facets of human life and society, and the incentive for expeditious commercialization of this technology is strong. However, factors, such as the long time lag between nanotechnology research and the development of commercial products, the large capital investment needed for a viable commercial venture, and financial/operational risks associated with commercial applications of nanotechnology, have impeded rapid adoption of this technology in the commercial domain.

Nanotechnology is a combination of science and technology and its commercialization requires the theoretical understanding of the underlying process, and specialized equipment for producing nanotechnology products through nanofabrication. Along the path to commercialization, nanotechnology's biggest liability is its novelty (Mazzola 2003). Inventions often attract attention because of their ingenuity, but a product must be useful and compelling. Undoubtedly, nanotechnology has solid commercial prospects but the process of converting basic inventions into marketable products is not easy.

There are two key steps to product development. The first step involves positioning the technology. That is, finding out what is nanotechnology's competitive edge. There is no simple answer to this question because of the wide range of products that can be developed using nanomaterials. The second step is to develop applications that leverage the unique aspects of nanoscale devices and systems. Currently, much of nanotechnology is at a stage where it requires significant incubation for application and product development.

At present, government funding is the key source of early support for nanotechnology research and development. As an example, the United States Federal Government has demonstrated its interest in the research, development, and commercialization of nanotechnology through an allocation of more than $3 billion since 2003 for programs in this arena. In the United States, the National Nanotechnology Initiative (NNI) consists of individual and cooperative nanotechnology-related activities of 25 federal agencies with a range of research and regulatory roles and responsibilities. The NNI as a program does not fund research; however, it informs and influences the federal budget and planning process through its member agencies. Federal research grants for nanotechnology are defined and awarded by individual U.S. government departments and agencies, in accordance with their respective missions. In addition to these grants, there are special programs designed to seed commercialization activity that stimulates economic growth. These programs support small business collaboration with universities and other research institutions.

Despite substantial support from government, the progress of nanotechnology commercialization has been slow so far (Maynard and Michelson 2007). Commercialization of nanotechnology products requires that they be produced in a predictable, reliable way, and in sufficient quantities. Until this is achieved, production will be limited to academia and R&D departments within industry. Academic institutions will continue to play a key role in our efforts to facilitate and promote the commercialization of nanotechnology.

17.2 Overview of Nanotechnology-Based Commercial Products

Nanotechnology may be defined as the engineering of functional systems at the molecular scale. Nanotechnology refers to the projected ability to construct items from the bottom up, using techniques and tools currently being developed to make highly advanced products. In its mature form, nanotechnology will have a significant impact on almost all types of industry. Like electricity or computers before it, nanotechnology will offer improved efficiency in every facet of life.

Nanotechnology covers processes associated with the creation and utilization of structures in the 1–100 nm range. Nanofabrication involves engineering at the atomic length scale. Engineering at this scale makes it feasible to create, atom by atom, fibers that are very small in diameter but extremely strong. In the health care domain, extremely minute probes can detect disease by examining individual strands of DNA. Nanofabrication makes it possible to manufacture capillary systems for providing nutrients to man-made replacement organs.

The nanofabrication process has been used for the creation of new chemical and biological substance detectors, which incorporate structures holding molecules that change their electrical conducting properties in the presence of the substances being detected. The development of a new class of nanoscale transistors and molecular electronics has also been made possible by the utilization of nanotechnology.

Nanotechnology is capable of producing products, materials, and devices that impact a wide spectrum of industries and consumer products. Therefore, it is pragmatic to view nanotechnology as a "platform technology" with applications in a number of industrial sectors, and with potential for producing a variety of products. The list of current and potential nanotechnology applications continues to grow. However, it would be appropriate to consider the following areas of nanotechnology application as the most promising beneficiaries (not ranked):

- Electronics and semiconductors
- Information technology (computing and telecommunication)
- Aerospace and automotive industries
- Chemical processes and engineering
- Agriculture
- Energy
- Disease diagnosis
- Health monitoring
- Drug delivery
- Food processing and storage
- Water treatment and air pollution control

The five top-ranking areas of nanotechnology product development are (McNeil et al. 2007):

- Semiconductors, nanowires, lithography, and printing products
- Nanostructures, nanotubes, and self-assembly
- Coatings, paints, thin films, and nanoparticles
- Environmental sensing and remediation
- Defense applications and protection gear

Although tremendous excitement has been stimulated by the potential applications of nanotechnology, the usage of "first-generation" passive nanomaterials accounts for most of the commercial applications of this technology currently. The products/applications include titanium dioxide nanoparticles in sunscreen, cosmetics, and food products (usage of nanoparticles as additives to existing consumer products); silver nanoparticles in food packaging, clothing, disinfectants and household appliances; zinc oxide nanoparticles in sunscreen and cosmetics, surface coatings, paints and outdoor furniture varnishes; and cerium oxide nanoparticles as a fuel catalyst (exploitation of surface characteristics of nanoparticles to improve chemical reactions and interfacial bonding). The relatively short list of actual applications of nanotechnology indicates that further research is needed to diversify the utilization of this technology.

17.3 Path to Nanotechnology Commercialization

Nanotechnology is an international phenomenon. It is not so much an industry as a collection of tools and approaches that will achieve significant commercial success only when compelling applications are developed and then adopted by consumers. In addition, nanotechnology has to overcome many barriers related to robust production and large-scale manufacturing.

Although, many new applications of nanotechnology originate from commercial laboratories, most of them will be developed in academic and government laboratories. These organizations are not well equipped to develop commercial products. However, over the years, academic institutions and laboratories have established mechanisms for transfer of newly developed technologies to the commercial sector. In the United States, the general approach for technology transfer begins with the filing of a patent. The academic institution or lab usually covers the expense associated with the filing of a patent application. The invention is then licensed either exclusively or nonexclusively, usually for some fee plus royalties based on product sales, to a commercial organization. The terms for such licenses vary considerably (Hornyak et al. 2009).

The commercial organization receiving a nanotechnology application license from an academic institution or a laboratory is expected to have the following key elements in place:

1. A suitable commercialization strategy based on market awareness
2. Appropriate level of investment in product research and development facilities
3. Appropriate level of investment in manufacturing facilities
4. Appropriately trained personnel in nanotechnology-based products for manufacturing and marketing
5. Knowledge of existing and potential customer base for nanotechnology-based products
6. Appropriate protection for intellectual property (IP)

The path to nanotechnology commercialization is not a linear path. This is because of the disruptive nature of nanotechnology and because we live in a world where change, challenge, and risk will always offer barriers to progress (Tolfree and Jackson 2008). The key steps in the nanotechnology commercialization ladder are as follows:

1. Proposing an idea/concept—Every invention or innovation is the result of an idea or a concept proposed by an individual or a group of individuals. An idea or a concept is driven by a specific need, sometimes by inspiration, and often by a desire to make money. Many times the idea for a new product results from an interconnection between existing products. There is obviously a significant gap between an idea and a proven concept. Advancing to a point where an idea is conceptually viable requires determination and focused research.
2. Design, modeling, and simulation—Once a proof of concept has been carried out to determine that the proposed idea is sound, a design study is the next step. The design is expected to satisfy a number of criteria. The knowledge and design tools must be available to carry out this step. As we move into the nanotechnology domain, new modeling software and design approaches are needed for atomic- and molecular-scale processes.
3. Prototyping—Prototyping, rapid prototyping, and preproduction prototyping are essential for volume manufacturing. In order to take a design directly to the preproduction prototyping level, modeling and simulation tools need to be integrated with fabrication processes. These tools need to function across multiple technology disciplines to enable nanotechnology system designers develop new products.
4. Packaging—This function is needed to interconnect, protect, and provide an interface to the macro world. Packaging includes cost, reliability, and accuracy. Packaging materials are of critical importance in obtaining the appropriate protection in any given environment.

5. Testing and reliability—Long-term reliability is of utmost importance for all miniaturized components to be embedded into products.
To achieve market credibility, a product must demonstrate total
product reliability. Design for reliability must identify and prevent
failures before a product advances to the manufacturing stage. All
possible failure modes in different environmental conditions need to
be tested to verify failure model predictions.

6. Final product realization and marketing—Usually, customers have
little interest in how a given product was manufactured or what kind
of technology was used to produce it. Buyers are, of course, keenly
interested in its performance and the cost advantage it has to offer.
When a company succeeds in achieving a marketable product, the
journey to commercialization is mostly over but the final destination
of penetration into a competitive market is still to be reached. In a
continuously changing global market, there are no certainties and
the success or failure of a product depends on the marketing capabilities of a company. Thus, it is very important for a company to
put in place an effective marketing strategy before it reaches the end-
product stage.

Each of the six above-mentioned steps can be subdivided into smaller tasks
depending on the nature of the product. Appropriate infrastructure should
be in place to carry out these steps. If such an infrastructure does not exist,
then companies have to look for prototyping or manufacturing facilities
outside their location and doing so can significantly add to the development
and production costs.

17.4 Barriers to Nanotechnology Commercialization

The commercialization of nanotechnology is presently slow paced due to a
number of barriers. These barriers may limit our ability to capture the full
potential of nanotechnology. The present and potential barriers to nanotechnology commercialization are listed as follows (McNeil et al. 2007):

1. Time between research and commercialization is estimated to be
3–10 years. Venture capitalists find this time factor to be a detriment.

2. There is a serious gap between research and commercialization that
must be addressed by government agencies and the venture capitalists. The scientists may publish their research results and not be
interested in commercialization. A common notion is that for every
dollar invested in basic research, an investment of almost $100 is
required to produce a competitive product.

3. Proper infrastructure (labs, equipment, measuring devices, etc.) availability is lacking. The availability of appropriate infrastructure is crucial for the growth of small companies that cannot afford the high cost of nanotechnology instrumentation, equipment, and facilities.

4. Small companies do not have the capability to produce products on a large scale.

5. There is a lack of effective mechanisms to facilitate the technology transfer from academic institutions and laboratories to the commercial sector.

6. In the United States, the Patent Office takes up to 36 months to respond to patent applicants.

7. Lack of standards is hindering nanotechnology advancements. There is an urgent need to develop standards for each aspect of the new nanotechnologies. These aspects include research and development, production and manufacturing, products, and waste disposal.

8. Efforts by national, state, and local economic development initiatives have not yet resulted in a remarkable increase in new nanotechnology commercial sector employment.

9. The public perception that nanotechnology products are unsafe must be challenged to make sure that general public has a complete awareness of the full potential of nanotechnology.

10. Successful commercialization of nanotechnology products requires unprecedented levels of collaboration across many different domains to effectively address the inherent complexities associated with the lifecycles of these products. At present, there is an acute shortage of such collaborations.

11. There is a serious need for development of methods to evaluate the toxicity of engineered nanomaterials.

12. Nanotechnology is a multidisciplinary field. Advances in nanotechnology will need the scientific and technical expertise provided by chemists, physicists, material scientists, biochemists, molecular biologists, engineers, toxicologists, and medical scientists working together. A potential barrier to nanotechnology growth in the United States is the lack of well-trained scientists, engineers, technicians, and researchers.

13. In the United States, the FDA and Patent Offices do not have enough qualified staff to evaluate new nanotechnology products.

Most of the above-mentioned barriers to nanotechnology commercialization can be significantly lowered if the following measures are adopted (McNeil et al. 2007):

1. Current funding provided by government and several other organizations for new nanotechnology ventures is focused on basic research. More funding is needed for the commercial development of nanotechnology-based products.
2. More sophisticated and early market research is needed to bridge the science culture-to-commercialization gap. Markets and future products must be identified early in the research and development stage, followed by periodic reviews as innovations are transformed into prototypes that are placed into production.
3. There is a need to establish new federal and state laws and regulations that encourage small and medium nanotechnology business growth, the expansion of more multidisciplinary research and development, and better IP protection.
4. Additional federal, state, and local government support is needed to encourage more networking, joint ventures, and strategic alliances to expand international collaboration involving universities, research and development organizations, investors, manufacturers, and product distributors.

17.5 Key Success Factors for Nanotechnology Commercialization

Every nanotech start-up company goes through three key stages after its formation:

1. Inception
2. Funding
3. Growth

During the inception stage, the key success factors are as follows (Waitz and Bokhari 2003):

1. A strong IP position at the inception of the nanotech start-up company.
2. Clear, concise, well thought-out, and compelling business plan. The critical components of this plan should focus on the issues that will enable the company to bring its products to market.
3. A well-balanced multidisciplinary team having skill sets needed to attain the business plan goals.

Common pitfalls during the inception stage of a nanotech start-up company are as follows:

1. Lack of a realistic and fundable manufacturing strategy
2. Failure in planning for the progress that an incumbent technology will make during the time it takes to develop the nano-based technology

Success factors during the funding stage of a nanotechnology start-up company include

1. For government funding, writing an effective funding proposal that satisfies the soliciting agency's requirements.
2. For funding from venture capitalists, having a strong "done it before" team addressing a large market opportunity is a good starting point.

Common pitfalls during the funding stage of a nanotechnology start-up company include

1. Lack of focus. This is an important issue with investors such as venture capitalists. It is important to note that venture capitalists do not generally like to invest in a project that has multiple disparate target markets.
2. Wrong kind of investor. If a nanotech company is too far away from a potential product, then the company should not approach venture capitalists. It should rather try to obtain funding from government sources.
3. Lack of technical understanding by venture capitalists. Many traditional venture capitalists may shy away from a nanotech venture due to a lack of understanding of the technical merit of the project.
4. Lack of business understanding by venture capitalists. Since the idea of a nanotech company is relatively new, there are not enough success business models for investors to compare with.

Finally, during the growth stage of a nanotech company, the single-most important success factor is that of having a management team that has a strong target market knowledge. There are examples of companies that have created a nano-based technology with a target market in mind to discover later that the product did not meet the needs of that market. This oversight was due to a lack of domain knowledge in the management team.

Common pitfalls nanotech companies fall into during the growth stage include

1. During transition from an academic lab to a commercial product, it is common for academic founders to underestimate the difficulty in commercializing a new technology.

2. Resistance to new approaches by incumbent markets. Many nanotech-based products are targeted at existing markets. To achieve success, these new products will need to displace the incumbents based on price or performance.

3. Lack of industry infrastructure. Since the nanotech companies are typically on the cutting edge of technology, there is not an existing well-developed infrastructure to leverage.

17.6 Conclusion

No single technology offers more economic and societal promise than nanotechnology. Nanotechnology holds the promise of both incremental improvements of existing products and the potential for revolutionary changes that could transform entire industries and create entirely new ones. However, many barriers slow down the commercialization of nanotechnology. These barriers include capital issues, market readiness, regulatory uncertainty, health and safety, workforce preparedness, public perceptions, infrastructure, standards, manufacturability, and nomenclature.

There are several ways to lower these barriers. Since many challenges exist in the transition from government research to commercial ventures, there is a definite need to establish a facilitating relationship between public/private funding and government information. More focused and early market research is needed to bridge the gap between science and commercialization. Target markets and future products need to be identified early in the research and development stage. New federal and state laws and regulations are needed to encourage growth of small and medium nanotechnology companies. Government support will be crucial for developing international collaboration involving academic institutions, research organizations, investors, manufacturers, and marketers.

References

Dhillon, H., Qazi, S., and Anwar, S. 2008. Mitigation of barriers to commercialization of nanotechnology: An overview of two successful university-based initiatives. *Proceedings of the ASEE 2008 Annual Conference and Exposition*, Pittsburgh, PA.

Hornyak, G. L., Moore, J. J., Tibbals, H. F., and Dutta, J. 2008. *Fundamentals of Nanotechnology*. Boca Raton, FL: Taylor & Francis Group/CRC Press.

Maynard, A. and Michelson, E. 2007. Project on emerging technologies. Report prepared for Woodrow Wilson International Center for Scholars, Washington, DC.

Mazzola, L. 2003. Commercializing nanotechnology. *Nature Biotechnology* 21(10): 1137–1143.

McNeil, R. D., Lowe, J., Mastroianni, T., Cronin, J., and Ferk, D. 2007. Barriers to nanotechnology commercialization. Report prepared for US Department of Commerce, Technology Administration, Washington, DC.

Tolfree, D. and Jackson, M. 2008. *Commercializing Micro-Nanotechnology Products.* Boca Raton, FL: Taylor & Francis Group/CRC Press.

Waitz, A. and Bokhari, W. 2003. Nanotechnology commercialization best practices. http://www.quantuminsight.com/papers/030915_commercialization.pdf

Glossary

Atomic layer deposition (ALD): ALD is a thin film deposition technique that is based on the sequential use of a gas phase chemical process. The majority of ALD reactions use two chemicals, typically called precursors. These precursors react with a surface one at a time in a sequential manner. By exposing the precursors to the growth surface repeatedly, a thin film is deposited. ALD is a self-limiting (the amount of film material deposited in each reaction cycle is constant), sequential surface chemistry that deposits conformal thin films of materials onto substrates of varying compositions. ALD is similar in chemistry to chemical vapor deposition (CVD), except that the ALD reaction breaks the CVD reaction into two half-reactions, keeping the precursor materials separate during the reaction. Due to the characteristics of self-limiting and surface reactions, ALD film growth makes atomic scale deposition control possible. By keeping the precursors separate throughout the coating process, atomic layer control of film growth can be obtained as fine as ~0.1 Å (10 pm) per monolayer. (Copied from Wikipedia, http://en.wikipedia.org/wiki/Atomic_layer_deposition.)

Azobenzene: A chemical compound composed of two phenyl rings linked by a N=N double bond. One of the most intriguing properties of azobenzene (and derivatives) is the photoisomerization of *trans* and *cis* isomers. The two isomers can be switched with particular wavelengths of light: ultraviolet light, which corresponds to the energy gap of the $\pi-\pi^*$ (S_2 state) transition, for *trans*-to-*cis* conversion, and blue light, which is equivalent to that of the $n-\pi^*$ (S_1 state) transition, for *cis*-to-*trans* isomerization.

Block cipher: In cryptography, a block cipher is a symmetric key cipher, which operates on fixed-length groups of bits, termed blocks, with an unvarying transformation. When encrypting, a block cipher might take, for example, a 128 bit block of plaintext as input, and output a corresponding 128 bit block of ciphertext. The exact transformation is controlled using a second input—the secret key. Decryption is similar: the decryption algorithm takes, in this example, a 128 bit block of ciphertext together with the secret key, and yields the original 128 bit block of plaintext.

Boundary element method (BEM): The BEM is a numerical computation technique for representing and solving partial differential equations as integral equations. BEM uses only 2D elements on the surfaces, which are the material interfaces or assigned boundary conditions.

Bragg's law: Describes how a light beam is reflected or diffracted in a crystal lattice given by the equation $n\lambda = 2d \sin \theta$, where n is an integer (diffraction order), λ is the wavelength of the incident beam, d is the spacing between crystal planes, and θ is the angle between the crystal plane and the incident beam (the Bragg angle).

Buckyball: A soccer-ball-shaped molecule made of 60 carbon atoms.

Carbon nanotube: A sheet of graphite rolled into a tube.

CMOS inverter: A CMOS (complementary metal-oxide-semiconductor) inverter is a pair of two complementary transistors: one of n-polarity with electrons as carriers of charge and another of p-polarity with holes as carriers of charge, with gates tied together for the input signal and drains tied together for the output signal, while the p-transistor source is set at a high voltage and the n-transistor source is grounded.

Colloid: A type of mixture where one substance (the internal phase) is dispersed evenly throughout another (the continuous phase). The particles of a dispersed substance are only suspended in the mixture, unlike a solution, within which they are completely dissolved. A colloidal system may be solid, liquid, or gaseous.

Computer branch exchange (CBX): Computerized or controlled branch exchange. A computer-controlled telephone switching system that supports such services as conference calling, least-cost routing, direct inward dialing, and automatic re-ringing of a busy line.

Computer teléphony integration (CTI): Also called computer–telephone integration, CTI is a technology that allows the blending and interaction of telephone and computer technology to be integrated or coordinated in a single application.

Coulomb blockade (QB): In physics, a QB, named after Charles-Augustin de Coulomb, is the increased resistance at small bias voltages of an electronic device comprising at least one low-capacitance tunnel junction.

Crystalline colloidal array (CCA): A three-dimensional crystal formed by self-assembly of highly charged, monodisperse colloidal spherical particles, in which the dielectric constant differences between the particles and surrounding medium are great enough to create a bandgap that does not allow passage of photons of a specific wavelength. A natural example of this ordering phenomenon can be found in precious opal. A polymerized CCA (PCCA) is a more rugged crystal formed by embedding the CCA within a hydrogel matrix.

Current saturation: The current saturation arises from the current remaining constant no matter what the voltage applied, setting a ceiling on the current driven in a nano- or microresistor.

Density matrix: In quantum mechanics, a density matrix is a self-adjoint (or Hermitian) positive-semidefinite matrix (possibly infinite dimensional) of trace 1, that describes the statistical state of a quantum system. The formalism was introduced by John von Neumann (and according to other sources, independently by Lev Landau and Felix Bloch) in 1927.

Diffraction: The bending and spreading out of waves around small obstacles. Similar effects are observed when there is an alteration in the properties of the medium through which the wave is traveling, for example, a variation in the refractive index for light waves or in the acoustic impedance for sound waves.

Digital switching: The process of sending and receiving signals from a communication network in which information is converted and processed in the form of distinct electronic pulses. A series of these processes in a network can be used to convey the information from the sender to the receiver.

Electroluminescence: The phenomenon in which a material emits light in response to an electric current passed through it, or to a strong electric field.

Embedded system: An embedded system is a special-purpose computer system dedicated to specific tasks which might have real-time computing constraints. It is usually embedded as a subsystem in a complete system including mechanical/chemical components.

Excitons: An electrically neutral excited state of an insulator or semiconductor, often regarded as a bound state of an electron and a hole.

Finite element method: The finite element method is a numerical computation technique for approximately solving partial differential equations (PDEs) and involves discretization of the domain of the function into elements. The function is approximated by a characteristic form on each element.

Finite volume method: The finite volume method is a numerical computation technique for representing and solving partial differential equations as algebraic equations.

Hamiltonian: In quantum mechanics, the Hamiltonian H is the observable corresponding to the total energy of the system. It is a Hermitian matrix, which, when multiplied by the column vector representing the state of the system, gives a vector representing the total energy of the system. As with all observables, the spectrum of the Hamiltonian is the set of possible outcomes when one measures the total energy of a system. Like any other self-adjoint operator, the spectrum of the Hamiltonian can be decomposed, via its spectral measures, into pure point, absolutely continuous, and singular parts.

Hardware/software codesign: Hardware/software codesign is a system design methodology aimed at meeting system-level objectives by exploiting the synergism of hardware and software through their concurrent design.

Hydrogel: A cross-linked network of polymer chains that are water-insoluble, sometimes found as a colloidal gel in which water is the dispersion medium. Hydrogels are highly absorbent (they can contain over 99% water) natural or synthetic polymers, and possess a degree of flexibility, very similar to natural tissue, due to their significant water content.

Light-emitting device: A device that emits light typically in response to electrical excitation, mostly used in optical communications and displays.

Lithography: Process of copying a pattern onto a surface using UV, electron beam, or x-rays.

Macromodels: Macromodels are compact mathematical expressions that are derived from the time- or frequency-domain responses of a device via suitable identification or approximation algorithms. A macromodel captures the essential static and dynamic behaviors of the device using a minimal set of equations in terms of the physical design parameters and the material properties.

Markov chain: In mathematics, a Markov chain, named after Andrey Markov, is a stochastic process with the Markov property. Having the Markov property means that future states depend only on the present state, and are independent of past states. In other words, the description of the present state fully captures all the information that could influence the future evolution of the process. Being a stochastic process means that all state transitions are probabilistic.

Master equation: In physics, a master equation is a phenomenological set of first-order differential equations describing the time evolution of the probability of a system to occupy each one of a discrete set of states. Many physical problems in classical, quantum mechanics, and problems in other sciences can be reduced to the form of a master equation, thereby performing a great simplification of the problem.

Model of computation: A model of computation defines the set of operations allowable in computation and their respective costs in terms of computational resources such as execution time, memory space, etc. A model of computation explains how the behavior of the whole system is the result of the behavior of each of its components.

Molecular beam epitaxy (MBE): MBE is one of several methods of depositing single crystals. It was invented in the late 1960s at Bell Telephone Laboratories by J. R. Arthur and Alfred Y. Cho. Molecular beam epitaxy takes place in high vacuum or ultrahigh vacuum (10^{-8} Pa). The most important aspect of MBE is the slow deposition rate (typically less than 1000 nm/h), which allows the films to grow epitaxially. (Copied from Wikipedia, http://en.wikipedia.org/wiki/Molecular_beam_epitaxy.)

Monodispersity: The state of uniform molecular weight of all molecules of a substance or of a polymer system. Monodisperse collections can be easily created through the use of template-based synthesis, a common method of synthesis in nanotechnology.

Monte Carlo method: Monte Carlo methods are a class of computational algorithms that rely on repeated random sampling to compute their results. Monte Carlo methods are often used when simulating physical and mathematical systems. Because of their reliance on repeated computation and random or pseudorandom numbers, Monte Carlo methods are most suited to calculation by a computer. Monte Carlo methods tend to be used when it is unfeasible or impossible to compute an exact result with a deterministic algorithm.

Multiquantum wells: Multiquantum wells is a combination of more than one potential well, which can confine the charge carriers, namely, holes and electrons.

Nanocircuit current divider: The familiar current division, as obtained from Ohm's law, is invalid as resistor length plays predominant role in the current division and the current saturates to a value that is proportional to the cross-sectional area or width of the resistor.

Nanocircuit voltage divider: The familiar voltage division as obtained from Ohm's law is invalid as resistor length plays predominant role in the voltage division.

Nanoelectromechanical systems (NEMS): Nanoscale electromechanical devices similar to microelectromechanical systems (MEMS) but at much smaller physical dimensions.

Nanoelectronic transport: Nanoelectronic transport is the drift of carriers (electrons and holes) in devices on the scale of a few nanometers.

Nano-emitter: Photonic devices, which, are in nanometers scale with sharp electron-emitting capabilities is known as nano-emitter.

Nanophotonics: Nanophotonics or nano-optics, a branch of optical engineering, is the study of the behavior of light with particles or substances at the nanometer scale.

Nanoscale silicon: Silicon structures in the size scale of a few nanometers. Silicon is the main semiconductor material used in electronics. Sometimes the term "nanoscale silicon" refers to silicon devices exhibiting quantum size effects.

Nanoshell: A nanoparticle composed of silica surrounded by a gold coating.

Nanotechnology: Manipulation of matter at the atomic and molecular scale, seeing matter at the atomic and molecular scale, and exploitation of the unique capabilities and properties of structures fabricated at the atomic and molecular scale.

Nanowire: Electrical conductors created at nanoscale for interconnecting devices or systems on an integrated chip.

National Nanotechnology Initiative (NNI): A program established in 2001. The NNI (www.nni.org) provides a vision of the long-term opportunities and benefits of nanotechnology.

Nonohmic conduction: Nonohmic conduction arises from the failure of the linear current–voltage relation, as given by Ohm's law, resulting in the current saturation.

Photoluminescence: Commonly known as PL, it is a mechanism in which an epi-structure (semiconductor layer by layer structure) absorbs photons (electromagnetic radiation) and then reradiates photons.

Photonic bandgap: It is a forbidden energy gap for photons, in a way analogous to the electronic bandgap for electrons occurring in semiconductors. It occurs in photonic crystals, which are periodic optical nanostructures that are designed to affect the motion of photons in a similar way that periodicity of a semiconductor crystal affects the motion of electrons. Although photonic crystals occur in nature and have been studied scientifically in various forms for the last 100 years, man-made photonic crystals have of late been receiving considerable attention for use in optoelectronics and optical communications.

Photovoltaic solar cell: A device that converts light directly into electricity by the photovoltaic effect.

Polarization: A property of electromagnetic radiations in which the direction and magnitude of the vibrating electric field are related in a specified way.

poly(*N*-isopropylacrylamide) (PNIPAM): Temperature-responsive polymer, which when heated in water above 33°C, undergoes a reversible volume phase transition from a swollen hydrated state to a shrunken dehydrated state, losing about 90% of its mass.

Quantum bit: A quantum bit, or quibit, is a unit of quantum information described by a state vector in a two-dimensional vector space over complex numbers. It is an abstract model of a two-level quantum-mechanical system (such as electron spin).

Quantum circuits: A quantum circuit is a series of quantum operations applied on one or multiple quantum bits. The initial state of these qubits is the input and the final state is the output. All the operations are unitary, and hence reversible, transformations.

Quantum communications: Quantum communication is the process of conveying classical or quantum information from a sender to a receiver with quantum bits as the underlying information carrier.

Quantum computing: A method of computing that makes direct use of quantum mechanical phenomena such as superposition and entanglement to perform operations on data on quantum binary digits (qubits).

Quantum dot (QD): A semiconductor nanocrystal whose electrons show discrete energy levels like an atom.

Quantum dot cellular automata (QCA): A proposal for implementing classical cellular automata by systems designed with quantum dots has been proposed under the name "quantum cellular automata" by Doug Tougaw and Craig Lent as a replacement for classical computation using CMOS technology. In order to better differentiate between this proposal and models of cellular automata which perform quantum computation, many authors working on this subject now refer to this as a quantum dot cellular automaton.

Quantum networks: A quantum network is a group of interconnected nodes each with the capabilities of sending, receiving, and performing quantum operations on the incoming quantum bits. With these capabilities, a quantum network can be used to convey quantum information from the sender to the receiver.

Quantum superposition: Quantum superposition is a phenomenon that an object, typically a quantum bit, exists in more than one state simultaneously. For example, according to quantum mechanics, the spin state of an electron can be up and down simultaneously, each with a probability of 50%.

Quantum switching: The technology of routing a quantum bit to its destination in a quantum network where information is carried by quantum bits. Unlike classical digital switching, the process has to be performed in such a way that the quantum state carried by the qubit is not destroyed or modified.

Qubits or Q-bits: The smallest unit of information in quantum computing. Qubits hold an exponentially larger amount of information than traditional bits and do not rely on the traditional binary nature of computing. The Qubit approach takes advantage of inherent probability distributions in digital data, combined with transformations from different mathematical spaces, to perform compression in a lossless, reversible manner; no bit is left behind.

Radiative interactions: A mechanism in which the recombination of charge carriers takes place in a certain medium, and as a result, photons are produced.

Raman scattering or Raman effect: Raman scattering refers to the inelastic scattering of light by matter. It was discovered by C. V. Raman in liquids and by Grigory Landsberg and Leonid Mandelstam in crystals. The scattered photons have a frequency somewhat shifted from the frequency of the incident light. This frequency shift is used to obtain information about the vibrational modes and/or the electronic properties of the material. Raman scattering is a very powerful tool in material characterization.

RC enhancement: *RC* time constant is higher than that expected from the application of Ohm's law in a circuit containing nanoscale resistor along with a capacitor due to its resistance blow-up as a digital signal rises from low to high or vice versa.

Resistance blow-up: The resistance as inverse slope of the sublinear current–voltage characteristics rises sharply (under dc bias conditions and much intensely for the signal propagating at a given dc bias point) as the saturation regime of the current is reached.

Saturation velocity: Saturation velocity, also called the intrinsic velocity, is the ultimate velocity that a carrier can have in a semiconducting sample with a value that is comparable to the thermal velocity under nondegenerate statistics and the Fermi velocity under degenerate statistics.

Self-assembly: The spontaneous aggregation of particles (atoms, molecules, colloids, micelles, etc.) into thermodynamically stable, structurally well-defined arrays without the influence of any external forces. Self-assembly is also emerging as a new strategy in chemical synthesis and nanotechnology.

Short-λ optical devices: Photonic devices, which can emit photons in the lower wavelength region of the solar spectrum.

Side-channel attack: In cryptography, a side-channel attack is any attack based on information gained from the physical implementation of a cryptosystem, rather than brute force or theoretical weaknesses in the algorithms (compare cryptanalysis). For example, timing information, power consumption, electromagnetic leaks, or even sound can provide an extra source of information, which can be exploited to break the system. Many side-channel attacks require considerable technical knowledge of the internal operation of the system on which the cryptography is implemented.

Silicon nanostructures: Basically the same as "nanoscale silicon." Silicon structures in the size scale of a few nanometers. Silicon is the main semiconductor material used in electronics. Sometimes the term "nanoscale silicon" refers to silicon devices exhibiting quantum size effects.

Silicon photonics: Photonic devices such as waveguides, optical filters, optical amplifiers, transmitters, etc., fabricated using silicon-based materials. The importance of photonic devices made out of silicon is that they can be integrated cost effectively with traditional electronics.

Space division switching: A digital switching technology that takes signals from multiple input ports and, via one of many physical resources, transmits the signals to the corresponding output ports.

Spiropyran: Photochromic compound that can undergo reversible structural transformation in response to external inputs such as light, protons, and metal ions. The spiropyran converts from closed form to open form (merocyanine) after ultraviolet light irradiation. The merocyanine form is thermally unstable and reverts back to closed form.

Superlattices: Superlattice is a periodic structure of repeating quantum wells that sets up a new set of selection rules which affects the conditions for charges to flow through the structure. This nanostructure consists of two different semiconductor materials, which are deposited alternately on each other to form a periodic structure in the growth direction. Since the first proposal of synthetic artificial superlattices by Esaky and Tsu in 1970, great advances in the physics of such ultrafine semiconductors, presently called quantum structures, have been made within the past two decades. The concept of quantum confinement has led to the observation of quantum size effects in isolated quantum well heterostructures and is closely related to superlattices through the tunneling phenomena. Therefore, these two ideas are often discussed on the same physical basis, but each field has its own intrigue and different physics useful for applications in many electric and optical devices. (Copied from Wikipedia, http://en.wikipedia.org/wiki/Superlattice.)

Surface-to-volume ratio: A figure of merit that determines the change in electronic, structural, optical, and acoustic properties of the material. The surface-to-volume ratio relationship allows the possibility of using size in nanometer regime to design and engineer materials with new and possibly technologically interesting properties.

System-on-chip (SoC): SoC refers to integrating all components of an embedded computer system on a single integrated circuit. The components might be heterogeneous and may contain digital, analog, mixed-signal, radio-frequency, micro/nano-sensor/actuator functions—all on a single microchip.

Telecommunications: The science and technology of communicating at a distance, especially the electronic transmission of signals, telecommunicating or the sending of signals representing voice, video, wire, wireless radio, optical, or other electromagnetic means.

The singularity: The postulated point or short period in our future when our self-guided evolutionary development accelerates enormously (powered by nanotechnology, neuroscience, AI, and perhaps, uploading, which, transfers the consciousness and mental structure of a person from a biological matrix to an electronic or informational matrix so that nothing beyond that time can reliably be conceived).

Time division switching: A digital switching technology that takes signals from multiple input ports and, in a preassigned time slot, transmits the signals to the corresponding output ports.

Transmission control protocol/Internet protocol (TCP/IP): A set of communications protocols that has evolved since the late 1970s, when it was first developed by the Department of Defense. Because programs supporting these protocols are available on so many different computer systems, they have become an excellent way to connect different types of computers over networks.

Transport equation: The generic scalar transport equation is a general partial differential equation that describes transport phenomena such as heat transfer, mass transfer, momentum transfer, etc.

Tunnel junction: A tunnel junction is any junction between two different materials, where electrons move through the junction by quantum tunneling. Tunnel junctions serve a wide variety of different purposes. A tunnel junction is, in its simplest form, a thin insulating barrier between two conducting electrodes. If the electrodes are superconducting, Cooper pairs with a charge of two elementary charges carry the current. In the case that the electrodes are normal conducting, i.e., neither superconducting nor semiconducting, electrons with a charge of one elementary charge carry the current.

Wide bandgap semiconductor: Wide bandgap semiconductors are semiconductors that have larger electronic bandgaps, may be one or two electron volts (eV).

Wireless communication: MEMS-based communications, RF MEMS.

Wireless sensor networks: Wireless integrated sensor networks are an emerging class of networked embedded systems that combine sensing, computation, and communication in an inexpensive and very small form factor device with limited energy. Their designs exploit recent advances in the micro- and nanotechnology and are meant to act as a bridge between the physical and the virtual worlds.

Workload balancing: In computer and telecommunications, load balancing is a technique to spread work between two or more computers, network links, CPUs, hard drives, or other resources, in order to achieve optimal resource utilization, maximize throughput, and minimize response time.

Index